DATA CONVERTERS

Data Converters

by

FRANCO MALOBERTI
Pavia University, Italy

 Springer

A C.I.P. Catalogue record for this book is available from the Library of Congress.

ISBN-13 978-1-4419-4087-2   ISBN-10 0-387-32486-0 (e-book)
                                       ISBN-13 978-0-387-32486-9 (e-book)

Published by Springer,
P.O. Box 17, 3300 AA Dordrecht, The Netherlands.

www.springer.com

Printed on acid-free paper

To Pina, Amélie, and Matteo

Contents

Preface

The purpose of this book is to help students, engineers, and scientists in the design and use of data converters, either as stand alone parts or used on a mixed analog-digital IC. For class use, each chapter includes many problems for illustrating the practical applications of the text. In addition, the book includes examples and insets outlining key aspects of the studied topics.

The contents of this book stem largely from a course on Data Converters given at the Texas A&M University, College Station, at the University of Texas, Dallas and more recently at the University of Pavia, Italy. The course assumes that the students have a working knowledge on analog circuit design, layout and are familiar with the principles of operation of transistors.

The book contains nine Chapters. Chapter 1 deals with the background knowledge necessary to properly understand and design data converters by providing the theoretical implications of data conversion and making the reader aware of the limits of the approximations used for studying a data converter. In addition, the chapter recalls the mathematical tools used for analysis and characterization of sampled-data systems.

Chapter 2 helps in a proper understanding of data converter specifications. The general information, the features and limits of the static and dynamic operation of a data converter are suitably presented for the evaluation and comparison of existing devices and for determining the specifications of new devices. The chapter also clarifies the definitions of technical terms used in manufacturer-supplied specifications.

The Nyquist-rate digital-to-analog architectures are studied in Chapter 3 that considers before the basic requirements on the voltage and the current references. Then, resistor based and capacitor-based architectures are discussed. Finally, the chapter studies the architectures that obtain the DAC function by steering unary or binary-weighted current sources and at the end briefly mentions special architectures.

Chapter 4 deals with the architecture, features and limits of Nyquist-rate analog/digital converters. It starts with the full-flash architecture capable of obtaining the conversion in only one clock period. Following this, studies the two-step solution whose algorithm requires at least two clock periods. Next, folding and interpolations method are discussed. Then, the chapter analyzes the interleaved and successive approximation algorithms before studying the pipeline architecture. Finally, it considers some techniques useful for special needs.

This book assumes that the reader is familiar with the features and design techniques of basic analog circuits; however, there are some special functions that are not studied in the proper details in analog courses. For this reason Chapter 5 concentrates on studying circuits that are specific to data converters starting from S&H circuits, realized in either bipolar or CMOS technology. After that, considers the clock boosting technique used to enhance (or make possible) the switching on of MOS transistors at very low supply voltages. Following this, the circuit techniques used in folding systems for current and voltage inputs are analyzed. Since various architectures use V/I converters, the chapter reviews some of the general V/I schemes. Finally, it describes the generation of overlapping and non-overlapping phases, as required by data converter controls.

Oversampling converters are increasing of importance in modern mixed analog-digital systems as they achieve high resolutions without requiring accurate technologies. Oversampling techniques are discussed in two chapters. Chapter 6 studies the basis of the methods starting from the plain oversampling method. Then, it discusses benefits from both noise-shaping and oversampling to give an optimum trade-off between speed and resolution. The chapter recalls the basic principles and discusses first and second order $\Sigma\Delta$ architectures providing the basis for the study, made in the successive Chapter 7, of high order modulators. In addition to single stage architectures Chapter 7 studies cascaded solutions normally named as MASH. Then it considers the continuous-time counterpart of the already studied sampled-data $\Sigma\Delta$ modulators before discussing band-pass implementations and, briefly $\Sigma\Delta$ DAC.

Chapter 8 deals with digital techniques used to enhance data converters performances. When the accuracy and component matching must be better than the one granted by technology it is necessary to calibrate the values or to correct the results using digital techniques. Therefore, various methods substantially assist the data converter designer in enhancing the expected performances. The chapter studies methods of error measurement, thus allowing for their correction or calibration in the analog or digital domain. The methods used can be either on-line (where the converter continues to function), or off-line with a dedicated operation for the error measurement or calibration. Correction techniques that enhance the spectral performances by a dynamic averaging of elements are also studied.

Chapter 9 deals with the methods used for testing and characterizing data converters. The chapter starts with the static method for testing DNL and INL. Following this, the testing of dynamic performance, namely settling time, glitch and distortion are discussed. Static ADC testing are also considered. The chapter studies the histogram method with different types of input and distortion and intermodulation test techniques. The use of sine wave and FFT in extracting part specifications is also considered.

A specific feature of this book is the extensive use of behavioral models in many of the examples provided. It is believed that the learning process is greatly enhanced by numerical verification of results or, for data converters, with a quantitative verification of behavior and performance by the use of suitably accurate models. For circuit design it is important to perform Spice simulations, but for data converters it is worth using a higher level of complexity. The choice of this book is to use Matlab and Simulink (© The MathWorks, Inc.) as basis for the behavioral simulations. The files of all the examples are available on the Web and can be downloaded. For information, please, visit http://ims.unipv.it/.

I am grateful to many students for the work done with behavioral simulations and for critically reading the manuscript. Among them, Edoardo Bonizzoni, Massimiliano Belloni, Cristina Della Fiore, Ivano Galdi. I am particularly grateful to Niall Duncan for his superb work in reviewing the technical content and improving the writing style of the entire book.

The material of this book has been greatly influenced by my association with Piero Malcovati, and I acknowledge his contributions.

Pavia, Italy, 2007

Franco Maloberti

Chapter 1

BACKGROUND ELEMENTS

In this chapter we shall deal with the background knowledge necessary to properly understand and design data converters. Data converters are used in electronic circuits at the interface between the analog and the digital world. Conceptually a data converter performs a transformation of signals: from continuous-time and continuous-amplitude to discrete-time and quantized-amplitude (and vice-versa). The transformation is inherently a non-linear operation. We shall see that data conversion affects the spectrum of the signal and can sometimes modify its information content. It is therefore important to be familiar with the theoretical implications of data conversion and to be aware of the limits of the approximations used for studying a data converter. In addition, proper knowledge of the mathematical tools used for analysis and characterization of sampled-data systems is necessary.

1.1   THE IDEAL DATA CONVERTER

The basic operation of an analog-to-digital (A/D) or a digital-to-analog converter (D/A) can be split into a sequence of simple elementary steps. Figure 1.1 (a) represents an A/D converter as the cascade of four functions: continuous-time anti-aliasing filtering, sampling, quantization and data coding. In the next sections we shall study why a continuous-time filter is necessary and what effects sampling and quantization have on signals. We shall then discuss the different coding schemes used to representing the signal in the digital domain.

1

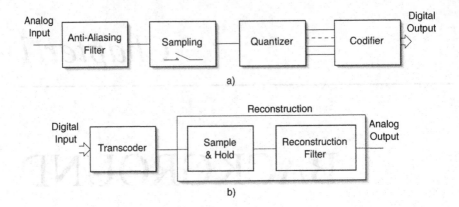

Figure 1.1.   Block diagram of the basic functions of an A/D (a) and a D/A (b) converter.

The D/A converter performs two basic functions: a transcoding stage, which converts the digital input into an equivalent analog signal, and a reconstruction stage. We shall see that the reconstruction is used to remove the high-frequency components of a sampled-data analog signal. As shown in Figure 1.1 (b) reconstruction is done in 2 steps: a sample-and-hold followed by a low-pass reconstruction filter. We shall study the relevant properties of these D/A functions in the later sections of this Chapter.

1.2   SAMPLING

A sampler transforms a continuous-time signal into its sampled-data equivalent. Ideally, a sampler yields a sequence of delta functions whose amplitude equals the signal at the sampling times. For uniform sampling with period T the output of a sampler is given by

$$x^*(t) = x^*(nT) = \sum x(t)\delta(t - nT). \tag{1.1}$$

Figure 1.2 shows the waveform of a continuous-time signal and the resulting sampled-data signal. The sampled data version is made, as stated by (1.1), by the superposition of weighted deltas. However, a practical circuit does not generate deltas but pulses with finite duration and amplitudes equal to the input at the sampling instances. Regardless of the pulse shape and duration, the pulses are intended to represent the input only at the exact sampling times, nT.

Equation (1.1) outlines the inherent non-linearity of the sampling operation: the input is multiplied by a sequence of delta functions (note that the multiplication is a non-linear operation). Therefore, as shown in Figure 1.3, sampling a signal is equivalent to the mixing of the signal with a train of deltas. Moreover,

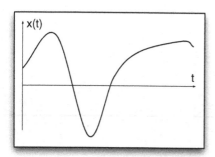 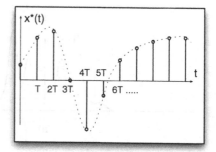

Figure 1.2. Continuous time signal (left) and its sampled data representation (right).

representing the sampler using a mixer is useful because it helps in the under-standing of a data converter when it is used in undersampling mode.

The Laplace transform of an infinite sequence of deltas is given by

$$\mathcal{L}\left[\sum_{-\infty}^{\infty} \delta(t - nT)\right] = \sum_{-\infty}^{\infty} e^{-nsT}. \tag{1.2}$$

The use of (1.1), (1.2) and the Laplace transform definition results in

$$\mathcal{L}\left[x^*(nT)\right] = \sum_{-\infty}^{\infty} (X(s - jn\omega_s) = \sum_{-\infty}^{\infty} x(nT)e^{-nsT}, \tag{1.3}$$

which provides two useful expressions for the Laplace transform of the sampled output. The right-hand equation will be used to discuss the relationships be-tween the s-plane and the z-plane. The other equation shows that the spectrum of $x^*(nT)$ is the superposition of infinite replicas of the input spectrum. These replicas are centered at multiples of the sampling frequency being shifted along

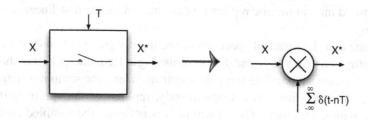

Figure 1.3. Ideal sampler and its non-linear equivalent processing.

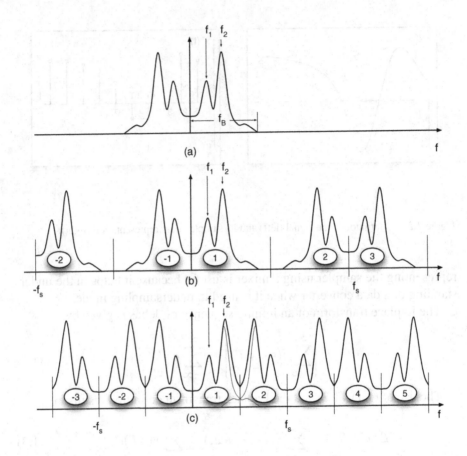

Figure 1.4. a) Bilateral spectrum of a continuous-time signal. b) Sampled spectrum by using $f_s/2 > f_B$. c) Sampled spectrum with $f_s/2 < f_B$.

the f axis by $nf_s(= n/T), n = 0, \pm 1, \pm 2, \cdots$. As a result, the spectrum is periodic with period f_s. Note that the transformation of the input spectrum from band-limited into an infinite replica reveals once more the non-linear nature of sampling.

Assume that the bilateral spectrum of the input signal is the one of Fig. 1.4 (a) showing two peaks at f_1 and f_2 and vanishing at frequencies higher than f_B. Fig. 1.4 (b) shows a possible sampled spectrum. Here, the sampling frequency is bigger than two times f_B. Consequently, replicas of the spectrum do not interfere with each other. This situation is beneficial: the sampled spectrum within the original signal bandwidth exactly equals Fig. 1.4 (a), thus making it feasible to return back to the continuous-time signal by filtering.

Fig. 1.4 (c) shows what happens when the sampling frequency is less than than twice the input bandwidth. The replicas partially overlap and modify the spectrum. Note that the spectrum of Fig. 1.4 (c) does not go to zero at $f_s/2$. Furthermore, the peak at f_2 has been shifted and its amplitude increased. Therefore, the spectrum alteration makes it impossible to preserve the continuous-time features.

The above discussion reminds us what is stated by the sampling theory:

A band limited signal, $x(t)$, whose Fourier spectrum, $X(j\omega)$, vanishes for angular frequencies $|\omega| > \omega_s/2$ is fully described by a uniform sampling $x(nT)$, where $T = 2\pi/\omega_s$. The band limited signal $x(t)$ is reconstructed by

$$x(t) = \sum_{-\infty}^{\infty} x(nT) \frac{sin(\omega_s(t - nT)/2)}{\omega_s(t - nT)/2}, \tag{1.4}$$

which is the convolution of the input with $sin(\omega_s t)/(\omega_s t)$, the inverse Laplace transform of a pulse of duration T.

Half the sampling frequency, $f_s/2 = 1/2T$, is often named the *Nyquist* frequency. The frequency interval $0 \cdots f_s/2$ is referred to as the Nyquist band (or band-base) while frequency intervals, $f_s/2 \cdots f_s, f_s \cdots 3f_s/2, \cdots$ are named the second and third Nyquist zones, and so forth. Since the spectrum in all Nyquist zones is the same it is sufficient to focus only on the band-base, $0 \cdots f_s/2$. When bilateral spectra are considered the frequency range of interest at becomes $-f_s/2, \cdots f_s/2$.

It was previously mentioned that if the sampling frequency is at least twice the bandwidth of the input then the replicas do not overlap. However, this condition must be verified not only for the signal, but also for noise and interferences. Noise has an unpredictable spectrum and can have components at any frequency. The same is true for interferences. Therefore, it is necessary to remove out of band interferences whose folding would corrupt the signal band. A filter placed in front of the sampler achieves this result. The frequency response of the filter must pass the signal band and reject the out-of-band interferences. This kind of filter, whose features are discussed shortly, is named *anti-aliasing* filter.

Remember!

The anti-aliasing filter protects the information content of the signal. Use an anti-aliasing filter in front of every quantizer to reject unwanted interferences outside of the band of the interest (either low pass or band-pass)!

Example 1.1

Verify the effects of sampling using a computer simulation. Perform the study in the sampled-data domain but use a very high sampling frequency. Operating in this way the Nyquist interval becomes very large with respect to the signal band and a sampled-data system works almost like a continuous-time system.

Solution

In order to investigate the effects of sampling we need a suitable input signal. It must be band-limited with spectrum that vanishes at a given frequency. A simple way to generate signals is to superpose sinewaves. However, a spectrum made by discrete lines is not good for this study. It is better to use signals whose spectrum spreads over a given range. For this reason this example uses the truncated $sin(\omega t)/(\omega t)$ function shown in Fig. 1.5. It lasts from 0 to 1 sec and vanishes for $t > 1$. Notice that the function is zero at both boundaries of the time-window. ωt ranges from -4π to 4π.

The simulation samples 2048 points in 1 sec resulting in a Nyquist frequency of 1024 Hz. Fig. 1.6 shows the spectrum of the truncated sine wave. Although the spectrum is irregular, it has a rectangle-like

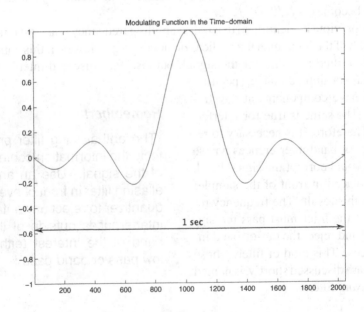

Figure 1.5.   Truncated sin(x)/x function (2048 points).

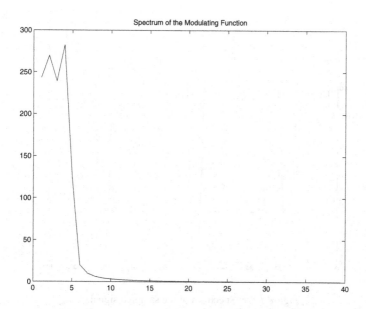

Figure 1.6. Spectrum of the truncated sin(x)/x function.

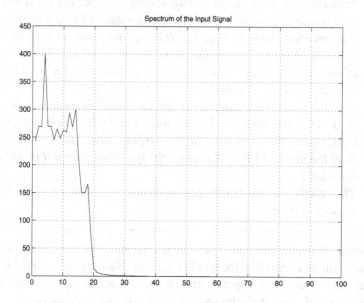

Figure 1.7. Spectrum of the input signal.

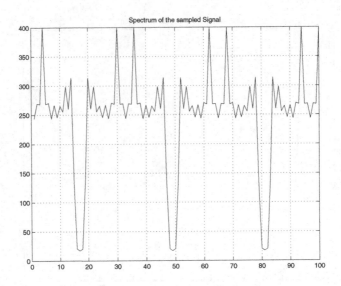

Figure 1.8.   Spectrum of the sampled signal.

shape which fades out at the 15 Hz (bin#16). This is the basis for building up a suitable input. What the Matlab code of Ex1_1 does is to modulate the truncated sine wave with pure sine waves. A weighted superposing of the four truncated and modulated sine waves leads to the spectrum shown in Fig. 1.7. It has a large tone at 3 Hz, two close tones at 11 Hz and 13 Hz and a small tone at 17 Hz. The bandwidth of the signal is 30 Hz, so according to the sampling theorem it should be sampled with $f_s > 60$ Hz
Fig. 1.8 shows what happens when the sampling frequency is as low as 32 Hz. The tone at 17 Hz is folded at 15 Hz but its effect is not easily seen. However, the tone at 13 Hz increases and becomes bigger than the one at 11 Hz. Moreover, the spectrum does not go to zero at Nyquist.
The reader can verify how the spectrum looks like when the Nyquist condition is satisfied.

We already discussed how using a continuous-time filter before the sampler avoids the folding of spurs into the band of interest. Namely, in order to preserve the signal the anti-aliasing response must be flat from 0 to f_B. Beyond this, it must reject critical spurs with the required anti-aliasing attenuation.

Fig. 1.9 (a) shows how the second Nyquist zone $f_s/2 \cdots f_s$ folds back into the first Nyquist zone. Since the bandwidth of the input signal is f_B, then spurs

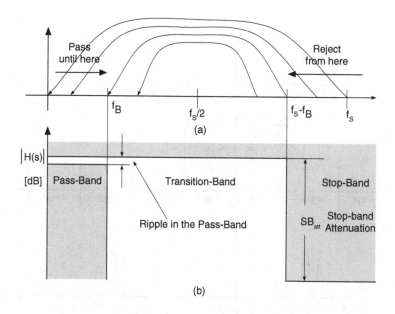

Figure 1.9. Aliasing effect (a), and anti-aliasing filter mask (b).

in the interval $f_s/2 \cdots (f_s - f_B)$ are not critical as their images are outside the $(0 \cdots f_B)$ signal range. In contrast, spurs inside the $(f_s - f_B) \cdots f_s$ range produce images inside the signal band. Therefore, the stop-band attenuation must be effective from $(f_s - f_B)$. Fig. 1.9 (b) shows a typical mask of the anti-aliasing filter. The transition-band is $f_B \cdots (f_s - f_B)$, within which the filter response must roll down such that the attenuation at the stop-band limit becomes the required A_{SB}.

The width of the transition-band and the required attenuation in the stop-band determine the order of the anti-aliasing filter. It is known that a pole yields a 20 *dB* per decade roll-off. Since the transition region is $log_{10}[(f_s - 2f_B)/f_B]$ decades, the order of the filter must be $A_{SB}|dB/\{20log_{10}[(f_s - 2f_B)/f_B]\}$ assuming it has all poles. If the transition region is 0.3 *decades* (one *octave*) an attenuation of 48 *dB* requires an 8-*th* order Butterworth filter, assuming that a flat response is necessary in the pass-band. However, if the transition band is one decade then an attenuation of 60 *dB* only requires a 3-*th* order filter. Therefore, large transition regions greatly simplify the anti-aliasing filter design. However, higher sampling frequencies that widen the transition region require fast circuits for sampling and for the necessary analog and digital signal processing.

The design of the anti-aliasing filter is critical as it is the first block of the conversion chain and limited performances at this stage affect the overall accuracy.

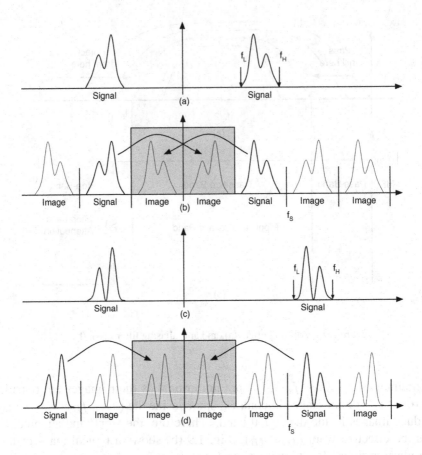

Figure 1.10.   Undersampling of a signal in the second Nyquist and the third Nyquist band.

It is therefore necessary to use anti-aliasing filters with low harmonic distortions, large output swing, and low-noise.

1.2.1   Undersampling

Undersampling exploits the folding into the band-base of signals whose frequency components are higher than the Nyquist limit, $f_s/2$. The technique is normally used to bring high-frequency spectra into the band-base. The method is also known as harmonic sampling, band-base sampling, *IF* sampling, and *IF*-to-digital conversion.

Fig. 1.10 illustrates the method with two examples. The spectrum of the signal in Fig. 1.10 (a) extends from f_L to f_H. If the higher limit, f_H, is bigger than twice the signal band, $(f_H - f_L)$ then it is possible to use a sampling frequency f_s such than $f_s > f_H$ and $f_s/2 < f_L$. This bounds the spectrum into the second Nyquist zone. Moreover, the aliasing brings an entire replica of the input spectrum into the base-band. Fig. 1.10 (b) shows that the replica in the band-base comes from the negative input spectrum shifted by f_s and the positive input spectrum shifted by $-f_s$ generates the negative frequencies image. Thus, the band-base becomes a mirrored version of the original spectrum.

Fig. 1.10 (c) shows another possible situation. The band of the input signal and the sampling frequency are such that the third Nyquist zone entirely bounds the input spectrum. The sampling moves a replica of the input spectrum into the band-base by a $-f_s$ shift, and the image in negative band-base, $-f_s/2 \cdots 0$, comes from an f_s shift of the negative input spectrum. Fig. 1.10 (d) shows the result after sampling. Since the in-band image comes from the third Nyquist zone the obtained input replica is not mirrored.

Both cases considered in Fig. 1.10 have an input that is band-limited and bounded inside one of the Nyquist zones. The various spectrum replicas do not overlap one another. Therefore, the information content of the input signal remains unchanged, albeit with the spectrum mirrored. This observation reveals an extended consequence of the sampling theorem: even if the signal band is in a high Nyquist zone, then the sampling theorem is still properly verified if the input bandwidth is less than half of the sampling frequency.

As is required for any sampled-data system, under-sampling converters must be preceded by an anti-aliasing filter. Undersampling uses aliasing to bring a spectrum from a high Nyquist zone into the base-band. However, interferences coming from other Nyquist zones can also be aliased to the base-band or, the base-band itself can contain unwanted spurs. The anti-aliasing filter must remove spurs from all the Nyquist zones except the one containing the signal band. For Fig. 1.10 (a) the anti-aliasing-filter must be band-pass from $f_s/2$ to f_s. For the case of Fig. 1.10 (c) the filter must reject spurs outside the interval $f_s \cdots 3f_s/2$. The complexity of the band-pass filter depends on the margins granted by the system design at both stop-band endings.

The inherent demodulation of the input makes undersampling methods attractive for communication systems. An intermediate frequency mixer is no longer necessary as the sampling itself performs the frequency translation. Thus, undersampling data converters produce a digital signal suitable for direct band-base digital processing. This is

> **Remember That**
>
> Under-sampling requires an anti-aliasing filter! This removes unwanted spurs, which can occur in the base-band or be aliased back from any other Nyquist zones. The anti-aliasing filter for under-sampled systems is band-pass around the signal band.

advantageous, however designing the analog sections of the data converter is difficult: bandwidth, distortion and jitter performances must meet the specifications not just at the sampling frequency but also at the signal frequency.

1.2.2   Sampling-time Jitter

Until now we have assumed that a sampled-data signal reproduces the input at the exact sampling-times. That is true for ideal sampling, but in a real situation sampling is affected by uncertainty in the clock. Also, the delay between the logic that generates the sampling phase and the effective sampling is in some extent unpredictable. The combination of the two effects determines *jitter* in the actual sampling instants. Fig. 1.11 shows an example of sampling with jitter. A positive sampling error, $\delta(0)$, influences the sampling at time $t = 0$. The signal is rising and the error, $\Delta X(0)$, is positive. At the next sample the sampling uncertainty $\delta(T)$ is again positive but the negative derivative of the input signal leads to a negative $\Delta X(T)$ error. The flat behaviour at time $2T$ leads to a negligible inaccuracy even if $\delta(2T)$ is significant. Finally, a negative $\delta(3T)$ and the rising input leads to a negative error at $3T$. Accordingly, the sampling jitter affects the value of the sampled signal by an error that depends on both the jitter, and the time derivative of the input.

For a sine wave $X_{in}(t) = A \cdot sin(\omega_{in}t)$, the error $\Delta X(nT)$ is given by

$$\Delta X(nT) = A \cdot \omega_{in} \cdot \delta(nT) \cos(\omega_{in}nT) \qquad (1.5)$$

which is a sampled-data quantity $\delta(nT)$ amplified by A and modulated by a cosine at the input frequency.

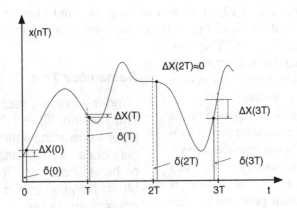

Figure 1.11.   Errors caused by sampling time jitter.

Assume that $\delta(nT)$ is the sampling of a random variable $\delta_{ji}(t)$; the error $\Delta X(nT)$ is the sampling of $x_{ji}(t) = \delta_{ji}(t)\omega_{in}X_{in}(t)$ (the 90° phase shift given by the cosine is disregarded). Furthermore, if the spectrum of $\delta_{ji}(t)$ is white then the spectrum of $x_{ji}(t)$ is also white as cosine modulation has no effect on white spectra. Therefore, this simple study leads to a model of the random jitter made by a white noise source $x_{ji}(t)$ added to the input before an ideal sampling.

> **Be aware!**
>
> *72 dB SNR* requires at least 2 psec jitter for sampling a 20 MHz sinewave. If the frequency augments by *10* the clock jitter must diminish by the same factor *10*. If the system requires *6 dB* more in the *SNR* the clock jitter must improve by a factor *2*.

The above is the first identified limits to the A/D converter accuracy. We shall see shortly that other noise sources (either fundamental or caused by non-idealities) further affect the data converters performance.

The power of the jitter error $x_{ji}(t)$ is

$$< x_{ji}(t)^2 > = < [A\omega_{in}\cos(\omega_{in}nT)]^2 > < \delta_{ij}(t)^2 > \qquad (1.6)$$

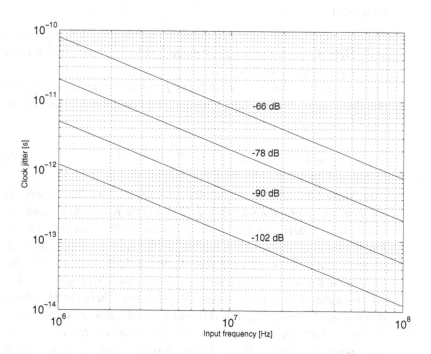

Figure 1.12. Clock jitter at different SNR and input frequencies.

giving rise to

$$< x_{ji}(t)^2 > = \frac{A^2 \omega_{in}{}^2}{2} < \delta_{ij}(t)^2 > . \qquad (1.7)$$

The power of the input sine wave is $A^2/2$. The consequent signal-to-noise ration (SNR) is

$$SNR_{ji,DB} = -20 \cdot \log\{< \delta_{ij}(t) > \omega_{in}\}. \qquad (1.8)$$

Fig. 1.12 plots the clock jitter versus the input frequency necessary to obtain a given SNR. Observe that for large SNR and high frequencies the clock jitter must be in the fraction of $psec$ range. For instance, achieving an SNR of $90\,dB$ with a $100\,MHz$ input sine wave requires a $50\,fsec$ clock jitter!

Example 1.2

A sampled-data system with 100 MHz clock frequency must obtain SNR = 80 dB. When $f_{in} = 20$ MHz 20% of the noise budget is assigned to the noise jitter. The input signal has a full-scale amplitude of 1 V. Estimate the degradation of the SNR when the input frequency extends into the second Nyquist zone.

Solution

The total noise power is $1/(2 \cdot 10^{SNR/10}) = 0.5 \cdot 10^{-8}\ V^2$. Therefore, at 20 MHz the power of the jitter noise is $0.1 \cdot 10^{-8}\ V^2$, which gives a clock jitter of $5 \cdot 10^{-13}\,s$.

The total power is made by a term independent on the frequency plus the jitter contribution. Accordingly, the noise as a function of the input frequency is

$$v_n^2 = 0.4 \cdot 10^{-8} + 0.1 \cdot 10^{-8} \left[\frac{f}{20 \cdot 10^6} \right]^2 . \qquad (1.9)$$

The Matlab file $Ex1\_2.m$ provides the required code to study the problem and to plot the SNR. Fig. 1.13 shows the result. At frequencies lower than 20 MHz the SNR is higher than 80 dB as there is a lower jitter noise contribution (it is almost zero at very low frequency). At 20 MHz the SNR is the expected 80 dB. At higher frequencies the jitter noise becomes dominant and at Nyquist the SNR loses 3 dB. The effect is more evident in the second Nyquist zone where the SNR drops down from 77 dB to 72.5 dB. Assuming that 1 dB loss is acceptable the usable input band is only 0.6 times the Nyquist interval

The above observation also indicates that this sampler is not suitable for under-sampling: the jitter noise is acceptable in the first Nyquist zone but becomes too high in other Nyquist zones.

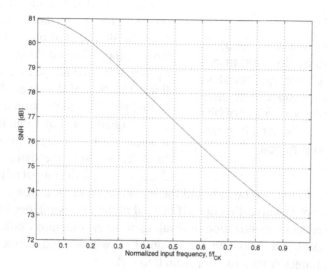

Figure 1.13. SNR degradation caused by the jitter noise.

1.3 AMPLITUDE QUANTIZATION

Amplitude quantization changes a sampled-data signal from continuous-level to discrete-level. The dynamic range of the quantizer is divided into a number of equal *quantization intervals*, each of which is represented by a given analog amplitude. The quantizer modifies the input amplitude into a value that represents which quantization interval it resides in. Often the value representing a quantization interval is the mid-point of the interval. In some cases either the upper or the lower bound represent the interval.

Assuming that $X_{FS} = X_{max} - X_{min}$ is the range of the quantizer and M is the number of quantization intervals, the amplitude of each quantization interval or *quantization step*, Δ, is

$$\Delta = \frac{X_{FS}}{M}. \tag{1.10}$$

Since the mid point of the *n-th* interval $X_{m,n} = (n + 1/2)\Delta$ represents all the interval amplitudes, quantizing an input level other than $X_{m,n}$ leads to an error. This called the *quantization error*, ϵ_Q. The output Y of a quantizer with input X_{in} is

$$Y = X_{in} + \epsilon_Q = (n + 1/2)\Delta; \quad n\Delta < X_{in} < (n+1)\Delta. \tag{1.11}$$

Fig. 1.14 (a) depicts the quantization process: the quantization error ϵ_Q is added to the input to obtain the quantized output. The addition is a linear operation but the added term is a non-linear function of the input. Fig. 1.14 b) that plots the quatization error for a $3 - bit$ quantizer shows that the quantization error ϵ_Q ranges from $-\Delta/2$ to $\Delta/2$. Outside the dynamic range $X_{min} \cdots X_{max}$ the output of the quantizer saturates to the two bounds and the quantization error increases linearly in the positive or negative direction. If instead of the mid-point, one of the two edges represents the quantization interval, then the ϵ_Q diagram would shift up or down by $\Delta/2$. The maximum variation of ϵ_Q within the dynamic range is still Δ but it ranges from 0 to Δ or from 0 to $-\Delta$.

Instead of identifying a quantization interval by its mid-point we can also represent the interval by a digital code. In such a case the output is not a discrete analog level but a digital word. Frequently the number of quantization intervals is a power of 2, 2^N: where N is the number of bits required to number the quantization steps. Thus, for example, for the case of Fig. 1.14 (b) we could use the analog value of the mid-point, a decimal symbol from 0 to 7 or a binary 3-*bit* code. All the representations are obviously equivalent.

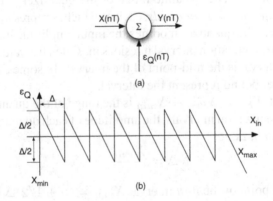

Figure 1.14.   (a) Quantization error added to the input signal obtains the quantized output. (b) Quantization error for a 3-bit data converter.

1.3.1 Quantization Noise

The previous sub-section showed that quantization corresponds to the addition of a quantization error to the input. A large quantization error leads to a reduced capability to preserve the signal features. This is what happens when electronic noise corrupts an analog system. As is known, the effect of noise is quantified by the signal-to-noise ratio (*SNR*) defined by

$$SNR|_{dB} = 10 \cdot \log \frac{P_{sign}}{P_{noise}}, \qquad (1.12)$$

where P_{sign} and P_{noise} are the power of the signal and the power of the noise in the band of interest.

When studying the effect of quantization it can be convenient using the *SNR* concept. This is only viable if the quantization error can be regarded as noise. However, this is not always possible as some kinds of inputs lead to a quantization error other than noise. For example, a *dc* input leads to a constant quantization error. Also, if the input remains confined inside the same quantization interval the quantization error is just a shifted version of the input thus leading to an error spectrum that differs from the signal spectrum by only a *dc* term. On the contrary, signals that cause recurrent crossing of the quantization thresholds work well. Frequent code transitions decorrelate successive samples of the quantization error, thus spreading the spectrum and making it like a noise. Therefore, signals with large busy amplitudes are promising candidates for the desired noise-like representation.

The above qualitative discussion is formally specified by the following set of necessary conditions:

- all the quantization levels are exercised with equal probability;
- a large number of quantization levels are used;
- the quantization steps are uniform;
- the quantization error is not correlated with the input.

A large input signal fulfils the first requirement. The second condition holds if the quantizer uses a large number of bits, that is true in many cases. However, an important category of data converters (the sigma-delta) employs very few quantization levels, often just two levels: 1-*bit*. Consequently, for sigma-delta architectures the second condition for describing the quantization as additive noise is not well founded. Nevertheless, the benefits resulting from the noise assumption are so relevant that, with a proper watchfulness, designers use the noise approximation anyway.

Most quantizers comply with the third requirement. Only a few data converter use a non-linear response (like the logarithmic response used in telephony for coding audio signals). Even the last rule is normally verified. However, if

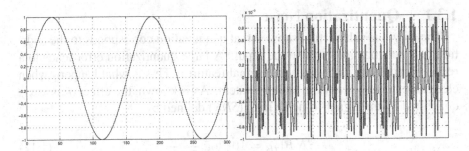

Figure 1.15. Quantized signal (left) and quantization error (right), $f_s = 150f_{in}$.

the data converter input is a sine wave, as is normally used for testing, an improper choice of frequency can be problematic: when the ratio between the sampling frequency and the input sine wave frequency is a rational number the quantization noise becomes correlated with the input.

Fig. 1.15 shows a 10-*bit* quantized signal and its quantization noise. The ratio f_s/f_{in} is 150. Observe that the signal swing is in the ± 1 range while the quantization error is $\pm 1/1024$. Also, the error shows a periodic pattern. The resulting correlation between error and input supports the recommendation of a proper selection of the sine wave frequency for doing the test. This point will be discussed in more detail shortly.

1.3.2   Properties of the Quantization Noise

Two key properties describe the features of any noise generator: the *time average power* and the *noise power spectrum*. The noise power spectrum depicts how the time average power spreads over the frequency. For quantization noise, being a sampled-data signal, the time average power is contained and the spectrum is meaningful only in the Nyquist interval.

Estimating the time average power of the quantization error assumes a constant probability distribution function $p(\epsilon_Q)$ in the range $-\Delta/2 \cdots + \Delta/2$; and that outside this range $p(\epsilon_Q)$ is zero. The first condition given in the previous sub-section justifies this assumption. Moreover, $|\epsilon_Q|$ is always less than $\Delta/2$. Since the integral of the probability distribution function over the infinite range $-\infty \cdots \infty$ is equal to one, it results

$$p(\epsilon_Q) = \frac{1}{\Delta} \ for \ \epsilon_Q \in -\Delta/2 \cdots \Delta/2$$
$$p(\epsilon_Q) = 0 \ otherwise. \tag{1.13}$$

The time average power of ϵ_Q is given by

$$P_Q = \int_{-\infty}^{\infty} \epsilon_Q^2 \cdot p(\epsilon_Q) d\epsilon_Q = \int_{-\Delta/2}^{\Delta/2} \frac{\epsilon_Q^2}{\Delta} d\epsilon_Q = \frac{\Delta^2}{12}. \qquad (1.14)$$

As expected, the time average power of the quantization noise decreases as the number of bits increases. Furthermore, it is proportional to the square of the quantization step. The use of (1.14) and the power of the signal permits us to calculate the SNR. We consider sine wave or triangular inputs. Since the maximum amplitude of a sine wave or a triangular wave within an X_{FS} dynamic range is $X_{FS}/2$, the power of a sine wave with maximum amplitude is

$$P_{sin} = \frac{1}{T} \int_0^T \frac{X_{FS}^2}{4} \sin^2(2\pi ft) dt = \frac{X_{FS}^2}{8} = \frac{(\Delta \cdot 2^n)^2}{8} \qquad (1.15)$$

and the power of a triangular wave with maximum amplitude is

$$P_{sin} = \frac{X_{FS}^2}{12} = \frac{(\Delta \cdot 2^n)^2}{12}. \qquad (1.16)$$

Therefore, the above equations and (1.14) lead to

$$SNR_{sine}|_{dB} = (6.02 \cdot n + 1.78) \, dB, \qquad (1.17)$$

$$SNR_{trian}|_{dB} = (6.02 \cdot n) \, dB. \qquad (1.18)$$

Equations (1.17) and (1.18) establish useful relationships between the maximum achievable SNR and the number of bits of a quantizer. Results show that every bit of resolution improves the SNR by $6.02 \, dB$. Also, the power of the quantization noise diminishes by a factor 4 for every additional bit.

The above SNR calculation only accounts for the quantization noise. In real circuits the electronic noise of passive and active components brings about additional noise. Moreover, the dynamic behavior of circuits deteriorates at high frequency. Speed limitations cause errors that, under some conditions, can be viewed as further noise. By contrast, as we will see shortly, many applications use a signal band that is much less than the Nyquist interval. Therefore, since the band of interest includes only a fraction of the total quantization noise power, the SNR can be made higher.

The above observations make the noise a more comprehensive limit affecting a real data converter. This is the reason why equation (1.17) and (1.18)

Keep Note

Every additional bit of resolution improves the SNR by 6.02 dB. Accordingly, the power of the quantization error decreases by a factor 4.

are often used more extensively for defining the *equivalent number of bits* (*ENB*). For sine wave or triangular inputs we have

$$ENB_{sin} = \frac{SNR_{tot}|_{dB} - 1.78}{6.02}, \tag{1.19}$$

$$ENB_{trinag} = \frac{SNR_{tot}|_{dB}}{6.02} \tag{1.20}$$

where SNR_{tot} is the signal-to-noise ratio accounting for the total noise that affects the signal band of the conversion system.

We already observed that the uncertainty in the sampling time can be described by white noise, thus increasing the total noise power. Therefore, jitter and other possible noise sources reduce the effective number of bits. Consider the effect of an *N-bit* quantizater and a δ_{ji} jitter in the sampling of an input sine wave at f_{in}. If the amplitude is $X_{FS}/2$, then $P_{ji} = (X_{FS} \cdot \pi f_{in}\delta_{ji})^2$. The equivalent number of bits becomes

$$ENB = \frac{10 \cdot log(\pi^2 f_{in}^2 \delta_{ji}^2 + 2^{-2N}/12) - 1.78}{6.02}. \tag{1.21}$$

Fig. 1.16 plots equation (1.21) for *sub-psec* jitter and $N = 12$ and $N = 14$. The signal frequencies go from 40 *MHz* up to 200 *MHz*. Similar to the results

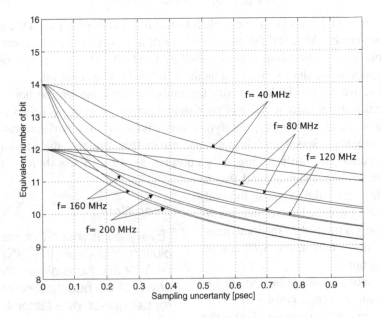

Figure 1.16.   Equivalent number of bit versus the sampling jitter (12 and 14 bit quantizer).

obtained in the Example 1.2 the *ENB* diminishes if the jitter increases. Note that the benefit of a higher resolution in the quantizer quickly vanishes. The *ENB* drops faster with $N = 14$ than with $N = 12$ until the two performances become almost equivalent for a large jitter. Thus the use of quantizers with many bits is not enough to obtain high resolutions: many bits must be accompanied by a careful control of all the noise sources.

The other key property of the quantization noise is the power spectrum. As already mentioned, it represents how the noise power spreads over the Nyquist interval. Recall that the power spectrum is the Laplace transform of the auto-correlation function. For sampled data signals

$$P_\epsilon(f) = \int_{-\infty}^{\infty} R_\epsilon(\tau) e^{-j2\pi f\tau} d\tau = \sum_{-\infty}^{\infty} R_\epsilon(nT) e^{-j2\pi fnT}. \qquad (1.22)$$

Unfortunately (1.22) does not help much as it is not easy to find an equation representing the auto correlation of the quantization error.

What we can do is make an approximate estimation of the auto correlation function. One of the conditions that justify the use of the quantization noise is that the input signal is busy. The same condition shows a limited relationship between successive samples of the quantization error. Therefore, we can assume that the auto correlation function, $R_e(nT)$, goes rapidly to zero for $|n| > 0$. Since $R_e(nT)$ is a sampled data function it is reasonable to account for $R_e(0)$ only: the auto correlation becomes a delta in the time domain and the spectrum becomes frequency independent as the Laplace transform of a delta is a constant. The power spectral density is therefore white with power $P_Q = \Delta^2/12$ spread uniformly over the unilateral Nyquist interval $0 \cdots f_s/2$. The unilateral power spectrum is

$$p_\epsilon(f) = \frac{\Delta^2}{6 \cdot f_s}; \quad \textit{meeting the condition} \quad \int_0^\infty p_\epsilon(f) df = \Delta^2/12. \quad (1.23)$$

For a bilateral spectrum the power P_Q spreads over a double frequency interval $-f_s/2 \cdots f_s/2$ and the $p_\epsilon(f)$ spectrum becomes $\Delta^2/(12f_s)$.

In summary, if the quantization fulfils given requisites, then the quantization process can be described using additive noise with a white spectrum. This approximation is useful as the addition to the input of a non-linear term shown in Fig. 1.14 becomes the addition of a linear term: noise with white spectrum. As shown in Fig. 1.17 the

> **Learn that**
>
> The spectrum of the quantization noise is white over the entire Nyquist interval. The amplitude for a unilateral representation is $\Delta^2/(6f_s)$. For bilateral representations is $p_\epsilon(f) = \Delta^2/(12f_s)$.

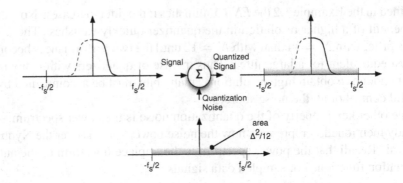

Figure 1.17. Modelling quantization by an additive white noise.

spectra of the input and the noise are superposed to give the spectrum of the quantized signal. Having a linear system enables the use of the Laplace or, for sampled-data systems, the $\mathcal{Z}$-transform.

1.4 kT/C NOISE

Quantization noise is a fundamental limit of data converters. Another unavoidable limit is the *kT/C* noise. It occurs in all real sampled-data systems as it is due to the unavoidable thermal noise associated with the sampling switch. Obviously, the *kT/C* noise only goes to zero for infinite sampling capacitance or zero temperature. This is why it is described as a fundamental limit of any real sampled-data system.

The operation of a sampler can be modeled by the simplified circuit shown in Fig. 1.18 (a): the input voltage V_{in} charges a sampling capacitor through a switch. At the sampling time the switch turns off, thus holding the value of the input voltage across the capacitor. The resistance R_s in Fig. 1.18 (b) represents the series of the on-resistance of the switch and the output resistance of the signal generator. Observe that the sampling operates correctly if the time constant $\tau_s = R_s C_s$ is negligible with respect to the sampling time (condition that enables the system to reach the time-invariant condition). Moreover, the bandwidth of the input signal must be much smaller than $1/\tau_s$.

Fig. 1.18 (b) provides the equivalent circuit for the noise estimation. The spectrum of the thermal noise contributed by R_s is white, $v_{n,R_s}^2 = 4kTR_s$. Moreover, the $R_s C_s$ network establishes a low-pass filtering that makes the noise spectrum across the capacitor colored. The spectrum of v_{n,C_s} is given by the $4kTR_s$ spectrum multiplied by the square of the transfer function of the $R_s C_s$ filter.

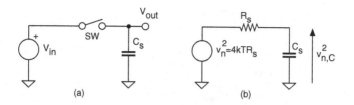

Figure 1.18. Simple model of a sampler and its noise equivalent circuit.

$$v_{n,C_s}^2(\omega) = \frac{4kTR_s}{1 + (\omega R_s C_s)^2}. \tag{1.24}$$

When the switch goes off the capacitor holds not just the input but also the noise. Since the noise spectrum across C_s is not band-limited the aliasing folds noise components from high Nyquist zones into the band-base. Since the corner frequency of the $R_s C_s$ low-pass filter is much larger than the Nyquist limit many bands of non-negligible power (both mirrored or non-mirrored) are superimposed giving rise to an almost white resulting spectrum.

The noise power in the band-base is given by the integral of the noise power of all the folded bands. Therefore, the total noise power stored on C_s when the switch goes off is

$$P_{n,C_s} = \int_0^\infty v_{n,out}(f)df = 4kTR_s \int_0^\infty \frac{df}{1 + (2\pi f R_s C_s)^2} = \frac{kT}{C_s}. \tag{1.25}$$

Observe that P_{n,C_s} does not depend on R_s. Increasing R_s raises the white noise floor but also improves the low-pass filtering action. The two effects compensate each other thus canceling the R_s dependency.

The *kT/C* noise added to other noise power components gives the total noise power used to obtain the *SNR* and, by turns, the equivalent number of bits. On the other way around, a given resolution and full scale reference determine the total noise power budget. A designer must then divide this noise budget between various sources. A fraction is assigned to quantization, another fraction to the jitter, and another accounts for the electronic and *kT/C* noise sources. The trade-off between various design constraints determines the fraction of the noise budget

Remember

The kT/C noise is a fundamental limit caused by sampling. Sampling any signal using 1 pF leads to 64.5 μV noise voltage. If the sampling capacitance increases by k the noise voltage diminishes by $\sqrt{k}$.

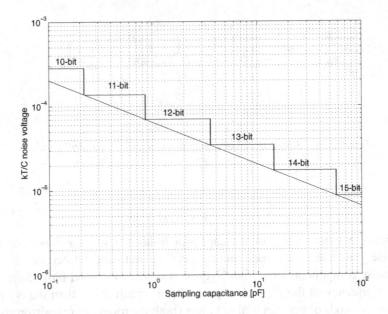

Figure 1.19. kT/C noise voltage versus the capacitance value. The staircase shows the quantization step for 1 V_{FS}.

that can be allocated for the kT/C sources, thus determining the minimum value of the sampling capacitance.

Fig. 1.19 plots the noise voltage $\sqrt{kT/C}$ for capacitors ranging from $0.1\,pF$ to $100\,pF$. The same figure shows the quantization noise voltages ($V_{Ref}/(2^N\sqrt{12})$ = 0.29Δ) for different numbers of bits and a *1 V* reference. Observe that when $\sqrt{kT/C}$ equals the quantization noise voltage then the total power noise increases by a factor of 2 and the *SNR* decreases by 3 *dB* decreasing the equivalent number of bits by $1/2$ *LSB*. For example if the kT/C noise of a $0.8\,pF$ sampling capacitance equals the quantization noise of a 12-*bit* converter with *1 V* reference, then the $0.8\,pF$ sampling element is therefore only suitable for less than 12-*bit* performances or for higher reference voltages. For 12-*bit* it is recommended to use a larger sampling capacitance like $1.6\,pF$ or better $3.2\,pF$. The total noise power would lead to 11.7-*bit* and 11.85-*bit* respectively.

Example 1.3

A pipeline data-converter uses a cascade of two sample-and-hold circuits in its first stage. The clock jitter is 1 psec. Determine the minimum sampling capacitance that enables 12 bit resolution. The full scale voltage is 1 V; the input frequency is 5 MHz.

Solution

Let us assume that an extra 50% noise is acceptable (the system would lose 1.76 dB, 0.29-bit). Since the quantization noise power is $\Delta^2/12$, the noise budget available for KT/C and jitter is

$$V_{n,budget}^2 = \frac{V_{FS}^2}{24 \cdot 2^{24}} = 2.48 \cdot 10^{-9} V^2. \tag{1.26}$$

Observe that the jitter only affects the signal in the first stage as the second stage samples an held signal generated by the first stage. Its jitter noise is

$$V_{n,ji}^2 = \{\frac{V_{FS}}{2} \cdot 2\pi f \delta_{ji}\}^2 = 2.47 \cdot 10^{-10} V^2. \tag{1.27}$$

Thus, the total noise power that the sampler can generate is $2.23 \cdot 10^{-9} V^2$. Assuming equal capacitance in both S&H circuits the noise for each is $V_{n,C} = 1.12 \cdot 10^{-9} V^2$, leading to a sampling capacitance

$$C_S = \frac{kT}{V_{n,C}^2} = \frac{4.14 \cdot 10^{-21}}{1.12 \cdot 10^{-9}} = 3.7 pF. \tag{1.28}$$

Observe that the noise jitter establishes a maximum achievable resolution that can not be exceeded even with very large sampling capacitances. This limit using the same 50% margin used above is for 1 psec jitter and 5 MHz input 15.3 bit.

1.5 DISCRETE AND FAST FOURIER TRANSFORMS

The spectrum of a sampled data signal can be estimated using equation (1.3) (given again below for the reader's convenience)

$$\mathcal{L}[x^*(nT)] = \sum_{-\infty}^{\infty} x(nT)e^{-nsT}, \tag{1.29}$$

or its Fourier counterpart

$$\mathcal{F}[x^*(nT)] = X^*(j\omega) = \sum_{-\infty}^{\infty} x(nT)e^{-j\omega nT}. \tag{1.30}$$

Unfortunately (1.30) requires an infinite number of samples that, of course, are not available in practical cases. What is available is typically a finite sequence, N, of samples starting at an initial sampling time, conventionally assumed $t=0 \cdot T$, and ending at the sampling time $(N-1)T$ with a final sample representing the signal in the interval $(N-1)T - nT$.

A convenient approximation of equation (1.30) is given by the Discrete Fourier Transform (*DFT*). The *DFT* extends *N-periodic* the input serie outside the range *0, (N-1)T*; that means to assume *x[(i+kN)T] = x(iT)* for *0 < i < (N-1)* and all *k*. Therefore, the series becomes periodic with period *NT*. The *N-periodic* extension leads to a spectrum made by *N* lines located at $f_k = k/[T(N-1)]$ for *0 < k < (N-1)*. The *DFT* results as

$$X(f_k) = \sum_{n=0}^{N-1} x(nT)e^{-j2\pi kn/(N-1)}. \tag{1.31}$$

Since the *DFT* is a complex function, we obtain the magnitude and the phase of the *DFT* using real and imaginary parts

$$|X(f_k)| = \sqrt{Real\,[X(f_k)]^2 + Im\,[X(f_k)]^2}; \tag{1.32}$$

$$Ph\{X(f_k)\} = \arctan\left[\frac{Im\{X(f_k)\}}{Real\{X(f_k)\}}\right]. \tag{1.33}$$

Since equation (1.31) requires N^2 computations, for long series the calculation time can become big. In order to avoid the limit the *DFT* can be calculated by another equation as established by the *Fast Fourier Transform* algorithm (*FFT*), proposed by Cooley-Tukey. The *FFT* algorithm reduces the number of computations from N^2 down to $N \cdot log_2(N)$. Therefore, for instance, with a series of 1024 points the computation time diminishes by a factor 10. However, the reduction is effective when the series is made by a power of 2 elements. Various software codes with different execution speeds are available for the *FFT* calculation. One of them is the *FFTW*, optimized for a fast and accurate implementation of the basic *FFT* algorithm.

1.5.1   Windowing

The *DFT* and the *FFT* assume that the input is *N-periodic*. In other words, it is assumed that the input samples are replicated over and over continuously. In reality, this feature is properly verified if the input is a repeating waveform and its frequency components are an integer multiple of the sample rate divided by the number of points of the sequence *(N)*.

Real signals are never periodic and the *N-periodic* assumption lead to discontinuity between the last and first samples of successive sequences. Therefore, even using pure sine waves can give discontinuity depending on the value of the sampling frequency, the signal frequency and the number of points of in the series.

Fig. 1.20 is a sequence of 256 sine-wave samples repeated four times. Each sequence contains 3.25 periods of the sine wave. It is evident that the *N-periodic* transformation leads to discontinuity at points 256, 512, 768, and so

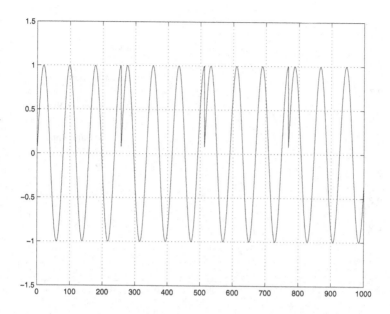

Figure 1.20. A signal that misses N-periodicity leads to discontinuities.

forth. Therefore, instead of giving a sharp tone at $3.25/256$ the spectrum will spread out indicating energy at frequencies other than the one of the signal.

Windowing alleviates this limit by tapering the two endings of the series, thus reducing the discontinuity. Windowing is expressed by

$$x_w(kT) = x(kT) \cdot W(k), \qquad (1.34)$$

where $W(k)$ is the windowing series made by N terms. The maximum value of $W(k)$ is typically 1. Observe that the signal shaping caused by windowing reduces the total power. This is not a serious problem as designers are principally interested in ratios between frequency components, and not in the absolute power associated with a given frequency component.

The simplest windowing function is a triangular shaping. Other windowing functions are also used. Fig. 1.21 compares four possible alternatives:

Useful Tips

When using FFT or DTF make sure that the sequence of samples is N-periodic. Otherwise use windowing. With sine wave inputs avoid repetitive patterns in the sequence: the ratio between the sine wave period and the sampling period should be a prime number.

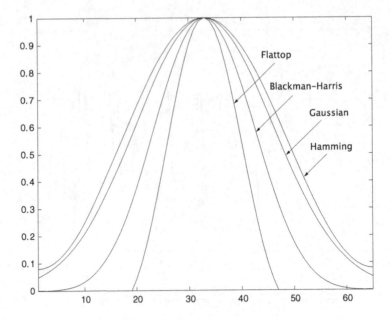

Figure 1.21. Windowing functions.

Blackman-Harris, Gaussian and *Hamming* and *flat-top*. The *Blackman-Harris* brings the signal to zero at the endings, thus making sure that $x_w(N + 1) = x_w(1)$. The others just attenuate the signal at the endings thus limiting the possible discontinuity. The various windowing functions differently shape the output spectrum. The *Hamming* window gives narrow spectral peaks. The *Blackman-Harris* gives wider peaks but with low spreading. The *Gaussian* is somewhere in between. Another possibility is the *flat-top*. It gives wide peaks, but the flatness of the peaks' top enables an accurate levels measure.

Even if windowing helps in improving the spectrum of the *DFT* or the *FFT* it is essentially an amplitude modulation of the input used to correct the limit of the *N-periodic* transformation and, for this, also source of undesired effects. For example, the features produced by a spike at the beginning or at the end of the sequence are completely masked and are windowed away almost completely.

One of the disadvantages of windowing is that it gives rise to spectral leakage. Although windowing gives accurate *SNR* measurements, if the input is a sine wave then the resulting spectrum may not be a pure tone represented by a spike in the spectrum as is expected. To avoid leakage it is convenient to use coherent sampling for which an integer number of clock cycles, k, fits into the sampling window. In addition, to avoid repetitive patterns in the output stream then k must be a prime number. A sequence of 2^N samples requires the use of sine wave frequencies f_{in} such that

$$f_{in} = \frac{k}{2^N} f_s, \qquad (1.35)$$

where k is a prime number, f_s is the sampling frequency and the input series is made by 2^N points.

Example 1.4

Study the effects of the Blackman window by computer simulations. Verify windowing in both the time domain and the frequency domain. Use two input sine waves with a fractional number of periods on a 1024 point series. The superposition of the sine waves must cause discontinuity.

Solution

The Matlab file Ex.1_4 is the basis for the behavioral simulations. The sine wave frequencies are 1/17 and 1/19. Accordingly, a series of 1024 samples contains 60.23 cycles of the first sine wave and 53.89 cycles of the second sine wave. Using amplitudes of 0.6 and 0.4 respectively for the two sinusoidal components leads to a discontinuity. It is easy to verify that the amplitudes of the first and last (# 1024) samples are 0.3393 and 0.0298. The value that should follow is 0.1899, therefore, the N-periodic transformation causes a 0.1899 $\cdots$ 0.3393 discontinuity. Fig. 1.22 shows the input signal before (a) and after (b) the Blackman window shaping. Observe that the Blackman window

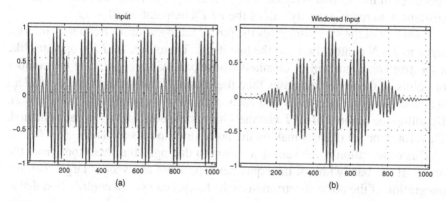

Figure 1.22. Time domain input of the example before (a) and after windowing (b).

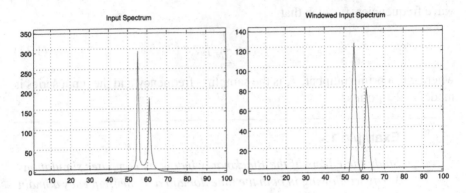

Figure 1.23. Spectra with and without windowing.

---

significantly attenuates the first and last part of the input series. Fur-thermore, the series goes to zero at the two endings. Fig. 1.23 shows the spectra of the two signals of Fig. 1.22. Observe that the discon-tinuity causes the spectrum to spread out and as a result it does not go to zero in the region between the two sine wave frequencies. The use of windowing better defines the two lines. Note however, in the windowed spectrum the absolute amplitude of the two sine waves is reduced by more than a factor 2 but the amplitude ratio is preserved.

The *FFT* of an N sample sequence leads to a spectrum made by N discrete lines. They are equally spaced in the frequency interval $0 - f_s$. Since the spectrum in the second Nyquist zone $(f_s/2 - f_s)$ mirrors the base-band, it is customary to represent only half of the *FFT* computation $(0 - f_s/2)$.

Each line of the spectrum gives the power falling within a frequency interval equal to f_s/N centered around the line itself. Therefore, the *FFT* operates like a spectrum analyzer with N channels whose bandwidth is f_s/N. If the number of points of the series increases then the channel bandwidth of the equivalent spectrum analyzer decreases and each channel will contain less noise power. Doubling the number of *FFT* channels halves the noise power in each channel. Therefore, the set of lines making the noise floor lowers by 3 *dB*.

This observation is relevant if we want to distinguish a small tone from the noise. If the bandwidth of the equivalent spectrum analyzer is large, then the integration of the noise spectrum in each channel can give a contribution that is larger than the power of the small tone. In short, the lines depicting the noise floor can mask the small tone. To observe the small tone over the noise floor it is necessary to reduce the channel width of the equivalent spectrum analyser.

Accordingly, it is necessary to increase (by multiples of powers of two) the number of points of the sequence used to calculate the *FFT*.

A quantizer with *M-bit* resolution generates a quantization noise equal to $V_{FS}^2/12 \cdot 2^{2M}$. The power of the full scale sine wave is $V_{FS}^2/8$. Therefore, the noise power is $3/2 \cdot 2^{2M}$ lower than the power of the full scale sine wave. If the *FFT* series is N points then the noise power is equally distrib-

> **Notice**
>
> If the length of the input series increases by a factor 2 the floor of the FFT noise spectrum diminish by 3 dB. Tones caused by harmonic distortion do not change. Only very long input series can reveal very small tones.

uted among $N/2$ spectral lines (unilateral representation). The *FFT* represents the quantization noise spectrum with a set of discrete lines whose amplitude is, on average, below full scale by a factor $3/2 \cdot 2^{2M}/N$. The result, expressed in *dB* yields

$$x_{noise}^2|_{dB} = P_{sign} - 1.78 - 6.02 \cdot M - 10 \cdot \log(N/2). \qquad (1.36)$$

The term $10 \cdot log(N/2)$ is called the processing gain of the *FFT*.

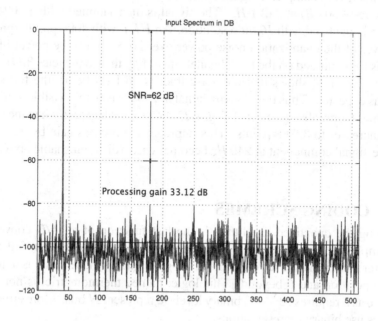

Figure 1.24. FFT of a 10-bit quantizer with a 4096 points sequence.

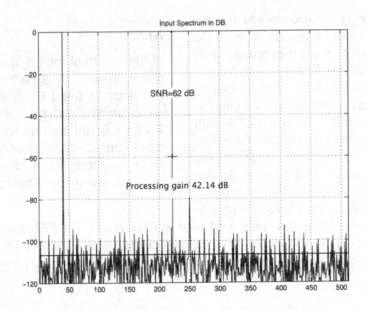

Figure 1.25. FFT of a 10-bit quantizer with a 32768 points sequence.

Consider, for example, a signal with a full scale sine wave at 39 Hz and a small tone ($-80 \, dB$) at 249.1 Hz. The signal is then sampled with a 1024 Hz clock and quantized with *10 bit* resolution. An *FFT* with 4096 points spreads the power of the quantization noise power over 2048 *lines*. The power of the signal is concentrated on the two channels including the frequencies 39 Hz and 249 Hz. Fig. 1.24 shows that the noise floor is 95.1 dB below the full scale signal as expected. Thus the $-80 \, dB$ signal at 249 Hz is barely visible amongst noise lines (some above $-85 \, dB$). Fig. 1.25 shows what happens when the input series increases to 32768 points. This improves the process gain by 9 dB and now the signal component at 249 Hz becomes easily distinguishable above the noise.

1.6 CODING SCHEMES

Coding the quantized amplitude is the last function of an *A/D* converter. The simplest scheme is the one generated by the comparators of a full flash converter which uses a set of $(2^N - 1)$ logic signals whose value is *1* up to a given position and *0* beyond. This logic, named thermometric, is not very effective as it requires $(2^N - 1)$ binary levels to represent N bits. More effective schemes use binary representations.

- **USB - Unipolar Straight Binary**: is the simplest binary scheme. It is used for unipolar signals. The USB represents the first quantization level,

$-V_{ref} + 1/2V_{LSB}$ with all zero's $(\cdots 0000)$. As the digital code increases, the analog input increases by one *LSB* at a time, and when the digital code is at the full scale $(\cdots 1111)$ the analog input is above the last quantization level $V_{ref} - 1/2V_{LSB}$. The quantization range is $-V_{ref} \cdots + V_{ref}$.

- **CSB - Complementary Straight Binary**: the opposite of the *USB*. *CSB* coding is also used for unipolar systems. The digital code $(\cdots 0000)$ represents the full scale while the code $(\cdots 1111)$ corresponds to the first quantization level.

- **BOB - Bipolar Offset Binary**: is a scheme suitable for bipolar systems (where the quantized inputs can be positive and negative). The most significant bit denotes the sign of the input: 1 for positive signals and 0 for negative signals. Therefore, $(\cdots 0000)$ represents the full negative scale. The zero crossing occurs at $(01 \cdots 111)$ and the digital code $(1 \cdots 1111)$ gives the full positive scale.

- **COB - Complementary Offset Binary**: this coding scheme is complementary to the BOB scheme. All the bits are complemented and the meaning remains the same. Therefore, since $(01 \cdots 111)$ denotes the zero crossing in the BOB scheme the zero crossing of COB is becomes $(10 \cdots 000)$.

- **BTC - Binary Two's Complement**: is one of the most used coding schemes. The bit in the *MSB* position indicates the sign in a complemented way: it is 0 for positive inputs and 1 for negative inputs. The zero crossing occurs at $(\cdots 0000)$. For positive signals the digital code increases normally for an increasing analog input. Thus, the positive full scale is $(0 \cdots 1111)$. For negative signals, the digital code is the two's complement of the positive counterpart. This leads $(1 \cdots 0000)$ to represent the negative full scale. The *BTC* coding system is suitable for microprocessor based systems or for the implementation of mathematical algorithms. It is also the standard for digital audio.

- **CTC - Complementary Two's Complement**: is the complementary code of *BTC*. All the bits are complemented and codes have the same meaning. The negative full scale is $(0 \cdots 1111)$; the positive full scale is $(1 \cdots 0000)$.

1.7 THE D/A CONVERTER

The transcoder is the first stage of the D/A transformation. It conceptually generates a sequence of pulses whose amplitude is the analog representation of the digital code. Then, the reconstruction process changes the sequence of pulses into a continuous-time signal. This operation is obtained by the cascade of a sample and hold $(S\&H)$ and a filter. Normally, a single circuit performs both transcoding and $S\&H$ functions. Fig. 1.26 show the stair-case

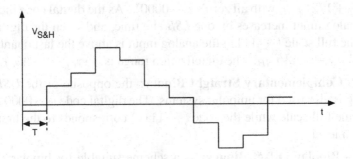

Figure 1.26. Signal waveform after the *S&H* of a *DAC* .

like waveform generated by the the *S&H*. The reconstruction filter smoothes the staircase-like waveform by removing high frequency components to obtain the analog result.

1.7.1   Ideal Reconstruction

The spectrum of a sequence of weighted deltas is an infinite replica of equal spectra (possibly altered by the *S&H*), however the desired reconstructed signal is contained within the band-base. Therefore, reconstruction should completely removes the replicas and leaves the band-base unchanged.

Ideally, reconstruction uses a filter with transfer function

$$H_{R,id}(f) = 1 \ \ for \ \ -\frac{f_s}{2} < f < \frac{f_s}{2}$$
$$H_{R,id}(f) = 0 \ \ otherwise. \tag{1.37}$$

This kind of filter is named an *ideal reconstruction filter*. It removes images and transforms the sampled-data into the continuos-time representation exactly. Unfortunately, the impulse response, $r(t)$, of the ideal reconstruction filter is

$$r(t) = \frac{sin(\omega_s t/2)}{(\omega_s t/2)} \tag{1.38}$$

and is not obtainable as it is non-recursive.

1.7.2   Real Reconstruction

A real reconstruction filter obtains an approximation of the ideal reconstruction response. The function, as already mentioned, is obtained by cascading a sample-and-hold $(S\&H)$ and a low pass (*reconstruction*) filter. The transfer function of a sample-and-hold is

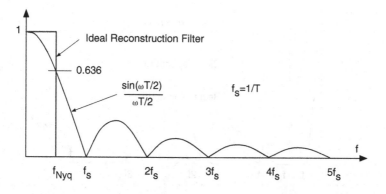

Figure 1.27. Amplitude response of the ideal and real sample-and-hold.

$$H_{S\&H}(s) = \frac{1 - e^{-sT}}{s\tau}, \qquad (1.39)$$

where τ is a suitable gain factor whose dimension is time for making $H_{S\&H}$ dimensionless.

Equation (1.39) on the $j\omega$ axis becomes

$$H_{S\&H}(j\omega) = j\frac{T}{\tau}e^{-j\omega T/2}\,\frac{sin(\omega T/2)}{\omega T/2}, \qquad (1.40)$$

showing a phase shift proportional to ω and an amplitude attenuation proportional to the *sinc (=sin(x)/x)* function.

Fig. 1.27 compares the magnitude response obtained by (1.40) with the magnitude response of the ideal reconstruction filter. Notice that the *S&H* filter causes signal attenuation in the Nyquist interval. At the Nyquist limit, $f_s/2$, the amplitude is 0.636 ($-3.9\ dB$). In addition, the attenuation in high Nyquist zones is not very large as the gain response of the *sinc* goes to zero at multiples of the sampling frequency but the relative maxima between two successive zeros only drops below 0.1 ($-20\ dB$) in the fifth Nyquist zone. In summary, an *S&H* causes an unwanted attenuation in the band-base and leaves a good portion of images in high Nyquist zones.

The phase response is an important issue for some applications. For this reason, in addition to the magnitude error the designer must often account for the phase error caused by the sample-and-hold in the signal band and correct it with suitable filters like an all-pass scheme.

The reconstruction filter corrects for the limitations caused by the sample-and-hold. The specifications of the reconstruction filter are similar to the ones discussed previously for the anti-aliasing filter. For anti-aliasing it was

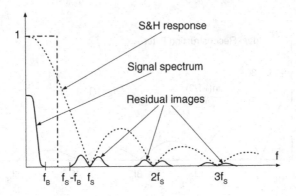

Figure 1.28.   Spectrum of a sampled-data signal passed through a sample-and-hold.

necessary to remove spurs from the high Nyquist zones to avoid possible folding into the band-base. Similarly, for the reconstruction filter, the signal images in the high Nyquist zones must be strongly attenuated. In addition, signal correction is required, with an emphasis on rectifying the unwanted attenuation in the band-base. Since the magnitude response of the *S&H* is $sin(x)/x$, then the ideal correction in the signal band is its inverse $x/sin(x)$. Observe that if the signal band is a small fraction of f_s then the attenuation caused by the *S&H* is negligible and compensation is often not necessary. Only when the signal band extends over a good fraction of the band-base can the in-band *sinc* attenuation become problematic thus needing a compensation filter that mimics in the signal band the $x/sin(x)$ response and drops down in the stop-band region.

Figure 1.28 depicts the typical spectrum of a transcoded signal followed by an *S&H*. Notice that the base-band spectrum occupies about half of the Nyquist interval. The *sinc* attenuation at the base-band limit is 0.9 (-0.91 *dB*).

Therefore, applications with medium accuracy should use magnitude correction in the signal band.

The spectrum in Fig. 1.28 drives the definition of the stop-band filter specifications. The transmission zeros of the *sinc* shape the images at multiples of f_s. The attenuation close to the zeros is very good. Only at frequencies far from the zeros do the images grow up and become significant. Therefore, since the *S&H* grants some attenuation at the signal band

---

Rule of thumb

The designer should consider using an in-band $x/sin(x)$ reconstructing compensation if the band of the signal occupies about a quarter of the Nyquist interval or more.

limit, $f_s - f_B$, this attenuation should be accounted for and subtracted from the one required by the system in specifying the stop band attenuation. If, for instance, the signal band is $1/20 \cdot f_s$ the value of $sin(x)/x$ at $f = 19/20 \cdot f_s$ is 0.0498 giving a 26 *dB* bonus to the stop-band attenuation of the reconstruction filter.

The transition region from pass-band to stop-band, $f_B, \cdots, f_s - f_B$, influences the reconstruction filter similarly to the anti-aliasing case. As it was observed for anti-aliasing, the order of the filter depends on the logarithm of the relative transition-band: $log_{10}\left[(f_s - f_B)/f_s\right]$. If the required stop-band attenuation in *dB* (including the benefit granted by the *S&H*) is $A_{SB}|_{dB}$, the order of a maximally flat reconstruction filter must be larger than $A_{SB}|_{dB}/\{20 \cdot log_{10}\left[(f_s - f_B)/f_s\right]\}$. Since the order of the filter decreases as the transition band increases, a suitable margin between f_B and f_s is important for ensuring an acceptable reconstruction filter specification.

Example 1.5

Verify by computer simulations the effect of the sample-and-hold on the signal images. Use a flat top input signal to mimic a band-limited input.

Solution

The Matlab file $Ex1\_5.m$ provides the basis for this study. The file defines a flat top function with 64 points. The obtained function is then interpolated over 1024 points to produce the input signal. Figure 1.29 shows the input signal and its spectrum. As expected the

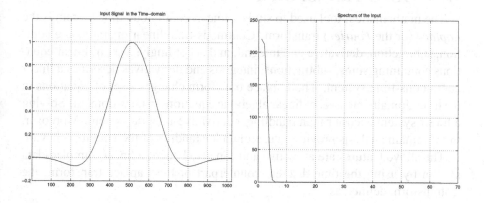

Figure 1.29. Input signal used in Example 1.5. Left: Time domain plot. Right: Obtained spectrum.

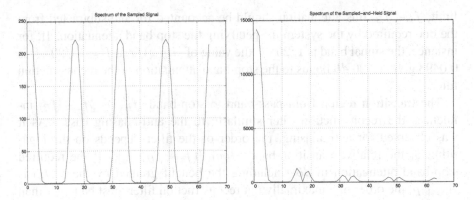

Figure 1.30. Spectrum of the sampled signal and its sampled-and-held version.

input spectrum is almost flat for a few frequency bins and vanishes at bin #6 corresponding to $5/1024 \cdot f_s$. Thus the bandwidth is a small fraction of the Nyquist interval.

The sampler and the sample-and-hold result in the spectra of Fig. 1.30. Since the sampling frequency is $f_s/64$, (bin#17) the sampled signal shows the band-base and its replicas located around multiples of $f_s/64$. The effect of the sample-and-hold is evident in the right hand plot. The sinc function produces zeros at $f_s/64$ and its multiples. Around the zeros the residual images pop up as expected.

1.8   THE Z-TRANSFORM

The mathematical tool used to study a linear, continuous-time system is the *Laplace* (or the *Fourier*) transform. Continous-time linear operators are addition, subtraction, derivative and integral; and accordingly, a set of linear equations containing sums, subtractions, integrals and derivatives describes a linear continuous-time system. The Laplace transform changes a set of equations in the time-domain into a set of linear algebraic equations in the s-domain. Solving a linear system is easier than handling a derivative-integral system. Moreover, the s-domain analysis provides a useful understanding of the circuits behavior.

The above features are attractive and can also benefit a linear, sampled-data system by using the time-discrete counterpart of the Laplace transform, the $\mathcal{Z}$-transform defined as

$$\mathcal{Z}\{x(nT)\} = \sum_{-\infty}^{\infty} x(nT)z^{-n}. \tag{1.41}$$

The above equations is a geometrical series that only converges in some regions of the z-plane. If the series converges then the $\mathcal{Z}$-transform is the analytical extension of the converged solution over the entire z-plane.

Discrete-time linear operators are: addition, subtraction and delay. The $\mathcal{Z}$-transform operator is also a linear operator. Thus, the $\mathcal{Z}$-transform of a linear combination of two or more discrete-time variables is the linear combination of their $\mathcal{Z}$-transforms:

$$\mathcal{Z}\{a_1 \cdot x_1(nT) + a_2 \cdot x_2(nT)\} = a_1 \cdot \mathcal{Z}\{x_1(nT)\} + a_2 \cdot \mathcal{Z}\{x_2(nT)\}. \quad (1.42)$$

Moreover, the $\mathcal{Z}$-transform of a delayed signal is

$$\mathcal{Z}\{x_1(nT - kT)\} = \sum_{-\infty}^{\infty} x(nT - kT)z^{-(n-k)}z^{-k} = X(z)z^{-k}, \quad (1.43)$$

showing that z^{-1} is the unity delay operator.

Equation (1.3) defined the Laplace transform of a sampled-data signal. Comparing that equation with the definition of the $\mathcal{Z}$-transforms leads to the relationship

$$z \rightarrow e^{sT}, \quad (1.44)$$

that establishes an important link between the complex plane s and the complex plane z. The relationship (1.44) is named *mapping* since it maps points on the z plane with points on the s plane (and vice-versa).

The use of the real and imaginary part of s ($s = \sigma + j\omega$) and the magnitude and phase of z ($z = |z|e^{j\Omega}$) leads to

$$|z| \rightarrow e^{\sigma T}; \quad \omega \rightarrow \Omega. \quad (1.45)$$

Thus, points on the imaginary axis of the s-plane ($\sigma = 0$) correspond to points on the unit circle of the z-plane ($|z| = 1$). Furthermore, all the points on the imaginary axis whose imaginary parts differ by $2n\pi$ map to the same point of the unit circle of the z-plane. Therefore, the mapping established by (1.44) is not biunivocal: a single point on the z-plane maps to an infinite number of points on the s-plane. The feature marks a key difference between continuous-time and sampled-data systems as it does the modification of a continuous-time spectrum into an infinite replica produced by sampling.

The s-plane z-plane transformation is useful for estimating the stability and frequency response of a sampled data system. The poles of a continuous-time system are in the left half s-plane. For a sampled-data system the poles must be inside the unit circle of the z-plane.

Since in a continuous time-system the frequency response is the transfer function evaluated on the $j\omega$-axis. The mapping between the s-plane and the

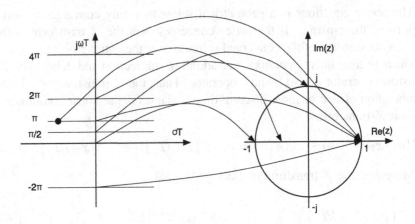

Figure 1.31. Mapping between points of the s plane (left) and the z plane (right).

z-plane tells us that the frequency response of a sampled-data system is the z-transfer function calculated on the unity circle.

Fig. 1.31 shows the mapping of relevant points of the s plane. The point $s = 0$ and all $s = \pm j2k\pi/T$ maps to $z = 1$. $sT = j\pi$ maps to $z = -1$ and $sT = j\pi/2$ maps to $z = j$. A point on the negative half s-plane $s = -\bar{\sigma} + j\omega T = \pi$ maps to a point on the real z-*plane* inside the $0 \cdots -1$ interval.

The above-described properties are useful for estimating the stability and frequency response of a sampled data system. Since the poles of a continuous-time system are in the left half s-plane, for a sampled-data system the poles must be inside the unit circle of the z-plane. The frequency response of the system is the z-transfer function calculated on the unity circle of the z-plane. The value at $z = -1$ (phase π) is the response at Nyquist.

Example 1.6

Determine the points of the s-plane which map to the following points of the z-plane: $z = -0.5$; $z = -0.5 + j\sqrt{3}/2$; $z = -0.8j$; $z = 1.2$. The sampling period is T_s.

Solution

Equation (1.44) gives the mapping from the plane s to z. Its inverse provides the reverse mapping from z to s

$$\sigma = \frac{1}{T_s}\log|z|; \quad \omega = \frac{1}{T_s}\{phase(z) \pm 2n\pi\} \qquad (1.46)$$

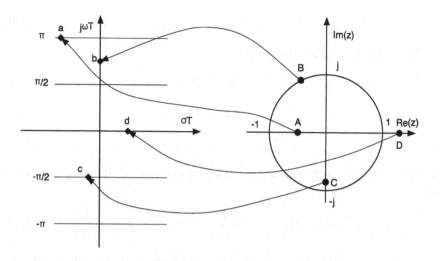

Figure 1.32. Mapping from the z plane (right) to the s plane (left).

which gives real and imaginary parts of $s = \sigma + j\omega$. Observe that the non biunivocal mapping comes from the $\pm 2n\pi$ term: n is any integer.

Figure 1.32 shows the mappings. The first point A is on the real axis with phase $\pi/2$. The corresponding point is on the left s half-plane as its magnitude is less than 1. The phase of the second point, B, is $2\pi/3$ and lies on the unity circle. The mapping point is on the imaginary axis at $j\omega_b = 2/3 \cdot \pi/T_s$. The point C is inside the unity circle with phase $3\pi/4 \rightarrow -\pi/2$. The point D is outside the unity circle with phase 0. The corresponding mapped points are on the left and right s half-plane with the proper imaginary parts.

The category of data converter called sigma-delta uses sampled-data processing for enhancing the benefit of a fast sampling. The used processing function is the integration that in the s domain is represented by the transfer function

$$H_I(s) = \frac{1}{s\tau}. \tag{1.47}$$

Since for a sampled-data quantity the signal is available only at discrete times, the integral $y(nT)$ of the variable $x(nT)$ can be only approximated. Possible expressions used are obtained by the approximate relationships

$$y(nT + T) = y(nT) + x(nT + T)T, \tag{1.48}$$

$$y(nT + T) = y(nT) + x(nT)T, \tag{1.49}$$

$$y(nT + T) = y(nT) + \frac{x(nT + T)x(nT)}{2}T. \tag{1.50}$$

The use of the $\mathcal{Z}$-transform obtains

$$zY = Y + zXT, \tag{1.51}$$

$$zY = Y + XT, \tag{1.52}$$

$$zY = Y + XT\frac{z+1}{2} \tag{1.53}$$

giving rise to three different approximate expressions of the integral transfer function

$$H_{I,F} = T\frac{z}{z-1}; \quad H_{I,B} = T\frac{1}{z-1}; \quad H_{I,Bil} = T\frac{z+1}{2(z-1)}, \tag{1.54}$$

where the indexes $"F"$, $"B"$ and $"Bil"$ indicate forward, backward and bilinear.

The three above expression are equivalent to the continuous-time integration through approximate mappings that, comparing (1.47) and (1.54) is given by

$$sT \rightarrow \frac{z-1}{Tz} \quad (Forward\ transformation), \tag{1.55}$$

$$sT \rightarrow \frac{z-1}{T} \quad (Backward\ transformation), \tag{1.56}$$

$$sT \rightarrow \frac{z-1}{2T(z+1)} \quad (Bilinear\ transformation), \tag{1.57}$$

where the sampling time T matches the dimensions on the left side of the equations.

Since the above mappings moves the poles differently than the ideal mapping $s \rightarrow ln(z)/T$, the trasformation of a stable continuous-time network into a sampled-data equivalent using circuits that implement the approximate mappings requires a further verification of stability and performance.

Indeed the Forward transformation is critical for stability, the Backward transformation is problematic for conserving performances, while the bilinear transformation is the most effective but not much used in data converter applications.

PROBLEMS

1.1   Repeat Example 1.1 but use a truncated sine wave. Build an input spectrum made by two main peaks, one at 10 kHz and the other at 60 kHz. Use 4096 points and $f_s = 4.096\,MHz$.

1.2   The bandwidth of an input signal is 22 kHz. Determine the mask of the anti-aliasing filter and estimate the order of the Butterworth filter required if the sampling frequency is 1 MHz, and the rejection of folded spurs must be at least 80 dB.

1.3   The bandwidth of an input signal ranges from 41 MHz to 42 MHz. Determine all the possible undersampling solutions that bring the signal in the interval $2\,MHz \cdots 3\,MHz$. Mirroring of the spectrum is not a relevant issue.

1.4   Repeat Example 1.2 but perform the verification in the frequency domain using the FFT. Verify that the spectrum of the signal folded from the second Nyquist zone does not depend on the phase difference between the two sine waves used.

1.5   Determine, by computer simulation, the spectrum of the quantization noise at the output of a 2-bit quantizer. The input is a sine wave with frequency $64/2048 \cdot f_s$ and amplitude $0.38\,V_{FS}$. Explain the cause of possible tones.

1.6   A processing system samples and holds a sine wave with amplitude $0.46\,V_{FS}$ and frequency $39/2048 \cdot f_s$. The result is quantized with 10-bit resolution. Use a behavioral simulator (like Matlab) to represent the output and determine the spectrum of the output signal.

1.7   What is the minimum time jitter that allows an SNR = 92 dB when sampling a $100\,MHz$ sine wave? What is the new jitter specification if the *SNR* is reduced to $80\,dB$?

1.8   Repeat Example 1.4 but use the Hamming window. Use a series made by 2048 points and three input frequencies with 30.3, 45.11 and 65.2 sine waves in the series. Choose amplitudes and determine the phase of the third sine wave required to eliminate the discontinuty.

1.9   Determine the USB coding of 0.367 V_{FS} and the BOB and the BTC coding of -0.763 V_{FS}. Assume that the quantizer has 10-bit of resolution.

1.10   Sample a 0.23456 MHz, 0.67 V_{FS} sinewave at 1 MHz. Hold the samples for an entire sampling period and re-sample the result at 10 MHz. Estimate the spectrum of the resulting signal.

1.11   Quantize a linear ramp, that goes from 0 to full scale (1 V) in 1 sec, with a 6-bit quantizer. Use a sampling frequency of 10 kHz. Plot the quantization error and explain the result.

1.12   Determine, by a computer simulation, the quantization error at the output of an 8-bit quantizer. Assume the input a 372.317 Hz sine wave sampled at 2 kHz. Estimate the distribution function of the quantization error. The accuracy of the distribution function must be better than 5%.

1.13   What is the noise floor established by a sampling capacitor of 3 pF. The sampling frequency is 58 MHz. What is the equivalent resistance leading to the same noise floor?

1.14   Determine (in dB) the gain emphasis required to compensate the sample-and-hold response at 0.36 f_s and the attenuation necessary to ensure an image rejection better than 75 dB at 1.5 f_s.

1.15   Generate with a computer simulation a new version of the diagram of Fig. 1.16. The resolutions must be 15-bit and 16-bit; the input frequencies are 50 MHz and 100 MHz. Examine the jitter range $0.01 \cdots 0.1$ psec.

1.16   Derive the two spectra given in Fig. 1.16. Compare the results with the spectrum of a sampled sine wave with $f_{in}/f_s = 6712/4096$. Re-evaluate the spectrum with $f_{in}/f_s = 6800/4096$. Repeat the simulations using 2048 and 8192 points.

1.17   Determine the points of the z-plane which map to the following points of the sT-plane: $sT = 2+j2$; $sT = -2+j2$; $sT = -2-j2$; $sT = -1 + j\pi/2$; $sT = -2 + j\pi/2$; $sT = -3 + j\pi/2$.

1.18   Calculate the processing gain of a 2^{14} point *FFT*. Assume the spectrum represents a 1 V_{peak} sine wave passed through a 10-bit quantizer. What is the minimum amplitude of harmonic components that can be observed?

1.19   The specifications of a given design require a very low harmonic distortion. The highest harmonic must be 98 dB below the full scale input. Estimate the length of the FFT sequence that pushes the noise floor below -105 dB_{FS}. Assume 10-bit and 12-bit quantization.

1.20   What is the distortion caused by a S&H to an input spectrum occupying 0.34 of the Nyquist interval? What is the phase shift at the upper limit of the input band?

1.21   Determine the value of the sampling capacitance and the clock jitter of a 14-bit, 65 MHz ADC. The full scale voltage is 1 V. The ENB must be 13.5-bit.

1.22   Calculate the z-transform of the following sampled-data expressions: $a \cdot x(nT) + b \cdot x(nT + 2T) - x(nT); x(nT - T) - x(nT)$.

1.23   Calculate, in the z-domain, the transfer function $Y(z)/X(z)$ determined by the relationship $y(nT + T) = y(nT) + x(nT)$

REFERENCES

Books and Monographs

L. R. Rabiner, and B. Gold: *Theory and application of digital signal processing* - Prentice-Hall, Inc., Englewood Cliffs, New Jersey, 1975.

S. W. Smith: *The Scientist & Engineer's Guide to Digital Signal Processing*. California Technical Publishing, San Diego, California, 1997.

A. V. Oppenheim and R. W. Schafer: *Discrete-Time Signal Processing*. Prentice-Hall, Inc., Englewood Cliffs, New Jersey, 1989.

R. Soin, F. Maloberti, and J. Franca: *Analogue-Digital ASICs*. Peter Peregrinus Ltd., IEE Press, London, 1991.

G. Doetsch: *Guide to the Applications of the Laplace and z-Transform*. Van Nostrand-Reinhold, London, 1971.

Journals and Conference Proceedings

J. W. Cooley and J. W. Tukey: *An algorithm for the machine computation of the complex fourier series*, Mathematics of Computation, vol. 19, pp. 297–301, April 1965.

R. E. Crochiere and L. R. Rabiner: *Interpolation and decimation of digital signals - A tutorial review*, Proceedings of the IEEE, vol. 69, pp. 300–331, 1981.

H. Sorensen, D. Jones, M. Heideman, and C. Burrus: *Real-valued fast Fourier transform algorithms*, IEEE Transactions on Acoustics, Speech, and Signal Processing, vol. 35, pp. 849–863, 1987.

P. Duhamel and M. Vetterli: *Fast fourier transforms: A tutorial review and a state of the art*, Signal Processing, vol. 19, pp. 259–299, 1990.

A. J. Jerri: *The Shannon Samplign Theorem - Its Various Extension and Appications: A Tutorial Review*, Proceedings IEEE, Vol. 65, pp. 1565–1596, 1997.

Internet Material

J. Albanus: *Coding schemes used with data converters*, http://focus.ti.com/docs/analog/analoghomepage.jhtm.

W. Kester: *Seminar material, high speed design techniques: High speed sampling and high speed ADC*. http://www.analog.com/support/Design_ Support.html.

EE241: Educational Material, Stanford University: *Waves*. http://sepwww.stanford.edu/ftp/prof/waves/toc_html.

Chapter 2

DATA
CONVERTERS
SPECIFICATIONS

Using or designing a data converter requires a proper understanding of its specifications. These give general information and describe the features and limits of the static and dynamic operation of a data converter. In this chapter we shall study the basic elements for evaluating and comparing existing devices, and we shall learn how to determine the specifications of new devices. Furthermore, the elements of this chapter help in the choice of the appropriate data converters for a given mixed system.

This chapter will discuss the definitions of technical terms used in manufacturer-supplied specifications and will clarify the terminology used. Testing methods for measuring the parameters are studied in a later Chapter.

2.1 TYPE OF CONVERTER

The first specification which defines a data converter is its type. The conversion algorithm normally provides this kind of information. For example, we have flash, sub-ranging, or sigma-delta converters. Furthermore, the types of converters are classified in two main categories: Nyquist-rate and over-sampling. This distinguishes between the following design strategies: using an input that occupies a large fraction of the available bandwidth or using an input-band that only occupies a small part of the Nyquist range. The ratio

47

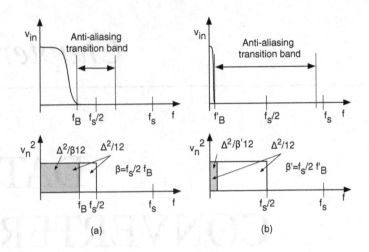

Figure 2.1. Comparing Nyquist-rate, (a), and oversampling, (b), strategies.

between the Nyquist limit and the signal band, $f_s/(2f_B)$, is called oversampling ratio (OSR). Converters with a large OSR are called oversampling converter, whereas Nyquist-rate converters have a small OSR, typically less than 8.

Fig. 2.1 illustrates the difference between Nyquist-rate and oversampling. In the former the transition region of the anti-aliasing filter is limited (leading to difficult specifications) and also a large fraction of the total quantization noise power is in the signal band. In contrast, the latter case has a large antialiasing transition region and only a small fraction of the total quantization noise occurs in the signal band. The sampling frequency for oversampling is, obviously, much larger than its Nyquist-rate counterpart, in some cases the OSR can be various hundreds.

2.2 CONDITIONS OF OPERATION

The behavior of a data converter strongly depends on the experimental set-up. Before studying the data converter's specifications, it is useful to examine how the operational environment can influence its performance.

Two important conditions of operation are the supply voltage and the temperature. A data-converter must comply with the requirements not only at the nominal supply voltage and room temperature but also within given ranges. The supply voltage should be allowed to fluctuate by 5% or even more; and the temperature range should be from $-20°C$ to $85°C$ (consumer applications) or $-55°C$ to $125°C$ (military applications). Specifications within a restricted range of supply voltages or temperatures don't properly represent the on-field use. Maintaining performance over a wide range of supply voltage or temperature is difficult especially for high resolution devices. For example, a *14-bit* converter requires accuracies as good as $600\,ppm/V$ (5V supply) or $0.3\,ppm/°C$ (consumer applications).

When measuring or using a data converter it is important to ensure that the printed circuit board (PCB) does not hamper the results. Power supply

> **Keep Note**
>
> The operational conditions of data converters are key for achieving (or measuring) specifications. Inaccurate set-up or *PCB* limits can totally mask the excellent performances of a device.

couplings and poor ground connections are critical issues. Often, high performance data-converters use separate pins for the analog and the digital supplies even though the pins are usually connected to a single supply generator on the *PCB*. This method exploits the bonding inductances to decouple the analog and digital internal supplies. Also, in order to obtain good V_{DD} or ground terminations, it is necessary to ensure a proper connection between the external supply generators and the pins. The length of the connecting lead must be as short as possible: they are equivalent to inductors. Ground loops between two sides of the *PCB*, especially at *RF* frequencies must be carefully avoided. The use of two-layer board is only suitable for low frequency measures. Multi-layer boards with separate ground and power planes are normally required to ensure high-level signal integrity at high frequency.

The master clock and reference voltages are other important signals that are supplied through the *PCB*. We know that the clock jitter degrades performances. Therefore, it is not just necessary to use signal generators with low jitter but it is also important to preserve this feature in the phase generator. The *PCB* traces leading the clock must be short with a solid ground plane underneath. This forms a microstrip transmission line and enables impedance matching. When low-speed data converters use external references it is necessary to utilize a clean voltage generator whose output impedance is low enough to avoid internal fluctuations greater than *1 LSB*. This is not easy as the discrete-time operation of the data-converter draws large pulses of current to charge and discharge capacitances.

Often, to avoid ambiguity and to help in the testing of parts, some manufacturers provide evaluation boards (or their layout) and give detailed guidelines explaining the evaluation procedures.

The set of parameters given in data sheets assume that their measure is done in optimum specified conditions. They should be very clear; however, when necessary, it may be necessary to clarify the features and accuracy of instruments to use in the experimental set-up for avoiding conflicts between part provider and customer.

2.3   CONVERTER SPECIFICATIONS

A large set of specifications describe the performance of data converters. Specifications are used to interpret and understand the material in Catalogues and to facilitate the use and characterization of products. Some specifications describe the features of either an *ADC* or a *DAC*, while others refer to the operation of both *ADC* and *DAC*. The specifications are divided into the following classes:

- General features.
- Static specifications.
- Dynamic specifications.
- Digital and switching specifications.

2.3.1   General Features

Most of the general features are self-explanatory. However, for the others it is worth recalling the definitions or providing some comments.

- **Type of Analog Signals**: the analog input or output of a data converter can be single-ended, pseudo-differential or differential. Single-ended analog signals are referred to a common ground that is connected to the analog ground of the converter. Pseudo-differential signals are symmetrical with respect to a fixed reference voltage that can differ from the analog ground of the converter. Differential signals are not necessarily symmetrical with respect to a fixed level: they are the difference between the inputs or outputs regardless of the common mode values.

- **Resolution**: is the number of bits that an *ADC* uses to represent its analog input or the number of bits that a *DAC* receives at its input to generate an analog output. The resolution, together with the reference voltage determines the minimum detectable voltage (for an *ADC*) or the minimum change in the output variable (for a *DAC*). This is also known as the quantization step.

- **Dynamic range**: is the ratio between the largest signal level the converter can handle and the noise level, expressed in *dB*. The dynamic range determines the maximum *SNR*.

- **Absolute maximum ratings**: are the limiting values that can be used, beyond which the service capability of the circuit may be impaired. The functional operation is not necessarily affected. However, exposure to absolute maximum rating conditions for an extended period of time influences the device reliability. Maximum ratings are divided into two categories: electrical and environmental. The environmental category includes the operating temperature range, the maximum chip temperature, the lead temperature, the maximum soldering time, the storage temperature range and, for airborne system applications, the vibration range.

- **ESD (electrostatic discharge) notice**: all *ICs* are sensitive to high electrostatic voltage. The human body and test equipment can store electrostatic charge at voltages as high as 4000 *V* which could discharge through the device. Although all *ICs* have protection circuitry, permanent damage may occur due to high-energy electrostatic shocks. Manufacturers always recommend proper *ESD* precautions to avoid loss of functionality.

- **Pin function descriptions and pin configuration**: this is a table with each pin's number, name and performed function. It is always provided along with the specifications. In addition, a drawing of the package also provides the pin configuration.

- **Warm-up time**: is the recommended amount of time required to stabilize the performances after powering up. The parameter accounts for the change of performances due to the temperature transient after powering the converter.

- **Drift**: is the change in a parameter (such as gain, offset and other static features) over a specified temperature range. Among them we have the drift temperature coefficient, normally specified in $ppm/^\circ C$, and the drift voltage coefficient specified in ppm/V. They can be calculated by measuring the parameter at the minimum and maximum operating range, and then dividing the parameter variation by the corresponding temperature range.

2.4 STATIC SPECIFICATIONS

The input-output transfer characteristic depicts the static behavior of a data converter. For an ideal case the input-output characteristic is a staircase with uniform steps over the entire dynamic range. Fig. 2.2 plots the initial part for a generic number of bits. If the first and last steps are $\Delta/2$ then the full-scale range is divided by $2^n - 1$ instead of 2^n to give Δ. Fig. 2.2 outlines that a quantization interval can be encoded using both digital code or midstep point. Also, Fig. 2.2 shows the quantization error. As known, the quantization error ranges between $\pm\Delta/2$ and is equal to zero at the midstep.

Deviations from the ideal transfer characteristic produce results like the

Notice

The static characteristic is the basis of all the static specifications; however, as studied in the last Chapter, the measure of some static parameters can use more convenient techniques.

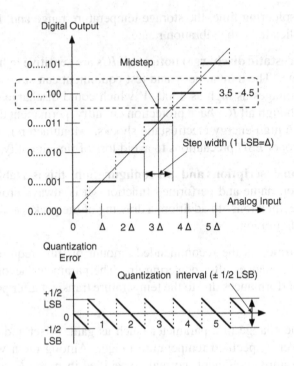

Figure 2.2. Ideal input-output transfer function of an A/D converter.

ones shown in Fig. 2.3. The curve of Fig. 2.3 (a) shows an almost random variation of the quantization intervals. There is no correlation between successive errors. The figure also shows the interpolating curve as a straight line running from the origin to the full scale. The characteristics of Fig. 2.3 (b) display small quantization intervals at the beginning and large quantization intervals at the end of the curve. As a result, the interpolating curve moves away from the straight line leading to a distorted response. These features are quantified by the *INL* and *DNL*, two of the static specifications defined below.

- **Analog Resolution**: is the smallest analog increment corresponding to a 1 *LSB* code change. For example, the resolution of a *16 bit* converter with $X_{FS} = 1$ is $15.26 \cdot 10^{-6} = 15.26\ \mu$.

- **Analog Input Range**: is the single ended or differential peak-to-peak signal (voltage or current) that must be applied to the *A/D* converter to generate a full-scale response. A peak differential signal is the difference between the two $180°$ out of phase signal terminals. Peak-to-peak differential is computed by rotating the inputs phase $180°$, taking the peak measurement again and subtracting it from the initial peak measurement.

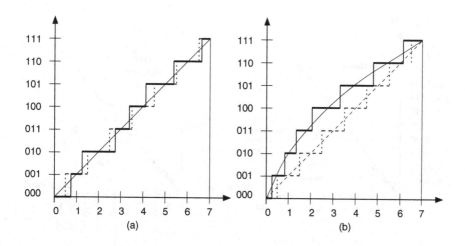

Figure 2.3. Input-output transfer characteristic of a real data converter.

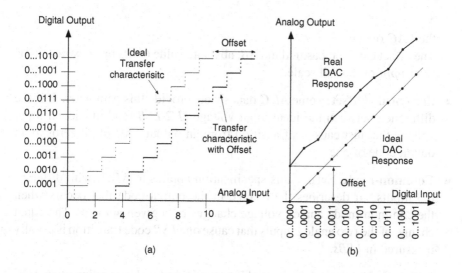

Figure 2.4. Offset error for an analog-to-digital (a) and a digital-to-analog (b) converter.

- **Offset**: the offset describes a shift for zero input. Offset is an error that can affect both an *ADC* or a *DAC*. Fig. 2.4 (a) compares the input-output transfer characteristics of a real and an ideal *ADC*. The offset changes the transfer characteristics so that all the quantization steps are shifted by the *ADC* offset. The offset of a *DAC* is defined by using the real response of Fig. 2.4 (b). The analog signal generated by the digital code $0 \cdots 0000$ is

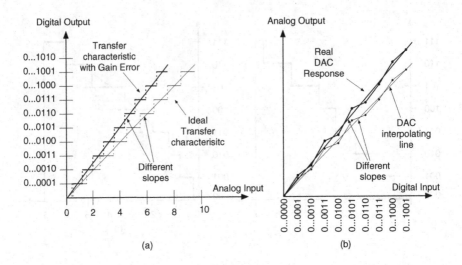

Figure 2.5. Gain error for an analog-to-digital (a) and a digital-to-analog (b) converter.

the *DAC* offset.

The offset can be measured in *LSB*, absolute value (*volts* or *amperes*), or as % or *ppm* of the full scale.

- **Zero Scale Offset**: some *ADC* data sheets provide this parameter. It is the difference between the ideal input voltage (*1/2 LSB*) and the actual input voltage that just causes a transition from an output code of all zeros to an output code of one.

- **Common-mode Error**: this specification applies to *ADCs* with differential inputs. It describes the change in the output code that occurs when the common-mode analog voltage changes by a given amount. The equal change of the two analog inputs that cause one *LSB* code transition is usually measured in *LSBs*.

- **Full-scale Error**: is a measure of how far the last code transition of an *ADC* is from the ideal top transition immediately below V_{Ref+}. It is normally measured in *LSB*.

- **Bipolar Zero Offset**: when a *DAC* is enabled for bipolar output and is loaded with $10 \cdots 000$ then the deviation of the analog output from the ideal midscale value is called the bipolar zero error.

- **Gain error**: is the error on the slope of the straight line interpolating the transfer curve. For an ideal converter the slope is D_{FS}/X_{FS}, where D_{FS} and X_{FS} are the full-scale digital code and full-scale analog range respectively. Since D_{FS} represents X_{FS}, we normally say that the ideal slope is one. The gain error defines the deviation of the slope of a data converter from the expected value. Fig. 2.5 shows the input-output diagrams for a real and an ideal *ADC* (a) and *DAC* (b).

  Another measure of the gain error is given by the difference between the input voltage causing a transition to the full scale and the reference (minus half *LSB*). When using this definition the gain error is known as the full scale error.

- **Differential non-linearity error (*DNL*)**: is the deviation of the step size of a real data converter from the ideal width of the bins Δ. Assuming that X_k is the transition point between successive codes $k-1$ and k, then the width of the bin k is $\Delta_r(k) = (X_{k+1} - X_k)$; the differential non-linearity is

$$DNL(k) = \frac{\Delta_r(k) - \Delta}{\Delta}. \tag{2.1}$$

This function is also known as the differential linearity error (*DLE*). Fig. 2.6 shows an example of *DNL* for a 12-*bit ADC*. The diagram shows the error to be within a $\pm 0.5\,LSB$ interval over the entire dynamic range. Fig. 2.6 measures the *DNL* in *LSB*. The *DNL* can be also measured in *Volts* (or *Amperes* when the input is a current) or as % or *p.p.m.* of the full scale. The

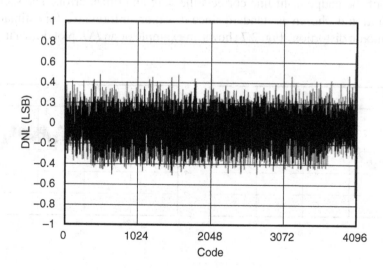

Figure 2.6. Differential non-linearity error (*DNL*) of a possible 12-bit ADC.

maximum differential nonlinearity is the maximum of $|DNL(k)|$ for all k. Often the maximum differential nonlinearity is simply referred to as *DNL*. An additional specification given by some data sheets is the root mean square (*RMS*) of the *DNL*.

$$DNL_{RMS} = \left\{ \frac{1}{2^N - 2} \sum_{1}^{2^N - 2} [DNL(k)]^2 \right\}^{1/2}. \tag{2.2}$$

- **Monotonicity**: is the *ADC* feature that produces output codes that are consistently increasing with increasing input signal and consistently decreasing with decreasing input signal. Therefore, the output code will always either remain constant or change in the same direction as the input.

- **Hysteresis**: is the limit that denotes a dependence of the output code on the direction of the input signal (i.e., increasing or decreasing signal). If this happens hysteresis is the maximum of such differences.

- **Missing code**: this denotes when digital codes are skipped or never appear at the *ADC* output. Since missing codes cannot be reached by any analog input the corresponding quantization interval is zero. Therefore, the *DNL* becomes -1.

- **Integral non-linearity (*INL*)**: is a measure of the deviation of the transfer function from the ideal interpolating line. Another definition of the integral non-linearity measures the deviation from the endpoint-fit line. The use of the endpoint-fit line corrects the gain and offset error. The second definition is chosen as standard since it is more informative for estimating harmonic distortion. Fig. 2.7 shows an example of an *INL* plot drawn using

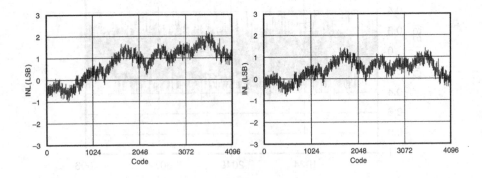

Figure 2.7.   INL obtained with the ideal interpolation line (left), or the endpoint-fit line (right).

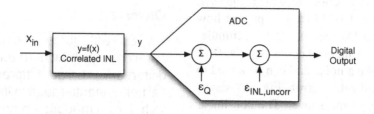

Figure 2.8. Equivalent system describing the *INL* error as pre-distortion effect.

the two definitions. The left curve does not start from zero and shows an *INL* climbing up. The right curve corrects the two limits and shows zeros at the two endings of the quantization range. The maximum of the *INL* is the maximum of $|INL(k)|$ for all k. Often, it is referred to as just *INL*. For the case illustrated in Fig. 2.7 the *INL* is more than 2 *LSB* using the first definition, and it is 1.3 *LSB* measuring the deviation from the endpoint-fit line. The *INL*, as for the *DNL*, is measured in *LSB*. It can also be measured using absolute value (*Volts* or *Amperes*), or as % or *p.p.m.* of the full scale.

Let us consider the endpoint-fit line which is the transfer curve with corrected offset and gain. An iterative use of (2.2) gives the transition point between codes after correction, $X'(k)$

$$X'(k) = \Delta' \left\{ k_{os} + \sum_1^k DNL(i) \right\}. \tag{2.3}$$

Where $\Delta' = \Delta(1 + G)$; G gain error; k_{os} is the offset measured in *LSB*. Since the offset compensated for the endpoint-fit line is $k_{os}\Delta'$, the *INL* in *LSB* becomes

$$INL(k) = \frac{X'(k) - k\Delta'}{\Delta} = (1 + G) \sum_{i=1}^k DNL(i). \tag{2.4}$$

Showing that the *INL* at bin k is the running sum of the *DNL* corrected by the gain error.

The *DNL* and *INL* provide information with different consequences on the noise spectrum. Assume the *DNL* is separated into its correlated and uncorrelated parts. The running sum of the correlated fraction is the main source of the *INL*. If the *INL* is few *LSB* over the entire range then the correlated part of the *DNL* is in the order of *INL* divided by the number of bins. It becomes a negligible fraction: looking at the *DNL* spectrum it is difficult to predict how large the *INL* can be. The accumulation of the uncorrelated part of the *DNL* looks like a noise and can be added to the quantization error. Fig. 2.8 shows a model of the above. The non-linear block $f(x)$ accounts for the running sum of the correlated part of the *DNL*. The quantization error and the uncorrelated part of the *INL* are treated as noise terms. Since the correlated part of the *DNL* is usually negligible we can view

> **Observe**
>
> ADCs with large integral non-linearity show harmonic distortion. Large differential non-linearities lead to INL with large random components. The error is equivalent to noise added to the quantization noise that degrades the SNR.

a large *DNL* as a source of extra noise. Its running sum is added to the quantization and degrades the *SNR*. A large *INL* means large deviation of the transfer curve from the straight line thus causing harmonic distortion. Harmonic terms affect the *SFDR* and *SNDR* (defined shortly).

Example 2.1

Evaluate the harmonic distortion caused by the INL. Use the model given in Fig. 2.8 and construct a 1.2 LSB INL response. The random variation of the INL is within the ± 0.45. Estimate the output spectrum due to an input sine wave whose amplitude approaches the full scale.

Solution

The Matlab code given in the file $Ex2\_1$ provides the basis for studying the problem. The INL is made by two terms, a random sequence of numbers whose variance is ± 0.45 and the polynomial function

$$y = x + ax^2 + bx^3 + cx^4; \quad x = (n - 2^{N-1})/2^N \tag{2.5}$$

where n is the running bin and N=12.
Using $a = -0.01$, $b = 0.01$ and $c = 0.02$ gives the INL plot of Fig. 2.9. The corresponding DNL is shown on the right hand side of Fig. 2.9. As expected the correlated term does not affect the DNL whose limits are ± 0.45.

The signal used is the 12-bit quantization of a 61 periods sine wave (the input sequence is made by 2^{12} samples). This produces the expected spectrum on the left of Fig. 2.10. The average of the spectral lines is the noise floor, -107.1 dB. It is equal to the sum of the $SNR = 74$ dB and the processing gain, 33.1 dB. The right side of

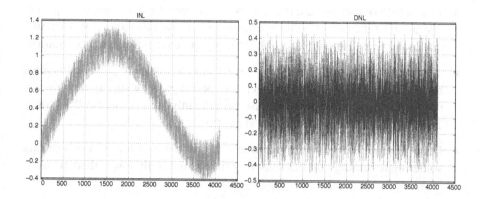

Figure 2.9. INL (left) and DNL (right).

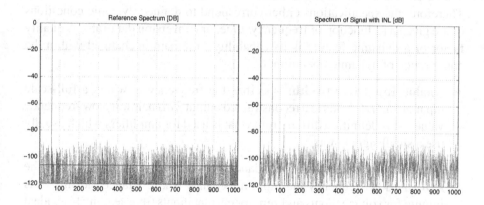

Figure 2.10. Normalized reference spectrum and spectrum of the signal affected by the *INL*.

Fig. 2.10 shows the effect of the INL on the signal spectrum. Since the uncorrelated part of the INL plot is $\pm 0.45LSB$, the noise floor increases by just fractions of dB. The correlated part of the INL causes tones. The large fraction of the INL with a parabolic shape denotes a dominant second order term. In fact, the biggest tone is the second harmonic with value $-76\ dB$.

- **Power Dissipation**: is the power consumed by the device during normal operation or during stand-by (or power-down) conditions.

- **Temperature ranges**: are the ranges of temperature that ensure the proper operation of the device. The operating range gives the limits of operation that preserve the functionality. The storage range gives the storing condition.

- **Thermal Resistance**: is the device capability to dissipate power consumed by the device itself. Some packages use special pads for supplying power to the *PCB*. The specification can also provide information on the *PCB* trace. The thermal resistance is measured in $°C/W$.

- **Lead Temperature**: is the maximum temperature of the *IC* leads during soldering assuming that the soldering time is normally lower than 10 *sec*.

2.5   DYNAMIC SPECIFICATIONS

The frequency response and speed of the analog components of a data converter determine its dynamic performance. Obviously, the performance becomes critical when the input bandwidth and the conversion-rate are high. Therefore, the specifications either correspond to defined dynamic conditions or are given as a function of frequency, time, or conversion data-rate. A quality factor of a dynamic feature is its capability to remain unchanged within the entire range of dynamic operation.

- **Analog Input Bandwidth**: specifies the frequency at which a full-scale input of an *ADC* leads to a reconstructed output 3 *dB* below its low frequency value. This definition differs from what is used for amplifiers which usually use a small signal input.

- **Input Impedance**: is the impedance between the input terminals of the *ADC*. At low frequency the input impedance is a resistance: ideally, it is infinite for voltage inputs and zero for current inputs (thus leading to an ideal measure of voltage or current.) At high frequency the input impedance is dominated by its capacitive component. Often, a switched capacitance structure performs the input sampling. In this case the specification provides the equivalent load at the input pin. At very high frequency the input impedance of the *ADC* must be the matched termination of the input connection.

- **Load Regulation or Output Impedance**: the load regulation measures the ability of the output stage of a *DAC* to maintain its rated voltage accuracy. It describes the change in output voltage per *mA* drained from the output terminal, and is expressed in *LSB/mA*. The output impedance is obtained from the load regulation by replacing the *LSB* in the *LSB/mA* ratio with the *LSB* value in *mV*. The result is in Ω.

- **Settling-time**: is the time at which the step response of a *DAC* enters and subsequently remains within a specified error band around its final value.

The input is a step signal applied at time $t = 0$. The final value is defined to occur a long time after the beginning of the step.

- **Cross-talk**: measures the energy that appears in a signal because of undesired coupling with other signals. In addition to coupling at the *IC* level a poor printed circuit board design can cause crosstalk. For instance, traces carrying critical signals running in parallel on the same layer of a *PCB* cause interferences.

- **Aperture uncertainty (Clock Jitter)**: is the standard deviation of the sampling time. It is also called aperture jitter or timing phase noise. It is normally assumed that clock jitter is like a noise with a white spectrum.

- **Digital to Analog Glitch Impulse**: is the amount of signal injected from the digital inputs to the analog output when the inputs change state. The maximum normally occurs at half scale when the *DAC* switches around the *MSB* and many switches change state, i.e., from $01 \cdots 11$ to $10 \cdots 00$. The parameter is the integral of the glitch area and is measured in *V-sec* or *A-sec*.

- **Glitch Power**: is similar to the previous *DAC* specification but its cause is more general. It can be due to a delay between bit controls or to timing mismatch in the analog sections. Normally its maximum occurs at half scale. Similarly to the previous specification it is the integral of the glitch area and is measured in *V-sec* or *A-sec*.

- **Equivalent input referred noise**: is a measure of the electronic noise produced by the circuits of the *ADC*. The result is that for a constant *dc* input the output is not fixed but there is a distribution of codes centered around the output code nominally encoding the input. With a large number of output samples the code histogram is approximately Gaussian. The standard deviation of the distribution defines the equivalent input referred noise. It

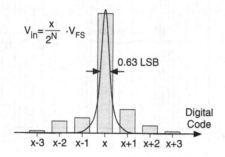

Figure 2.11. Estimation of the input referred noise using the histogram method for a dc input.

is normally expressed in terms of *LSB*s or *rms* voltage. Fig. 2.11 shows the histogram at the output of a possible data converter affected by a 0.63 *LSB* noise.

- **Signal-to-Noise Ratio (*SNR*)**: is the ratio between the power of the signal (normally a sinewave) and the total noise produced by quantization and the noise of the circuit. The *SNR* accounts for the noise in the entire Nyquist interval. The *SNR* can depend on the frequency of the input signal and it decreases proportional to the input amplitude. Fig. 2.12 shows the *SNR* of a hypothetical 12-*bit* data converter with 50 *MHz* sampling frequency. The *SNR* for a −0.5 *dB* input is 67 *dB*. The loss in the *SNR* shows that the noise caused by the electronics is larger than the quantization noise. When the input signal is −20 *dB* then, as expected, the *SNR* is 48 *dB*. Observe that the *SNR* performances versus frequency are good: the *SNR* is almost constant in the entire Nyquist range. Also, it only drops a few *dB* for frequencies in the second Nyquist zone. Therefore, the hypothetical converter of Fig. 2.12 is suitable for under-sampling a signal whose spectrum is in the second Nyquist zone.

- **Signal-to-Noise-and-Distortion Ratio (*SINAD or SNDR*)**: is similar in definition to the *SNR* except that non linear distortion terms, generated by the input sine wave, are also accounted for. The *SINAD* is the ratio between the root-mean-square of the signal and the root-sum-square of the harmonic components plus noise (excluding *dc*). Since static and dynamic limitations cause a non-linear response the *SINAD* is dependent on both the amplitude and frequency of the input sine wave. Fig. 2.13 shows the *SINAD* for the

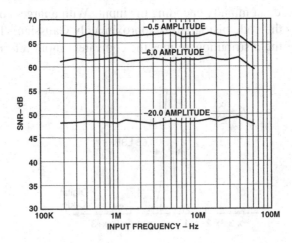

Figure 2.12. Possible SNR versus the input frequency at different input levels.

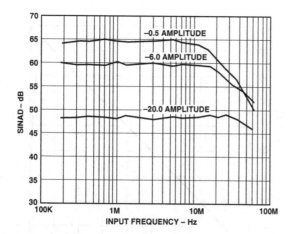

Figure 2.13. Signal-to-Noise-and Distortion Ratio versus input amplitude and input frequency ($f_{ck} = 50\,MHz$).

same hypothetical converter of Fig. 2.12. The *SINAD* shows that harmonic terms are negligible if the input is $-20\,dB_{FS}$ or less. Larger input amplitudes bring about distortion especially at high frequencies. Notice that the *SINAD* significantly degrades in the second Nyquist zone; thus, the use of this particular converter could be problematic when used in the second Nyquist zone and high-linearity is required.

- **Dynamic Range**: is the value of the input signal at which the *SNR* (or the *SINAD*) is 0 *dB*. The parameter is useful for some types of data converters that do not obtain their maximum *SNR* (or SINAD) at $0\,dB_{FS}$ input. This typically happens in sigma-delta converters. Fig. 2.14 shows a typical plot of the *SNR* versus the input amplitude for a sigma-delta *ADC*. The peak of the *SNR* is at 74 *dB* while the dynamic range is 80 *dB*. Therefore, the peak of the *SNR* occurs approximately at $-6\,dB_{FS}$.

- **Effective-Number-of-Bits (*ENOB*)**: measures the signal-to-noise and distortion ratio using bits. *SINAD* in *dB* and *ENOB* are linked by

$$ENOB = \frac{SINAD_{dB} - 1.76}{6.02}. \tag{2.6}$$

- **Harmonic Distortion (*HD*)**: is the ratio between the root-mean-square of the signal and the root-mean-square of harmonic components including aliased terms. Unless otherwise specified the *HD* accounts for the second through tenth harmonics: it is normally assumed that harmonic terms higher than the tenth have negligible effects. If f_{in} is the frequency of

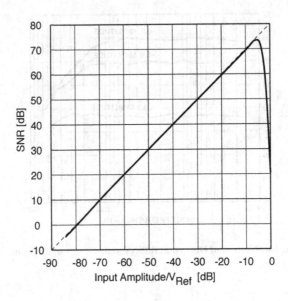

Figure 2.14. Typical SNR versus the input amplitude for sigma-delta converters.

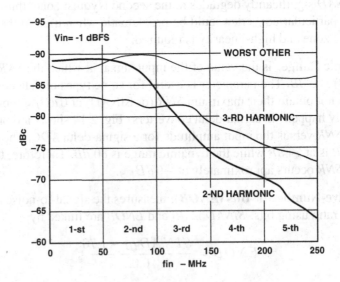

Figure 2.15. Harmonic components as a function of the input frequency.

the input signal and f_s is the sampling frequency, then the *n-th* harmonic component is at frequency $| \pm n f_{in} \pm k f_s |$, where k is a suitable number that folds the harmonic term into the first Nyquist zone. At high input amplitudes and huge frequency the largest terms are the second and the third harmonic. Some data sheets provide a plot of their amplitude as a function of the input frequency in *dBc* (*dB* below carrier). Fig. 2.15 shows the harmonics of a hypothetical 100 *MHz* clock, 12 *bit ADC*. The figure is for input frequencies up to 250 *MHz* (5*th* Nyquist zone). A fully differential system makes the second harmonic negligible in the first two Nyquist zones. At high frequency the benefit of the fully-differential architecture vanishes and the second harmonic distortion becomes dominant.

Example 2.2

Calculate the frequency of the first ten harmonic tones in the output spectrum of a non-linear data converter. The input is a single tone sine wave. Use computer simulations to verify the achieved results.

Solution

This example uses an input sequence with 2^{14} points. Assuming that the sequence lasts 1 sec the sampling frequency is 16.4 kHz. With an

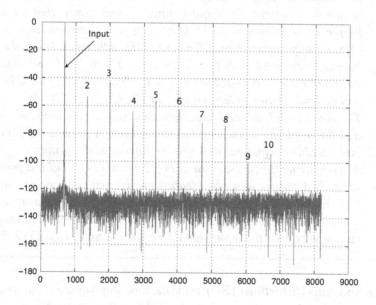

Figure 2.16. Harmonic tones for f_{in}=671 kHz (f_s=16.38 kHz).

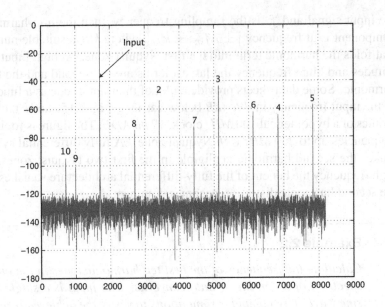

Figure 2.17.   Harmonic tones for f_{in}=1.711 kHz (f_s=16.38 kHz).

input frequency equal to 671 Hz the tenth harmonic is at 6.71 kHz, just below the Nyquist limit. All the harmonic tones will show up sequentially within the Nyquist range. Assume now that the input frequency increases to 1.711 kHz. The fifth harmonic is at 8.555 kHz, slightly higher than $f_s/2$=8.192 kHz. Aliasing mirrors the fifth harmonic down to f_s-5·f_{in}=7.829 kHz. The 6th, 7th, 8th and 9th harmonics are also folded once. The tenth term is at 17.11 kHz, and will be folded twice to leave a tone at $10 \cdot f_{in} - f_s$=726 Hz.

The codes Ex2_2.mdl and Ex2_2launch.m provide the Simulink model and the m-file for launching the simulation. A polynomial function and suitable coefficients describe the harmonic distortion.

By comparing the two spectra in Fig. 2.16 and Fig. 2.17 it is possible to verify the above described features with input frequencies equal to 671 Hz and 1.711 kHz. The diagrams show tones well separated from the input. However, if f_{in} is close to f_s/n (with n integer) harmonics can be very near the input signal (try, for example, f_{in}=2339 Hz).

- **Total Spurious Distortion (TSD):** is the the root-sum-square of the spurious components in the spectral output of the *ADC*. The input is a pure sine wave input of specified amplitude and frequency. *TSD* is often expressed in *dB* using, as a reference, the root-mean-square amplitude of the output component at the input frequency.

■ **Spurious Free Dynamic Range (SFDR):** is the ratio of the root-mean-square signal amplitude to the root-mean-square value of the highest spurious spectral component in the first Nyquist zone. The *SFDR* provides information similar to the total harmonic distortion but focuses on the worst tone. The *SFDR* depends on the input amplitude. With large input signals the highest tone is given by one of the harmonics of the signal. For input amplitudes well below the full scale the distortion caused by the signal becomes negligible and other tones not caused by the input become dominant due to the non-linear nature of the converter.

The *SFDR* is important for communication systems. Often it is necessary to perform analog-to-digital conversion on a small signal representing a channel that the antenna receives together with other big channels. It may happen that a high spur generated by a big channel falls very close to the small channel thus masking the associated information. Fig. 2.18 illustrates the problem. The input signal has two channels, a big one 0(*dB*) at around 6.72 *MHz*, and a small one (−90 *dB*) at around 3.8 *MHz*. The sampling frequency is 16.4 *MHz*. The big component generates a big third harmonic at 20.16 *MHz* which gets folded down to 3.76 *MHz*, just 40 *kHz* from the

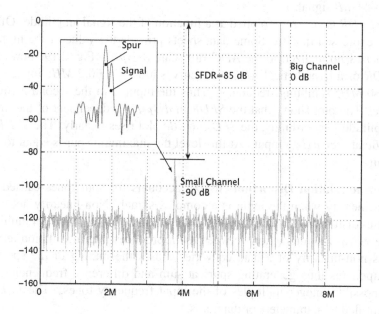

Figure 2.18. Spectrum of a small channel corrupted by the bigger spur due to a large channel.

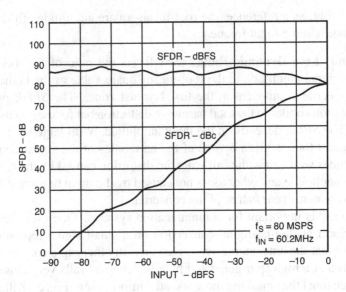

Figure 2.19. SFDR plot expressed in terms of dBc and dBFS.

small channel. Even if the *SFDR* is 85 *dB* the spur almost completely masks the -90 *dB* signal.

The *SFDR* is generally plotted as a function of the signal amplitude. Often it is expressed in *dBc*. Some data sheets provide the value of the highest spur *rms* normalized to the *ADC* full scale (dB_{FS}). Fig. 2.19 shows the *SFDR* of a hypothetical *ADC*. The curves are for a 60.2 *MHz* input with an 80 *MHz* sampling frequency. Thus the input is in the second Nyquist zone. The plot shows that the *SFDR* in dB_{FS} is independent of the input amplitude. Accordingly, the *SFDR* in *dBc* plot rises linearly. The *SFDR* is 0 *dBc* at $-86\, dB_{FS}$ input. At this level the maximum spur is equal to the input.

- **Intermodulation Distortion (IMD)**: accounts for spur tones caused by non-linearity when the input is a complex signal. Non-linearity not only causes distortion of a pure tone; but also, when the input is made of multiple sine waves the interaction between them produces intermodulation terms. This non-linearity of a data converter causes the mixing of the spectral components thus generating spurs at sum and difference frequencies for all possible integer multiples of the input frequency tones. The *IMD* is quantified by parameters or diagrams.

- **Two tone Intermodulation Distortion (IMD2)**: is the ratio of the *rms* value of either input tone to the rms value of the worst third order intermodulation product reported in *dBc*. The input is made by two closely spaced tones f_1 and f_2. Often the specification accounts for the third order spurs only, which occur at $(2f_1 - f_2)$, $(2f_2 - f_1)$. The reason for considering third order terms only is that they are close to the input frequencies $f_1 \cong f_2$. Other intermodulation terms are far away from the input and can be filtered out in the digital domain.

Example 2.3

Determine with computer simulations the IMD2 tones caused in a 14-bit ADC. The distortion is described by a non-linear block before the quantization $y = x + 10 - 4x^3$.

Solution

The code Ex2_3.mod and the corresponding launcher Ex2_3launch.m provide the basis for the computer solution. The sampling frequency is 16.38 kHz. The two tones at 1.169 kHz and 1.231 kHz have an amplitude of 1. The full scale is $\pm 2\,V$. The quantization step is $4V_{FS}/2^{14} = 1/2^{12}$. A non-linear block models the distortion of the transfer characteristic of the converter.

Fig. 2.20 gives the output spectrum in the frequency interval ranging from 500 Hz to 4 KHz. The figure outlines four intermodulation products $(2f_1 - f_2)$, $(2f_2 - f_1)$, $(2f_1 + f_2)$, $(2f_2 + f_1)$. The first two tones are the closest to the input frequencies. The figure also shows the effect of the third order non-linearity that produces two tones at $3f_1$

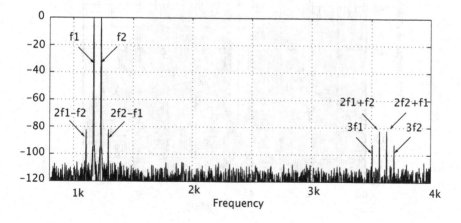

Figure 2.20. Spectrum of the output signal with IND2 tones.

and $3f_2$. The amplitude of pairs of intermodulation tones is the same because the sine wave components have the same amplitude. The simple use of trigonometric equations gives the first pair amplitudes equal to $3A_1A_2/4$. The reader can verify the effect of different sine wave amplitudes on the intermodulation tones.

- **Multi-Tone Power Ratio (MTPR)**: is specific for data converters used in communication systems. It defines the distortion of multi-tone transmission systems. The parameter is measured using a sequence of tones with equal amplitude A_0 placed at frequencies that are multiples of a fundamental frequency f_0. Few tones are left out. The harmonic distortion produces interference signals in the position of the missing tones. The *MTPR* is defined as the ratio of the *rms* signal amplitude A_0 to the *rms* value of the tones at the missing tones frequencies.

- **Noise-power ratio (NPR)**: similar to the *MTPR* it describes the linear performances of an *ADC* used in frequency division multiplexed (*FDM*) links. *NPR* is a parameter which is usually used to describe power amplifiers but the same concept is also used for data converters. In an *FDM* system the

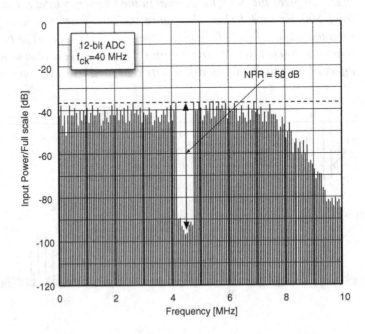

Figure 2.21. Typical output spectrum used to measure the NPR.

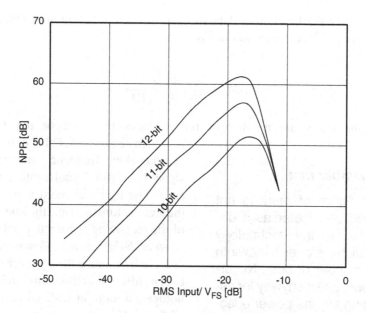

Figure 2.22. NPR for different ADC resolutions versus the RMS input amplitude.

signal is made of many carriers with different amplitude and phase. The signal looks like white noise passed through a band-pass filter. If one of the channels is removed then the spectrum will show a deep notch at the frequency of the missing channel. If we use the obtained signal to excite the *ADC* then the converter noise and the *IMD* will tend to fill in the notch. The depth of the notch after the *ADC* gives the *NPR*. Fig. 2.21 shows a typical spectrum used to determine the *NPR*.

The *NPR* is dependent on the *rms* value of the input. For low input levels the notch is mainly filled by the quantization noise and the thermal noise contributions which are almost independent of the input power. For high inputs the *ADC* saturation and distortion terms become dominant causing a quick drop of the *NPR* (Fig. 2.22).

- **Effective Resolution Bandwidth (ERBW)**: is defined as the analog input frequency at which the *SINAD* drops by 3 *dB* compared to its low frequency value. The *ERBW* gives the maximum signal bandwidth that the converter can handle. The *ERBW* should be well above the Nyquist limit.

- **Figure of Merit (FoM)**: is a parameter used to measure the power effectiveness of an *ADC*. It assumes that the total power is consumed mainly because of the bandwidth of the converted signal (*BW*) and the equivalent number of bits (*ENB*). Publications or data-sheets use different definitions of the *FoM*. The basis of all these definitions is

$$FoM = \frac{P_{Tot}}{2^{ENB}2 \cdot BW}. \qquad (2.7)$$

In some cases the number of bits replaces the *ENB* or the *ERBW* is used instead of *BW*. Other definitions use the clock frequency and not the signal band (for Nyquist-rate converters.) The figure of merit depends on the architecture used and the line-width of the technology. Effective solutions show an *FoM* below 1*pJ/conv-step*. In order to harmonize the comparison between different architecture and technologies a more articulated definition of the *FoM* with parameters that depend on technological values and the operation conditions can be used.

> **Remember that**
>
> The figure of merit is not a solid parameter as it depends on the technology line-width, signal bandwidth and number of bits. Nevertheless, use it anyway for assessing the the power effectiveness of the converter.

2.6　DIGITAL AND SWITCHING SPECIFICATIONS

The digital features are also specified by a given set of parameters. These ensure the proper interfacing with internal or external circuits and are useful for the synchronization of logic signals within the data converter. The specifications given below are the ones most commonly used in commercial data sheets.

- **Logic levels**: are the set of non overlapping ranges of amplitudes used to represent the logic state. The used logic levels ensure the compatibility with the defined logic standard (like the *CMOS* or *TTL*.)

- **Encode or clock rate**: is the range of possible encode rates that ensures the performances of the specifications. It can vary by one decade or more. It is best to operate a data converter using a maximum clock rate of about 25% of the maximum guaranteed by the specification.

- **Clock timing**: specifies the features of the clock. The information is normally given using a diagram. The external clock is normally regenerated inside the integrated circuit with edge-triggered flip-flops that latch the input on the rising or the falling edge. The clock duty cycle can be chosen arbitrarily under some constraints. A 50% duty cycle is normally the best for optimum dynamic performances.

- **Clock Source**: the clock signal specifies the timing operation of the converter. Circuits requiring a very low jitter generate the clock using a differential input sine wave. A crystal clock oscillator (with or without external filters) obtains the input sine wave. This ensures sine wave purity and provides accurate zero-crossing times. Internal amplifiers, under saturation, are used to square the input sine wave and thus generate the internal clock.

- **Sleep Mode**: specifies a power-down mode that turns off the main bias currents and minimizes the power consumption. Applying a logic level to a proper pin activates the power-down mode. The power-up and the power-down activation times depend on the time constant associated with the sleep circuit. It may take a few μs to go into the sleep mode and a few ms to power the circuit back.

PROBLEMS

2.1   Search the Web-sites of key data-converter manufacturers and list the various types of data converters produced. Draft a table with type, resolution and maximum clock-rates.

2.2   Plot the input-output transfer curve of an 8-bit data converter. Use a computer simulation and assume a random DNL with maximum variance 0.4 LSB. The interpolating curve is $y = x + 0.01(x - 0.5)^3; 0 > x > 1$.

2.3   The INL of a 12-bit data converter measured using the ideal interpolation line is -0.4 LSB and 1.3 LSB at the first and last code. Its maximum occurs at 2/3 of the full scale and is 2.1 LSB. Estimate the offset, gain error and the maximum of the endpoint-fit line INL.

2.4   Make a search on the Web to find the type of A/D converters that provide more than 14-bit. Consider the following input bandwidths: 44 kHz,150 kHz, 2 MHz, 20 MHz and 80 MHz.

2.5   Download the data-sheet of a 12-bit ADC from the Web. Estimate the interpolating curve that causes the INL. Verify the effect of the distortion on the output spectrum for a full scale sine wave.

2.6   Repeat Example 2.1 with the following distortion coefficients: $a = 0.02, b = -0.005 \ c = 0.01$. Determine the harmonic distortion for an input sine wave. Change the amplitude from full scale to -20 dB_{FS}. Keep the parameter a constant and change the values of b and c so that the harmonic distortion is optimised for an input full scale sine wave. Plot, using the optimum values, the distortion as a function of the input amplitude.

2.7   Perform a computer simulation with 5000 samples and plot the histogram of the output of 12-bit data converter. Use a constant input amplitude of 0.372 V_{FS} with a 20 mV_{FS} random noise component.

2.8 Repeat Example 2.2 using 3.7 kHz and 4.2 kHz input sine waves. Determine the output spectra for input square waves at the same frequency. Justify the results obtained.

2.9 Make a search on the Web for an ADC with 100 MS/s conversion rate and 14-bit. Compare the harmonic distortion provided data of the found parts and make an assessment of the cost/performances (if the cost of the parts are available).

2.10 Repeat Example 2.3 by using the following transfer response: $y = x + 10^{-4}x^2$. Use two sine waves one with half-scale and the other with a quarter of full-scale amplitude. The frequencies are 1.23 kHz $\pm$ 50 Hz.

2.11 The INL of a 10 bit ADC has a sawtooth like behavior with a jump of 1.6 LSB at half scale and 0.8 LSB and 1.1 LSB at a quater and three quarter of the full scale. Estimate the DNL and draft the input-output transfer characteristic around the same points with the given INL values and their opposite.

2.12 Search on the Web published data on power for ADC with $f_B = $ 2MHz and 10-bit resolution and compare the FoM. Find a possible dependence on the line-width of the used technologies.

REFERENCES

Books and Monographs

M. Gustavson, J. J. Wikner, and N. N. Tan: *CMOS Data Converters for Communications*, Kluwer Academic Publishers, Dordrecht, The Netherland, 2000.

Working Group of the IEEE Instrumentation and Measurement Society: *IEEE Std 1241-2000: IEEE Standards for Terminology and Test Method for Analog-to-Digital Converters*. IEEE, 2001.

W. Kester (ed.): *The Data Converter Handbook*. Elsevier's Science & Technology, Burlington, MA, 2005.

Journals and Conference Proceedings

F. Maloberti: *High-speed data converters for communication systems*, in IEEE Circuits and Systems Magazine, vol. 1, no. 1, pp. 26–36, 2001.

H. J. Casier: *Requirements for embedded data converters in an ADSL communication system*, in International Conference on Electronics, Circuits and Systems, ICECS 2001, vol. 1, pp. 489–492, 2001.

F. H. Irons, K. Riley, D. Hummels, and G. Friel: *The noise power ratio-theory and adc testing*, IEEE Transactions on Instrumentation and Measurement, vol. 49, no. 3, pp. 659–665, 2000.

M. Vogels and G. Gielen: *Architectural selection of A/D converters*, in Proceedings of Design Automation Conference, pp. 974–977, 2003.

Internet Material

Analog Devices: *Data Converters information and data sheets*.
http://www.analog.com.

Maxim: *Data Converters information and data sheets*.
http://www.maxim-ic.com.

ST Microelectronics: *Data Converters information and data sheets*.
http://www.st.com.

Texas Instruments: *Data Converters information and data sheets*.
http://www.ti.com.

Chapter 3

NYQUIST-RATE DIGITAL *to* ANALOG CONVERTERS

This chapter studies digital-to-analog architectures used in integrated circuits. We shall first consider the basic requirements on the voltage and the current references: the accuracy of any DAC (and also any ADC) critically depends on the quality of its references. Then, we shall study resistor based and capacitor based architectures. These architectures provide an attenuation of the reference voltage under the control of the digital input. Finally, we shall study the architectures that obtain the DAC function by steering unary or binary-weighted current sources.

3.1 INTRODUCTION

The input of a *DAC* is a multi-bit digital signal while the output is a voltage or a current capable of driving an external load. To meet the driving requirements many *DAC*s use active elements like op-amps or *OTA*s whose performances are, obviously, crucial for meeting the *DAC* specifications. Because of this we shall analyze the possible limits caused by active blocks but the study will be at the system level and will not go into design details. It is assumed that the reader already knows enough on operational amplifier or *OTA* design. Moreover, it is assumed that the reader has a basic knowledge of integrated circuit technologies and layout.

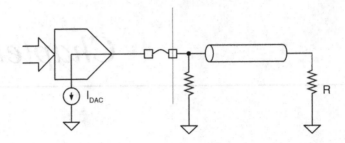

Figure 3.1. Use of a DAC with current output to drive a double terminated coaxial cable.

The output of a DAC can be a voltage or a current. In the latter case the current is normally used to drive an off-chip coaxial cable to yield a voltage across the terminations. The cable should be properly terminated on both sides (Fig. 3.1) to preserve the high frequency features of off-chip wire-transmitted signals.

The first Chapter assumed that the output of a *DAC* is a reconstructed waveform; i.e., a continuous-time signal whose bandwidth is completely within the first Nyquist zone of the sampled-data digital signal. However, many commercial integrated circuits don't provide a reconstructed signal but only its sampled and held *DAC* representation. It is assumed that a separate circuit, off-chip, realizes the reconstruction function. Furthermore, when the *DAC* is employed in an *ADC* then the analog waveform must be in sampled-data form, as the *DAC* signal is typically combined with other analog signals to implement some sampled-data algorithm.

Many *DACs* use integrated resistances or capacitors to attenuate (or amplify) the reference. Since the attenuation (or the gain) factor depends on the ratio between passive elements then having good matching and temperature independency is very important. A careful design of the layout (with inter-digitized or common centroid structures), the use of dummy elements, the matching of parasitic capacitances and the equalization of resistive-metal contacts leads to matching accuracies in the order of $0.02 - 0.1\%$. Thus, resolutions and linearitiy of up to $60 - 70\,dB$ are possible without trimming of analog passive elements and without using digital correction or digital calibration.

Various *DACs* (and also *ADCs*) use analog *CMOS* switches for the conversion algorithm implementation. They utilize the fact that *CMOS* technology easily obtains switches: a *MOS* transistor is an off-switch if its driving voltage is in the sub-threshold region while it becomes a low resistance (on-switch) if its driving voltage well exceeds the threshold and V_{DS} is small. The design of switches and their control is an importan part of the data converter design because it is necessary to obtain low on-resistance, high speed of switching and minimum side effects.

Since the on-resistance is inversely proportional to the product of the *W/L* aspect ratio and the applied overdrive voltage $(V_{GS} - V_{Th})$ a single *MOS* transistor can be enough for switching nodes towards fixed voltages. For switching nodes that experience some swing it is better to use complementary transistors since when the node voltage increases the n-channel on-resistance also increases, but the on-resistance of the p-channel decreases. Some implementations prefer to use a sigle n-channel transistor whose gate control is a shifted-up replica of the signal to be switched. The obtained control voltage typically exceeds the supply voltage and the circuit used to generate it is called clock booster.

An important limit when using analog switches comes from the clock-feedthrough: this is the injection of the *MOS* channel charge into source and drain when the gate control goes off. A charge injection into a low-impedance node will only cause a glitch whose duration depends on the value of the node impedance. However, a charge injection into a high impedance or capacitive node causes an offset that can be problematic, especially when the injected charge is signal dependent in a non-linear manner. The problem of clock-feedthrough is tackled with suitable techniques which are well described in basic textbooks and recalled in a successive chapter.

3.1.1 DAC Applications

Many high speed *DAC* specifications are generated by the video market needs. Video signals from a digital source, such as a computer or a *DVD*, are often displayed on analog monitors, thus requiring high-speed *DAC*s. High performance *DAC*s are also used for *TV*: *HDTV* quality pictures require *HDTV* quality displays. An analog *NTSC TV* displays 525 lines each frame while a good *HDTV* monitor displays more than 1000 lines per frame. In addition to the higher resolution a *HDTV* monitor also provides a higher contrast ratio and a more detailed color range. Unlike an analog *TV* which has about 8-*bit* contrast ratio, a digital light processing (*DLP*) or plasma *TV* provide contrast ratios that exceed 11-*bits*. *DNL*, conversion-rate, glitch energy and minimum chip area are key parameters for such video *DAC*s. To maximize the viewing quality on a state-of-art display technology, 12-*bit* and 150 *MSPS* conversion rate are therefore often necessary.

*DAC*s are also used for wired and wireless communications that use a variety of modulation and encoding schemes to exchange digital information. *DAC*s for *ADSL* or *ADLS2+* must handle signal bands of 1.1 *MHz* or 2.2 *MHz* with 12-*bit* of resolution. Wireless applications such as *UMTS*, *CDMA2000*, and *GSM/EDGE* require both high conversion rates and high resolution especially when multiple carriers are used instead of a single signal generating source.

UMTS uses up to four carriers per transmitter, and *GSM/EDGE* and *CDMA2000* applications may employ four to eight carriers for a single transmitter. Thus, the generation of a complex modulation waveform involves complex digital processing but also requires high performance *DAC*s. Conversion rates of 200 *MSpS* - 1 *GSpS* and resolutions from 12-*bit* to 16-*bit* can be necessary.

An analogue signal processor can use *DAC*s to replace potentiometers. A digitally controlled resistance allows functions like digital gain and offset cancellation, programmable voltage and current sources, and programmable attenuation. The solution is a resistive divider *DAC* whose wiper position is selected via an n-bit register value. The setting is updated infrequently and often the *DAC* uses a slow serial interface with volatile or non-volatile memory to store the setting.

A signal from a microphone or other sound source is converted to a digital signal for storage in a computer, where it can be edited if necessary or placed in some storage devices. Headphones or speakers playback the sound after the *DAC* conversion. The resolution used is high (16-*bit* or more) with a typical sampling-rate of 44 *kSpS*. Since the signal bandwidth is relatively low, *DAC*s used for audio applications normally exploit oversampling.

3.1.2   Voltage and Current References

The dynamic range of a *DAC* is established by the voltage (or current) reference which can be either generated inside the chip or provided via an external pin. In both cases it is necessary to ensure a very high level of accuracy as any error affecting the reference limits the overall system performance. Therefore, it is always necessary that references are set accurately and kept at a constant value independent of load changes, temperature, input supply voltage, and time.

The use of an external reference enables flexibility and permits a careful control of errors. However, series or shunting impedances in the interconnections can cause a static or dynamic error. Fig. 3.2 shows simple equivalent circuits of

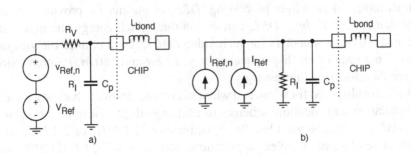

Figure 3.2.   Equivalent model of external references: (a) voltage and (b) current.

the external reference connections. The resistances R_V and R_I are the internal resistance of the references. C_p is a parasitic capacitance accounting for all the parasitics at the input pin. The connection of the external reference to the inside of the circuit is through the bonding, modeled by the inductance L_{bond} whose value is about 1 *nH* per *mm* of bonding connection.

Static errors within the reference corresponds to a fixed value change, they are normally irrelevant as only cause a gain error. By contrast, dynamic errors can be much more problematic because they affect the speed and the *SNR* of the circuit. A dynamic error mainly occurs because the reference is required to charge or discharge one or more capacitors. The voltage across the capacitors experiences a transient that must settle within half an *LSB* (or better) in a fraction of the conversion period. Thus, the time constant of the charging network must be much lower than $1/f_{ck}$. In addition, because of the bonding inductance the pin voltage also experiences a ringing transient at the natural frequency of the *LC* network thus requiring the use of series or parallel resistors which dampen the ringing to a level well below half *LSB* within a fraction of the clock period.

Fig. 3.2 also includes the noise of the references, $v^2_{Ref,n}$ or $i^2_{Ref,n}$. Since these noise sources can be viewed as an almost unchanged output additive term, their spectral level must be well below the quantization floor; therefore, for *n-bits* with dynamic range V_{FS} (or I_{FS}) and sampling frequency f_s it is necessary to have

$$v^2_{Ref,n} \ll \frac{V^2_{FS}}{6 \cdot 2^{2n} \cdot f_s} \quad or \quad i^2_{Ref,n} \ll \frac{I^2_{FS}}{6 \cdot 2^{2n} \cdot f_s}. \quad (3.1)$$

For example, a 14-*bit DAC* with 400-*MSPS* and $I_{FS} = 20\,mA$ needs a current reference whose noise spectrum is lower than $0.79\,nA\sqrt{Hz}$. Such a noise current injected into $50\,\Omega$ yields a voltage spectrum of $39.5\,nV\sqrt{Hz}$, figure that is a fairly low noise level. Remind that the spectrum of the input referred noise generator of an *MOS* transistor is $v^2_n = (8/3)kT/g_m$; therefore, just a single *MOS* transistor which transconductance is equal to $0.2\,mA/V$ gives to a noise spectrum as large as $v^2_n = 7.4\,nV\sqrt{Hz}$.

Remember

The noise of reference generators must be lower than the expected quantization noise floor. The request becomes very challenging for resolutions above 14-bit.

3.2 TYPES OF CONVERTERS

The basic components used in a *DAC* architecture normally classify the *DAC*. We distinguish between:

■ Resistor based architectures.

- Capacitor based architectures.
- Current source based architectures.

3.3 RESISTOR BASED ARCHITECTURES

The resistor based architectures include the resistive divider (or *Kelvin* divider), the digital potentiometer, and the *R-2R* resistive ladder.

Silicon technology realizes resistors using strips of resistive layers with a given specific resistance, $\rho_\square$. This is the resistance in $\Omega/\square$ of a square of the strip measured between two opposite sides of the square. The effective number of squares $(L/W)_{eff}$ and the contact resistance R_{cont} give the total resistive value

$$R = \rho_\square (L/W)_{eff} + 2 \cdot R_{cont}. \tag{3.2}$$

Integrated technologies provide resistive layers whose specific resistance ranges from a few $\Omega/\square$ to $k\Omega/\square$. An accurate value of number of squares requires using widths of the strip that are not at the minimum allowed by the technology; also, the selection of the layer must account for the linearity requirements that depends on temperature and voltage coefficients and must be such that the number of squares is not too low so that a good accuracy is ensured or too high so that the silicon area is relatively small even if the desired resistance is tens or hundred of $k\Omega$.

The absolute value of the resistances normally only matters for controlling the power consumption or for enhancing the speed of the circuit. Therefore, absolute accuracy is not particularly important. However, matching accuracy is much more significant. Matching errors cause gain error and in some situations harmonic distortion. In order to have good matching it is necessary to use resistive layers with well contolled fabrication steps and to draw a good layout. Obviously, when it is required to ensure matching between resistances they must use the same resistive layer and they must be at the same temperature. Moreover, the terminating resistances must be proportional to the resistance value. If the ratio between resistances is an integer number it is worth using a unity element to construct the various elements by series connections. The layout must also ensure that the orientation of matched elements and the direction of the current flow must also be the same.

Fig. 3.3 (a) shows an example of layout of 6 squares of a given resistive layer. Fig. 3.3 (b) shows three of such unity resistances forming a matched *R-2R* pair. The two unity elements of the *2R* resistance are on either side of the unity element *R*. This arrangement leads to the same centroid for *R* and *2R* thus compensating for any possible vertical gradient of the specific resistance. The connections are such that the current flows in the same direction in all the unity resistances thus avoiding possible errors due to anisotropy.

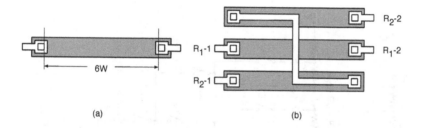

Figure 3.3. (a) resistance with L/W=6. (b) Two matched resistances with ratio 2.

The layout of Fig. 3.3 (b) does not account for boundary effects. Precision applications use dummy elements around the layout to make the boundary of all the resistors equal. Also, the metal connection establishes a series resistance whose value is tens of $m\Omega/\square$. If necessary, this parasitic effect is equalized by matching the number of squares of metal connection.

3.3.1 Resistive Divider

A resistive string connected across the positive and negative references obtains multiple voltages whose digitally controlled selection realizes a very simple *DAC*. This is known as a Kelvin divider as it was invented by Lord Kelvin in the 18th century. Fig. 3.4 (a) shows a *3-bit* resistive divider connected between ground and a unipolar reference V_{Ref}. The divider is made by 2^3 equal resistors, R_U, that generate 8 discrete analog voltages. It is easy to verify that the generated voltages are

$$V_i = V_{Ref}\frac{i}{8} \quad i = 0\cdots7. \tag{3.3}$$

The resistive divider of Fig. 3.4 (b) shifts the generated voltages by half an *LSB* ($V_{Ref}/2^{n+1}$) by moving a half unity resistance, $R_U/2$, from the top to the bottom of the resistive divider.

The selection of a voltage generated by the divider is done by switches whose *on* or *off* condition is controlled by the digital input. The selected voltage is then used at the input of a buffer capable of providing a very high input impedance so as to not disturb the divider (thus performing a volt-metric measurement). The buffer also provides a low output impedance for properly driving the *DAC* load.

Fig. 3.4 (b) shows also the simplest selection method (*3-bits*). It is a tree of switches directly controlled by the digital bits (or their inverse). The method does not require any substantial digital processing; however, there is the *on-*resistance of 3 switches (that would become n switches for an *n-bit DAC*) along the connection to the buffer.

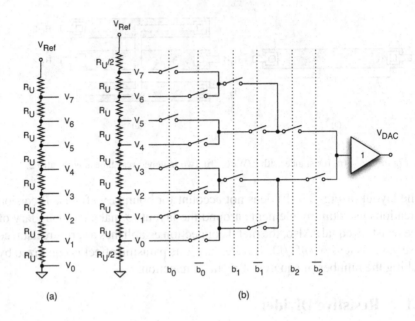

Figure 3.4. (a) Resistive divider. (b) Resistive divider with 1/2 LSB offset, Switch tree selector and output buffer.

The value of the switches on-resistances and the parasitic capacitances can impede the fast operation of a *DAC*. Thus, some architectures use a control logic to generate *DAC* register signals which are used to individually control 2^n switches. The switches connect all the taps of the resistive divider to a common line connected to the buffer. Fig. 3.5 shows a typical architecture that realizes

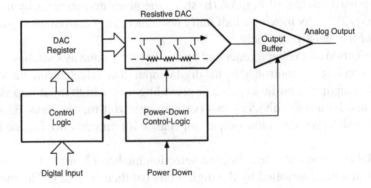

Figure 3.5. Typical architecture of a resistive divider DAC.

this kind of solution. The diagram includes the power down control, useful for reducing the power consumption in the idle state. Other possible functions to be incorporated in the circuit are a serial interface (like the I^2C) for input data and the power-on reset of the *DAC* register that sets the output voltage to zero at power-on.

Trade-off

The decoding method used to select the Kelvin divider voltages depends on a trade-off between speed, complexity, and power consumption.

3.3.2 X-Y Selection

A limit to the resistive divider architecture is that the number of switches or the number of control lines increases exponentially with the number of bits. For instance, an 8-*bit DAC* must use a tree of 510 switches or must generate 256 logic signals driving switches every conversion rate period. Moreover, the parasitic load due to the *off* switches connected to the common line at the buffer can be very large, thus reducing the speed or requiring an increase in the power used in the output buffer.

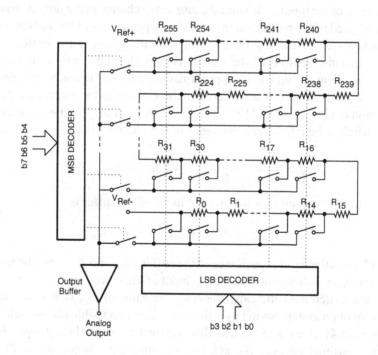

Figure 3.6. Resistive DAC with X-Y addressing scheme.

A solution that simplifies the selection and also saves power is the one shown in Fig. 3.6. The scheme, implemented for an 8-*bit* Kelvin divider, uses an *X-Y* selection. The 256 unity resistances, divided into 16 lines, are arranged in a serpentine fashion connected between bipolar positive and negative references, V_{Ref+} and V_{Ref-}. Each line represents one of the *4 MSBs* of the digital input. Therefore, the *MSB* decoder selects a line while the *LSB* decoder picks up an element of the line. A given pair: line selection *X* and column selection *Y*, identifies one and only one voltage of the divider. The first tap of the odd lines and the last tap of the even lines, next to R_0, R_{16}, R_{32}, $\cdots$ identifies the voltage of the first *LSBs* of that line. Therefore, the *LSB* decoder must swap the *LSBs* selection direction depending on whether the *MSB* identifies an even or an odd line of resistors.

The benefit of the *X-Y* selection is that the decoding signals are only $2 \cdot 2^{n/2}$ instead of 2^n. The layout is compact and minimizes the grading error. However, since the connection to the output buffer is through two switches instead of only one, the time constant of the *RC* network between resistive divider and buffer is doubled.

3.3.3   Settling of the Output Voltage

A change of the input code should cause a step change in the output, however the finite speed of the buffer and the parasitic capacitances of the Kelvin divider smooth this step transition. First assume that the speed of the buffer is very large so that the settling of the output voltage depends only on the divider whose equivalent circuit can be approximated by an *RC* network. For an *n-bit* divider the equivalent resistance between the *k-th* tap and ground is the parallel connection of $(k-1)R_U$ and $(2^n - k + 1)R_U$. Accounting for the on-resistance of the switches the equivalent resistance of the *RC* network is

$$R_{eq} = \frac{(k-1)(2^n - k + 1)}{2^n} R_U + N_{on} \cdot R_{on}. \tag{3.4}$$

The parasitic capacitance loading the input of the buffer is

$$C_{in} = C_{in,B} + N_{on} \cdot C_{p,on} + N_{off} \cdot C_{p,off}, \tag{3.5}$$

where $C_{in,B}$ is the input capacitance of the buffer, N_{on} and N_{off} are the number of *on* and *off* switches connected to the input of the buffer, and $C_{p,on} \simeq C_{p,off}$ are the associated parasitic capacitances. The value of C_{in} is almost constant as the sum of *on* and *off* switches is the same for any possible divider selection. However, (3.4) gives a resistance that depends on the selected tap. R_{eq} is parabolic starting from $N_{on} \cdot R_{on}$ at $k = 1$ reaching its maximum at $k = 2^{n-1} + 1$ (with value $2^{n-2}R_U + N_{on} \cdot R_{on}$) and going down to $N_{on} \cdot R_{on} + R_U$ at the last tap. Therefore, the variation of the time constant $R_{eq}C_{in}$ with the selected tap affects the settling in a non-linear manner: the response is fastest if the beginning or the end of the resistive string are selected and slower in the middle of the range.

Even if the output of the buffer passes through the reconstruction filter only spurs at frequencies higher than the Nyquist limit are filtered out. Harmonics of low-frequency signals and the intermodulation products coming from multi-tone inputs produce spurs

Observation

A code dependent settling of the output voltage causes distortion. High *SFDR* requires low resistances at every node. The variation of the settling time must be much smaller than the hold period.

inside the first Nyquist zone. Thus, a spur free output spectrum requires a settling time well below the *D/A* conversion period making it is necessary to use a very low unity resistance that minimizes the time constant.

Very low unity resistances reduce the time constant but the practical implementation can be problematic; an alternative solution is to use shunt resistances R_s across sub-strings of k unity elements. The resistance of the parallel connection becomes

$$R_{eq} = \frac{kR_U R_s}{kR_U + R_s}. \tag{3.6}$$

If the shunt resistance is lower than the total sub-string then the shunting element dominates the value of the sub-section resistance. Thus, the time constant is reduced without using low resistance unity elements. For example, if a shunt resistance $R_s = R_U$ is placed across each line in the X-Y arrangement of Fig. 3.6 (Fig. 3.7) the resistance of each line reduces by about a factor 16 as each line is the parallel connection of R_U and $16R_U$. The time constant can also be reduced by using minimum area switches that reduce the input capacitance or by using an active compensation of the capacitance at the input of the buffer.

Let us now consider the effect of the buffer. Its dynamic response is limited by the slew-rate, SR, and the gain-bandwidth product, f_T. Assume that the buffer input is a step voltage (i.e, the Kelvin divider does not limit the speed now). The step starts at $t = 0$ swinging from $V_{in}(0^-)$ by $\Delta V_{in}(0)$ and lasts for the conversion period, T. The output of the buffer slews with a constant rate, $dV_{out}/dt = SR$, until t_{slew} at which the output differs from the input by a given value, ΔV; then, the output settles exponentially to the final value. The value of ΔV is such that, at the first approximation, the derivative of the output voltage is continuous at the transition point. The time constant, τ, equals $1/(\beta 2\pi f_T)$ (where β is the feedback factor and f_T is the gain-banwidth product of the op-amp). Summing up, the transient is described by

$$V_{out}(t) = V_{in}(0^-) + SR \cdot t \qquad\qquad for \ \ t < t_{slew}$$

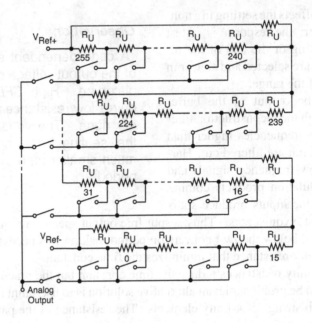

Figure 3.7. Resistive DAC with X-Y addressing scheme and shunt resistances.

$$V_{Out}(t) = V_{in}(0^-) + \Delta V_{in}(0) - \Delta V \cdot e^{-t/\tau} \quad for \ t > t_{slew} \qquad (3.7)$$

$$\Delta V = SR \cdot \tau; \qquad t_{slew} = \frac{\Delta V_{in}(0)}{SR} - \tau,$$

showing a non linear combination of a ramp and an exponential. A large step gives a long slewing period and reduces the time left for the exponential settling.

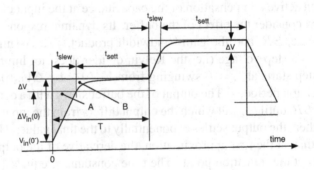

Figure 3.8. Response of a unity gain buffer with finite slew-rate and bandwidth.

A small step leads to a minimum (it may be zero) slewing time and a longer period for settling (Fig. 3.8).

Since the reconstruction filter obtains more or less the average over time of the waveform generated by the buffer, the average of the non-linear error denotes distortion. If the time allowed for settling is long enough the output reaches the final value with a good accuracy. The integrated error during the slewing period (area A of Fig. 3.8) is $\Delta V_{in}^2 / SR - SR \cdot \tau^2$ while the integrated error during the settling period (area B of Fig. 3.8) is $\Delta V \cdot \tau$. If $\Delta V_{in} < \Delta V$ there is no slewing phase and the integrated error is $\Delta V_{in} \cdot \tau$. Accordingly if the voltage step is small then the integrated error is proportional to the step size, while for large step, the error increases with the square of the step.

The above considerations are useful when making a qualitative estimation of the speed limits of the buffer. A quantitative assessment of performance degradation must be done with simulations at the transistor level.

3.3.4   Segmented Architectures

The use of shunting resistors across each line of an *X-Y* arrangement (Fig. 3.9) makes the shunting elements themselves an auxiliary divider that generates the voltages of an *n/2-bit DAC*. This observation is the basis of the segmentation technique: segmentation obtains a high-resolution *DAC* by combining the operation of two or more *DAC*s together.

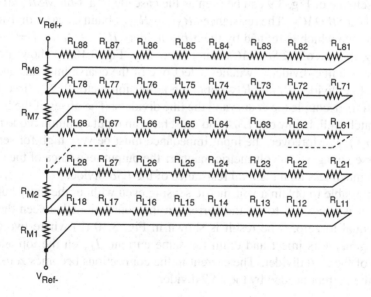

Figure 3.9.   3+3 bit X-Y network with shunting resistors.

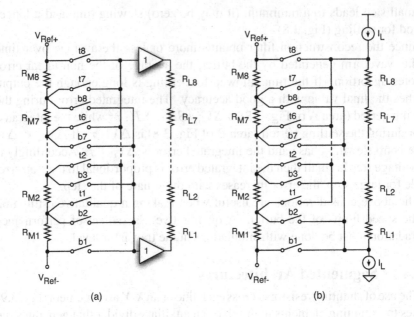

Figure 3.10. (a) Cascade of two DACs with decoupling buffers. (b) Volt-metric interconnection of the LSB DAC.

The scheme of Fig. 3.9 can be seen as the cascade of a *3-bit MSB DAC* and eight *3-bit LSB DAC*s. The resistances $R_{M1} - R_{M8}$ obtain a coarse division of the reference which is refined by the *3-bit* dividers $R_{Li1} - R_{Li8}$, $i = 1 \cdots 8$. Actually, the use of 8 *LSB DAC*s is not necessary. Fig. 3.10 (a) shows how to use only one fine divider. Switches select one of the coarse intervals and use it in the fine divider $R_{L1} - R_{L8}$ obtaining a division factor 2^{LSB}. Two unity gain buffers decouple the coarse and the fine dividers. The offset of the buffers must match well, below one *LSB*, and must have an input common mode range equal to V_{ref}. Moreover, the input impedance must be very high for sensing the coarse voltages in a volt-metric manner: the output resistance of the buffer must be much less than the total resistance of the *LSB* divider.

It is possible to obtain a volt-metric sensing even without the unity buffers: for this it is necessary to set the current in the connections between the two *DAC*s equal to zero. The result is shown in Fig. 3.10 (b) where two equal current generators inject and drain the same current, I_L, on the top and the bottom of the *LSB* divider. The current in the connections becomes zero if I_L equals the current needed by the *LSB* divider

$$I_L = \frac{\Delta V_{LSB}}{2^{n_{LSB}} R_L} = \frac{V_{Ref+} - V_{Ref-}}{2^{n_{MSB}} \cdot 2^{n_{LSB}} R_L}. \tag{3.8}$$

The static current is zero; however, every time the *MSB* changes the parasitic capacitances of the *LSB* nodes must charge or discharge thus draining current from the *MSB* divider and causing glitches in the output voltage.

3.3.5 Effect of the Mismatch

The resistances of a real implementation differ from the nominal value, R_u, because of two effects: a global error and a local fluctuation of parameters. The *i-th* resistance can be expressed as

$$R_i = R_u(1 + \epsilon_a) \cdot (1 + \epsilon_{r,i}), \qquad (3.9)$$

where ϵ_a is the absolute error and $\epsilon_{r,i}$ accounts for the relative mismatch. The voltage at the tap k of an *n-bit* divider biased between V_{Ref} and ground is

$$V_{out}(k) = V_{Ref} \frac{\sum_0^k R_i}{\sum_0^{2^n-1} R_i}; \quad k = 0, \cdots, 2^n - 1. \qquad (3.10)$$

The use of (3.9) in (3.10) shows, as expected, that the output voltages do not depend on the absolute error. The output voltages become

$$V_{out}(k) = V_{Ref} \frac{k + \sum_0^k \epsilon_{r,i}}{2^n - 1 + \sum_0^{2^n-1} \epsilon_{r,i}}; \quad k = 0, \cdots, 2^n - 1. \qquad (3.11)$$

Thus, the error depends on the accumulation of mismatches and is zero at the two endings of the string. A critical case is when the errors are correlated so that their accumulation causes a large *INL*. Consider, for example, a straight string with unity elements spaced by ΔX and gradient α in their relative values. The *k-th* resistor is

$$R_k = R_0(1 + k \cdot \alpha \Delta X); \quad k = 0, \cdots, 2^n - 1 \qquad (3.12)$$

and the output at the tap k becomes

$$V_{out}(k) = V_{Ref} \frac{k + \alpha \Delta X \cdot k(k+1)/2}{2^n - 1 + \alpha \Delta X \cdot (2^n - 1)2^n/2} \qquad (3.13)$$

which is parabolic with initial value 0 and final value V_{Ref}. Recall that the *INL* is $V_{out}(k) - \overline{V_{out}}(k)$. It is plotted in Fig. 3.11 for an 8-*bit* DAC with $\alpha \Delta X = \pm 10^{-4}$ (curves (a) and (b). Notice that the maximum value of the *INL* is at half scale and for the studied case is ± 0.8 *LSB*.

If the resistive divider is folded around its mid point and the only error is a gradient then the unity resistance increases (or decreases) up to the folding point and then returns to the initial value at the end of the string. Since the resistance of the first half equals that of the second half then the midpoint voltage is

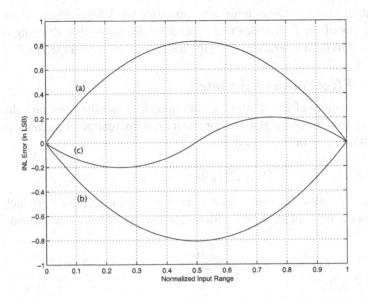

Figure 3.11. INL caused by a gradient in the specific resistance for a linear and a folded (curve (c) resistive string.

correct, thus making the *INL* at the folding point equal to zero. Fig. 3.11 (curve *c*) shows that the folded arrangement with the same $\alpha\Delta X$ used for the other curves reduces the maximum *INL* by a factor 4 ($\pm0.2\%$ *LSB*). Better results are obtained with multiple folding and, more in general, with suitably designed layouts.

Example 3.1

Determine, by computer simulations, the harmonic distortion caused by a $2.5 \cdot 10^{-5}/\mu$ gradient (α) in the resistivity of a straight string of resistors spaced by 4μ. The DAC is an 8-bit Kelvin divider connected between 0 V and 1 V.
Can the result be predicted by an approximate analysis? What resistivity gradient gives a $-95\,dB_{FS}$ spur?

Solution

To estimate the harmonic distortion the Matlab file $Ex3\_1$ uses an input sequence made by 2^{12} points. The noise floor of the fft of an 8-bit DAC output (quantization noise plus process gain) is

$$P_Q = -1.78 - 6.02 \cdot 8 - 10 \cdot log(2048) = -83\,dB;$$

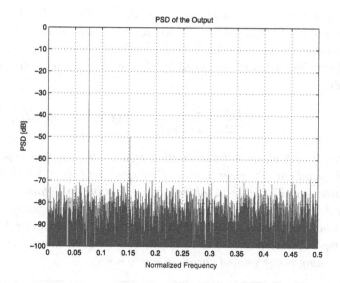

Figure 3.12. Power spectral density of the DAC output.

so any spurs higher than, say, -65 dB (18 dB above the noise floor) are well visible. Spurs at lower levels would require longer sequences. The input signal is a full-scale sine wave with full range amplitude 0.5V and frequency such that a prime integer number (311) of sine waveforms fits into the 2048 sequence. This obtains clean spectral lines without using windowing (although windowing is still recommended).

The function calcSNR calculates the power of the noise and SNDR; results show that, as expected, zero distortion leads to a SNR and SNDR equal to -49.8 dB. Fig. 3.12 shows that the linear graded resistors cause a second harmonic term -50 dB below the input (whose amplitude is normalized to 0 dB). Also, the second harmonic causes a loss of 3 dB in the SNDR. Observe that a gradient $\alpha \cdot \Delta X = 10^{-4}$ makes the resistance of the last element bigger than the first one by only 5%. Despite this small change the obtained harmonic distortion is not negligible.

For an approximate study we neglect the effect of quantization by using the approximation $k \cdot V_{Ref} \simeq V_{in} \cdot 2^n$ in the equation (3.13)

$$V_{out} \simeq \frac{2^n \left[V_{in}(1 + \alpha\Delta X/2) + V_{in}^2 \alpha\Delta X 2^n/(2 \cdot V_{Ref}) \right]}{2^n - 1 + \alpha\Delta X \cdot (2^n - 1)2^n/2}.$$

The use of the trigonometric relationship $sin^2 x = 1/2 \cdot (1 - cos2x)$ in the above equation leads, for an input sine wave, to a second harmonic with amplitude $V_{in}^2 \alpha \Delta X \cdot 2^n / (4 \cdot V_{Ref})$ (the value of the denominator is approximated by 2^n). With a full-scale sine wave it results

$$\frac{A_{spur}}{A_{in}} = \frac{V_{Ref}}{2} \frac{\alpha \Delta X \cdot 2^n}{4 \cdot V_{Ref}} = 0.0032 \rightarrow -49.9\,dB;$$

this is very close to the simulation result. The equation used to find the gradient that gives a $-95\,dB_{FS}$ spur yields $\alpha = 1.39 \cdot 10^{-7}/\mu$. However this value is, obviously, impossible to obtain. For such a demanding specification it is necessary to use common centroid structure for canceling the first order correlated errors or, better, to employ randomized series connections of unity elements to transform the long range graded correlation into noise.

3.3.6   Trimming and Calibration

In addition to a systematic variation (described in the first order by a gradient) the value of unity resistances can be affected by random errors. They are studied by using statistical tools and/or by repeated simulations using to the Monte Carlo technique that foresees a series of simulations with different values of of the mismatched elements, collects the results, and creates a histogram for providing the variance of the output error. The values used in the simulation reflect the statistical distribution of the random error (often a Gaussian).

The local fluctuations of the resistive parameters are unpredictable and cannot be compensated for with layout strategies. Also, the effect of the systematic error cannot be completely cancelled so the resulting global accuracy can end up being inadequate for the application. The solution for these situations is the trimming, or electronic calibration of resistors. The thin film resistor technology is particularly suitable for resistor calibration as the resistors are realized on the top of the passivation layer of the integrated circuit and so their value can be precisely adjusted at the wafer level using a laser. The quality of thin film resistors is good: the temperature coefficient is $< 20\,ppm/°C$ and the matching accuracy is $0.05 - 0.1\%$; with trimming it is possible to improve the matching up to 0.005%.

Note that:

The calibration by trimming or by using fuses (or anti-fuses) is permanent and static as it corrects the mismatches due to the fabrication process inaccuracies.

Unfortunately, the thin film technology is not always available and the trimming process is expensive. An alternative approach to the use of thin film laser trimming is to employ fuses or anti-fuses for opening or closing the interconnections of a network of resistive elements. If for example a large resistance, R_{cal} is in parallel with a unity element, R'_U, through a fuse, then blowing that fuse increases the value of the resistance from $R'_U \cdot R_{cal}/(R'_U + R_{cal})$ thus providing correction capability. Note that correction with fuses or anti-fuses is done during testing (either before or after packaging) and is permanent.

Example 3.2

Estimate the harmonic distortion caused by a random error in the unity resistances of a 10-bit resistive divider. Assume a standard deviation of 0.05 and normal distribution

Solution

The Matlab code Ex3_2 obtains the solution. The simulation uses a 2^{12} sequence that accommodates 91 periods of sine wave. Note that repeated use of the code gives different results because the random errors change every run. Fig. 3.13 shows a plot of the resistor values versus the value of bin. The figure also represents the DNL, which is the quantization step equal to the product of the resistance of the bin by the current of the string. In some cases the value of a resistance can change by two or more σ but the spectrum as given in Fig. 3.14 has

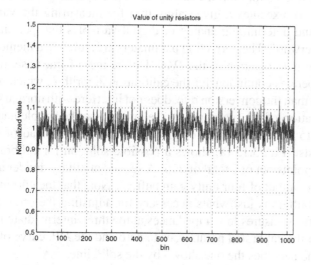

Figure 3.13. Value of the unity resistances versus the position bin.

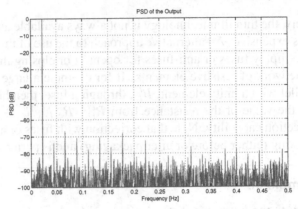

Figure 3.14.   Spectrum of the output signal with the mismatch shown in Fig. 3.13.

tones below -65 dBFS. The INL remains small as it is the accumulation of non-systematic errors.

In addition to trimming and the use of fuses or anti-fuses it is possible to achieve calibration by digital controlled circuits whose setting is defined at power-on (*off-line* calibration) or during the normal operation of the converter (*on-line* calibration). Since this error correction is not permanent the electronic calibration also compensates for slow drifts, like aging or (for off-line) temperature effects.

During off-line calibration the device does not operate as a *DAC*: the calibration network executes a given algorithm for measuring the value of the mismatches and determining analog or digital variables that permit the error correction afterward. The correction parameters are stored in a memory for use during the normal operation of the *DAC*. Unlike the off-line case, the on-line calibration operates continuously: the calibration algorithm, which may run at a lower speed than the conversion rate, does not interfere with normal operation.

The calibration of a large number of elements can be problematic. Often it is enough to match the value of groups of elements instead of the entire set of elements. Fig. 3.15 (a) shows a possible solution for calibrating three intermediate points out of a total of 256. A second auxiliary divider adjusts the voltages at the output of four unity gain buffers. Note that the current required by the unity buffers is just whats necessary for adjusting the errors; also, the use of the buffers ensures low impedances at the three intermediate nodes.

The *INL* at the calibrated points goes to zero and the *INL* curve of Fig. 3.15 (b), dotted line becomes the one shown by the solid line.

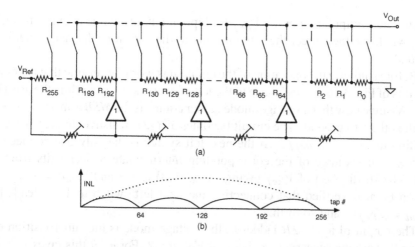

Figure 3.15. (a) Three points calibration of an 8-bit resistive divider. (b) INL improvement.

3.3.7 Digital Potentiometer

A slight variation of the resistive divider *DAC* realizes a digital potentiometer. This has the same functionality as a conventional potentiometer, except that the wiper terminal is controlled by a digital signal, so only discrete steps are allowed. The potentiometer has three accessible terminals: the high side terminal, the wiper terminal, and the low-side terminal. This type of digital potentiometer has 2^n equal resistors with a tap-point between each element for connecting to the wiper. The selection of the wiper position is controlled via an n-bit register value. Communication and control of the device can be supported by a parallel or serial interface.

Furthermore, non-volatile logic can be used to retain the wiper setting when the device is turned off. For devices with volatile logic the wiper is normally set at the mid-tap at power-up.

3.3.8 R–2R Resistor Ladder DAC

A limit of the resistive divider is that the number of resistors increases exponentially with the number of bits. This problem can be solved if an R–$2R$ ladder is used. The R–$2R$ ladder uses an R–$2R$ cell per bit plus a $2R$ termination hence reducing the total number of resistors from 2^n to about $3n$. The ladder network can generate either a voltage or a current as an output variable. Fig. 3.16 (a) and Fig. 3.16 (b) distinguish between the voltage mode and the current mode by using different reference and output connections while the control bits switches every "rung" (or arm) of the ladder toward two common lines. In voltage-mode the arms switch between reference voltage V_{Ref} and ground (assuming a unipolar output, $-V_{Ref}$ for bipolar output); while in current-mode the switches either direct the arm currents to the output or discharge them to ground.

Before studying the operation of the circuit it is worth observing that for both circuits of Fig. 3.16 the resistance to the left of every node (excluding the arm) is $2R$. Assume now that a voltage-mode ladder connects the *MSB* arm to V_{Ref} and all the others to ground. The circuit becomes a R–$2R$ resistive divider of V_{Ref} leading to $V_{Out} = V_{Ref}/2$. If the next left switch is the only one connected to V_{Ref}, the voltage of the corresponding intermediate node results from a $2R - 6/5R$ division of V_{Ref} giving $V_{n-1} = 3V_{Ref}/8$ and $V_{Out} = V_{Ref}/4$. It can be also verified that connecting the next left switch to V_{Ref} leads to $V_{Out} = V_{Ref}/8$ and so forth.

The output of an R–$2R$ ladder in the voltage mode is the superposition of terms that are the successive division of V_{Ref} by 2. For *n-bit* this gives

$$V_{Out} = \frac{V_{Ref}}{2}b_{n-1} + \frac{V_{Ref}}{4}b_{n-2} + \cdots + \frac{V_{Ref}}{2^{n-1}}b_1 + \frac{V_{Ref}}{2^n}b_0, \qquad (3.14)$$

which is the *DAC* conversion of a digital input $\{b_{n-1}, b_{n-2}, \cdots, b_1, b_0\}$.

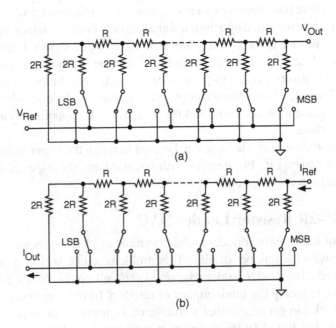

Figure 3.16. (a) Voltage mode R-2R ladder network. (b) R-2R current mode ladder networks.

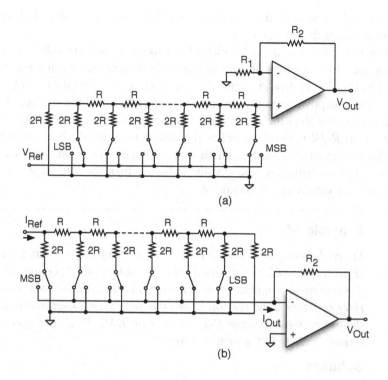

Figure 3.17. (a) Use of the voltage-mode R-2R ladder network. (b) Use of the current-mode R-2R ladder network in an output voltage DAC.

However, the output generated by an R–$2R$ ladder working in voltage-mode is not suitable for direct use: the ladder network requires a volt metric measure that is only made possible by using a buffer or an amplifier with high input impedance. It is also necessary to ensure a low output impedance for driving the *DAC* load.

Note that the output impedance of

Be Aware

A code dependent load of a voltage source causes harmonic distortion. Use reference generators whose output impedance is much lower than the load in any conditions!

the voltage-mode ladder is constant and this is good for stabilizing the amplifier (Fig. 3.17 (a)) or the unity gain buffer used. However, the reference voltage must supply a load that varies with the digital code and this is a drawback: the reference generator must provide an accurate voltage over a wide range of resistive loads. Moreover, some switches connect the nodes of the ladder toward V_{Ref} and others to ground. A low impedance of the reference generator and a low and linear on-resistance of the switches must be capable of charging the parasitic capacitances quickly.

A limit of the *R-2R* ladder network (either voltage or current mode) is that the input-output characteristics are not intrinsically monotonic. Such a feature is granted by the Kelvin divider: when the code increases by one the tap selection moves from one position to the one immediately upper and the voltage of that tap cannot be less than the lower one. On the contrary a code that increases by one in an *R-2R* network switches the connections of all the arms whose bits change. Because of the mismatch it may happen that an increase by one switches off a contribution whose value is higher than the total switched on. The worst case occurs at the mid-point.

Example 3.3

Study, by computer simulations, the effect of the mismatch in the resistors of a ladder network used in the voltage-mode configuration. Assume random mismatch with normal distribution for both arm and riser resistors. Find the conditions that lead to non-monotonicity and determine the maximum INL for an 8-bit DAC. Study the effect of a linear gradient in the arm resistors.

Solution

The file $Ex3\_3$ provides a Matlab code describing the circuit of Fig. 3.18 at the behavioral level for any number of bits. The parameter n (number of bits) is set for this example to 8. The nominal values of arm and riser resistors are 2 Ω and 1 Ω respectively. The parameter "errandom" defines the random error set for this simulation equal to $1/2^n$. The same is for the parameter "errorgradlad" that describes the linear gradient in the arm resistors.

The code estimates the resistance seen at the right and left side of all the nodes of the ladder. The nominal right side values (in Ω) are

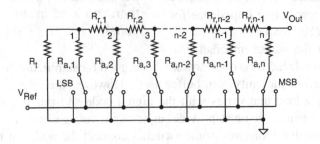

Figure 3.18. Ladder network used in this Example.

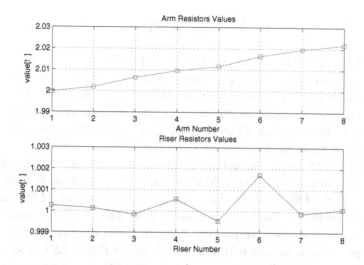

Figure 3.19. Normalized value of the arm and riser resistors.

$$2.0002 \quad 2.0007 \quad 2.0029 \quad 2.0118 \quad 2.0476 \quad 2.2000 \quad 3.0000 \quad Inf$$

The nominal left side values are 2 Ω. The parallel resistance of right and left resistors, $R_{par}(i)$, and the arm resistance $R_a(i)$ give the arm node voltage due to a bit 1 on the same arm

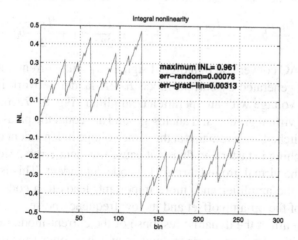

Figure 3.20. INL with the mismatch between resistors given in Fig. 3.19.

$$V_i = V_{Ref} \frac{R_{par}(i)}{R_{par}(i) + R_a(i)}.$$

The current that flows on the right side is $V_i/R_{r,i}$. It is again and again divided by the next arm connections and only a fraction of it reaches $R_{a,n}$ to give the voltage contributed by the i-th bit, $V_{DAC}(i)$. The Matlab code uses all the $V_{DAC}(i)$ voltages to calculate the output. Observe that different simulations yield different results since the random numbers change every run.

Fig. 3.19 shows the normalized resistor values used in one of the simulations while Fig. 3.20 shows that the corresponding maximum INL, 0.6 LSB, is at the mid-scale. An INL lower than 1 LSB denotes monotonicity; however the random error used is not low enough to ensure monotonic responses. Repetitive runs identify some random error combinations that give an INL larger than 1 LSB. Also, positive linear gradients $errorgradlad \geq 0.8/2^n$ give non monotonicity while negative gradients can cause missing steps.

The current-mode circuit performs a successive division by two of the reference current I_{Ref} provided that the voltage at the output node is ground. The right-most node of Fig. 3.16 (b) has two paths for the current: one is through the $2R$ rung, the other is through the rest of the network, whose resistance is also $2R$. The same holds for the next left node and again until the last cell divides the remaining current by two using the $2R$ termination resistance. The superposition of the currents selected by the switches forms the output current

$$I_{Out} = \frac{I_{Ref}}{2} b_{n-1} + \frac{I_{Ref}}{4} b_{n-2} + \cdots + \frac{I_{Ref}}{2^{n-1}} b_1 + \frac{I_{Ref}}{2^n} b_0 \qquad (3.15)$$

which is the DAC conversion of a digital signal $\{b\}$ with current output.

The current generated by current-mode R–$2R$ ladder network is normally converted to a voltage with an operational amplifier (Fig. 3.17 (b)). Since the voltages of the virtual ground and analog ground are equal the parasitic capacitance of the switched node remains at the same voltage independent of the code. Thus, the switching is faster and does not cause transient errors. However, the impedance at the virtual ground terminal is code dependent. This is a problem for stabilizing the amplifier and, more important, leads to a code dependent amplification of the op-amp offset and its low frequency noise.

Glitches can affect the dynamic response of the current-mode implementation. The transition from one code to another switches some arms from ground to the output or vice-versa. This current switching cannot be instantaneous and it may happen that some large terms continue to flow into the virtual ground while the others are already switched on. This momentary current exceeds the expected value causing a positive glitch in the output. Similarly, a negative glitch may occur.

Mismatch is also problematic for the current mode circuit: a code that increment by one can switch off a current whose value is higher than the current switched off causing non monotonicity. Consider a current-mode R-$2R$ network nominally switching at the mid-scale from $I_{ref}(1/2 - 1/2^n)$ to $I_{ref}/2$. If the mismatch errors are such that the MSB current is $I_{ref}(1 - \epsilon)/2$ the above transition becomes

$$I_{ref}(1 + \epsilon)(1/2 - 1/2^n) \rightarrow I_{ref}(1 - \epsilon)/2; \qquad (3.16)$$

the step amplitude is

$$\Delta I \simeq I_{ref}(1/2^n - \epsilon). \qquad (3.17)$$

Thus, if $\epsilon > 1/2^{n+1}$ the step amplitude is negative and the transfer characteristics become non monotonic.

The layout of resistors consumes a relatively large area if the value of unity resistance is large and the used layer has a low specific resistance (like low-resistive poly). For medium accuracy it is possible to save silicon area by replacing the resistors with MOS transistors. Fig. 3.21 shows a possible current-mode implementation. Since all the transistors have a *p-channel* equal aspect ratio the single resistance is made by one element while the series of two transistors achieves the double R and the termination resistance. Observe that two equal parallel MOS structures divide the input current into equal parts regardless of the non-linear response of the single element: the only requirement is that the two parts operate at the same non-linear point.

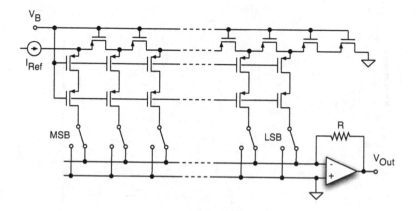

Figure 3.21. R-2R architecture with MOS transistors replacing the resistors.

Example 3.4

The resistances used in an 8-bit current-mode R-2R DAC have a normal distribution with standard deviation $2/2^8$. Determine by computer simulations the INL and the output spectrum with a full-scale input sine wave. Use various sets of random mismatches and observe the effect on the INL.

Solution

In current-mode the reference current is successively divided by two with a sequence of cells of the $R - 2R$ ladder. The mismatch between resistors changes the resistance at the right of each splitting node making the current in the arm and the one delivered to the right side of the $R - 2R$ network different.

The file Ex3_4 is the basis of this solution. The code generates a random or a gradient parameter in the resistors. In addition, it enables the setting of a specific value of an arm or riser element. Fig. 3.22 shows the normalized resistances obtained in a specifically selected run. The values range between 0.994 and 1.007. The obtained INL plot shown in Fig. 3.23 has minor variation until the big jump by about 1 LSB that occurs at the mid-range. The corresponding spectrum of a 4096 sequence with 31 sine wave periods is shown in Fig. 3.24. The

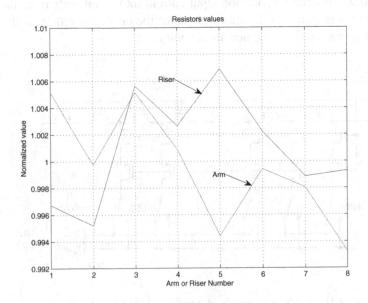

Figure 3.22. Normalized value of the R-2R resistors used in this Example.

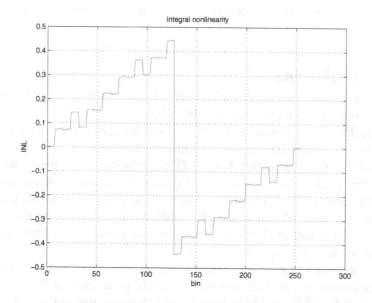

Figure 3.23. INL caused by the resistance value of Fig. 3.22.

spectrum contains third, fifth and seventh visible harmonics. Their amplitude is between -55 dB and -68 dB.
Observe that the simulation uses a -0.1 dB input followed by an

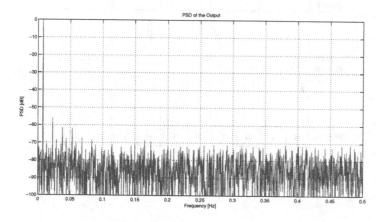

Figure 3.24. Spectrum of the output current.

8-bit quantizer because the use of a full-scale sine wave can generate overrange codes.

3.3.9 Deglitching

The non-synchronous switching in the $R-2R$ network produces glitches that can cause non-linearity and harmonic distortion.

The deglitching technique is a method used to improve the circuit performances by employing a track-and-hold after the *DAC*. Fig. 3.25 illustrates the method. The voltage at the output of the *DAC* is replicated by a track-and-hold (*T&H*). The track phase starts when the output of the *DAC* is stable and the possible glitches are already settled, while the hold phase starts before the new data enters the *DAC*. Therefore the *DAC* glitches that occur during the transitions of the input data are isolated from the output throughout the time period during which they are expected to occur.

The generation of the phases requires special care. Moreover, the possible distortion and noise of the *T&H* may affect the *DAC* performances. The linearity and noise of the *T&H* must be at least 10 *dB* better than that of the *DAC*.

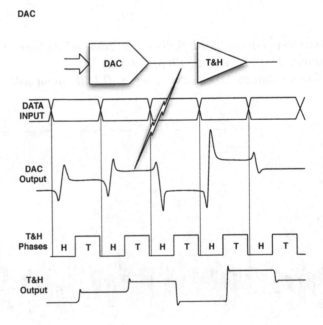

Figure 3.25. Deglitch circuit and possible voltages.

Audio *DAC*s require high-linearity for low-level signals that, unfortunately, occur around the mid-point where the number of bits that changes is at a maximum and the glitch is large. The use of a track and hold partially solves the problem although the linearity requirements become challenging. A possible alternative solution to deglitching consists of the use of a digital offset that moves the low-level range away from the mid-point. The mid-scale distortion remains the same but it affects medium-level signals. However, the addition of a *dc* term that is not important for audio applications reduces the usable range of the *DAC*.

3.4 CAPACITOR BASED ARCHITECTURES

The capacitor based architectures realize the *DAC* function by using an op-amp or an *OTA* with a capacitive divider or a capacitive *MDAC*. This section discusses these two different approaches in detail.

3.4.1 Capacitive Divider DAC

The series of two capacitors, initially discharged, connected between V_{Ref} and ground yields a middle point voltage given by (Fig. 3.26 (a)

$$V_{Out} = V_{Ref} \frac{C_1}{C_1 + C_2}.$$ (3.18)

As shown in Fig. 3.26 (b) the capacitors used are multiples of a unity element (C_U) and have total value $2^n C_U$. C_1 is made by k unity elements and C_2 is the rest of the 2^n thus leading to a voltage equal to $V_{Out} = V_{Ref} \cdot k/2^n$ as required by a *DAC*. Often the two capacitors make an array of binary weighted elements ($C_U, C_U, 2C_U, 4C_U, \cdots 2^{n-1}C_U$) that sums up $2^n C_U$ (Fig. 3.26 (c). The combination of the elements realizes the two capacitors by connecting the

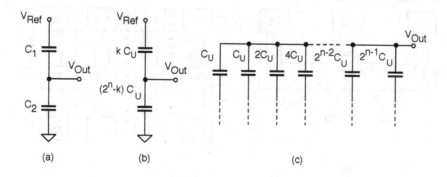

Figure 3.26. (a) and (b) Simple capacitor divider. (c) Array of binary weighted capacitors.

bottom terminals to ground or to V_{Ref} depending on the binary combination required to obtain k. If the full scale voltage is $(2^n - 1)V_{Ref}/2^n$, then one unity capacitance is always connected to ground, but if the transfer characteristic is shifted by half LSB then a half unity element is connected to ground and half to V_{Ref} resulting in a full scale voltage of $(2^n - 1/2)V_{Ref}/2^n$.

Integrated capacitances are made from two conductive plates separated by a thin layer of oxide and located a few microns or less from the substrate. There are two main configurations: with plates one on the top of the other (the capacitor can be poly-oxide-poly or Metal-Insulator-Metal, *MIM*, Fig. 3.27 (a)) or with plates made of side by side fingers of metals and vias (Metal-Metal Comb Capacitor, *MMCC*) as shown in 3.27 (b). Since the structures are close to the substrate in both cases, the parasitic from the plates to the substrate is not negligible. The poly-poly and the *MIM* have a bottom plate that partially shields the top one thus giving a top plate parasitic that is smaller than that of the bottom plate. The two parasitic capacitances of the *MMCC* structures are almost equal. Also, for *MIM* and *MMCC* the use of high level metals leads to a small parasitic.

The parasitic of the plate connected to V_{Ref} or ground does not affect the capacitive divider as it receives the required charge by low impedance nodes. Instead, the parasitic capacitances connected to the output node affect the capacitive division between output and ground. The output voltage for a binary weighted array of capacitors becomes

$$V_{Out} = V_{Ref} \frac{\sum_1^n b_i C_i}{\sum_0^n C_i + \sum_0^n C_{p,i}}, \qquad (3.19)$$

where b_i are the bits of the digital control and C_i and $C_{p,i}$ are the value and parasitic associated with the unity elements of the array. If the parasitic

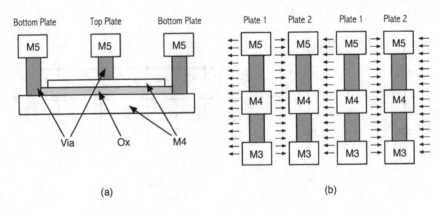

(a)   (b)

Figure 3.27.   (a) MIM capacitance. (b) MMCC capacitance using three metal levels.

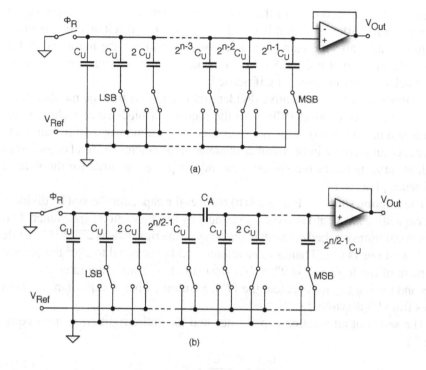

Figure 3.28. (a) n-bit Capacitive divider DAC. (b) The use of an attenuator in the middle of the array reduces the capacitance spread.

capacitances are independent of the output voltage equation (3.19) leads to a gain error only. However, non-linear parasitics that change with the output voltage cause harmonic distortion.

Any resistance loading the output will drain current and thus absorb the charge stored on the capacitors. Therefore, the voltage of the capacitive divider must be measured by a very high resistance, typically a capacitive probe realized by the input *MOS* transistor of an operational amplifier or an *OTA*.

Fig. 3.28 (a) shows the implementation of an *n-bit DAC*. The conversion is preceded by a reset phase Φ_R during which the array is discharged by setting all the switches such that the bottom plates are grounded. During the conversion phase the reset switch opens and the bottom plates of the binary weighted capacitors are connected to

> **Remember!**
>
> Capacitor based DAC must avoid discharging at the high impedance output node. Measure the output with a buffer or an amplifier with infinite input resistance.

V_{Ref} or ground according to the value of the control bits. The input capacitance of the buffer is in parallel with the parasitic capacitance of the array, thereby it increases the gain error and, being non-linear, the harmonic distortion. The op-amp of Fig. 3.28 (a) is in the unity gain configuration: a suitable feedback network obtains gain, if needed.

A drawback of the capacitive divider architectures is that the number of elements increases exponentially with the number of bits. Since technological limits and matching requirements determine the minimum size of the unity capacitance an increase in the number of elements augments the total capacitance and, in turn, increases the silicon area and the power required for the voltage reference generator.

The solution shown in Fig. 3.28 (b) reduces the capacitors' count by dividing the capacitive array into two parts separated by an attenuation capacitance, C_A. The maximum switched element in the array of the right side is $2^{n/2-1}C_U$ while the lowest one is C_U. Thanks to an attenuation factor equal to $2^{n/2}$ the biggest element of the left array is $2^{n/2-1}C_U$ instead of $\frac{1}{2}C_u$ allowing to convert $n/2$ bits and having C_U to represent the *LSB*. Summing up, the capacitances count goes from 2^n down to $2 \cdot 2^{n/2} - 1$.

The series of attenuation capacitance and the entire right array must equal the C_U

$$\frac{C_A \cdot 2^{n/2}C_U}{C_A + 2^{n/2}C_U} = C_u \tag{3.20}$$

which yields the value of C_A

$$C_A = \frac{2^{n/2}}{2^{n/2} - 1}C_U. \tag{3.21}$$

Unfortunately the value of C_A is a fraction of C_U: obtaining the necessary accuracy requires care in the layout.

3.4.2   Capacitive MDAC

The buffer used in the capacitive divider must allow an input dynamic range equal to the *DAC* reference interval and, at the same time, must ensure high linearity. The requirements are not easy for many operational amplifiers or *OTA*s which usually have their non-inverting terminal connected to a fixed voltage. A solution that avoids the swing of the input common-mode is the capacitive *MDAC* whose circuit schematic is shown in Fig. 3.29.

The circuit obtains the output voltage with a programmable gain amplifier made of an op-amp and capacitors. The input capacitance is a binary weighted array of capacitors connected under the control of the digital input to the reference; the feedback capacitor 2^nC_U. If the selected input elements are $k \cdot C_U$ the output voltage, with an ideal op-amp, is $-V_{Ref} \cdot k/2^n$.

The circuit of Fig. 3.29 also performs other functions, namely it resets the array and performs offset cancellation. During the phase Φ_R the op-amp is in the unity gain configuration and makes the offset available at the inverting terminal. The input array and the $2^n C_U$ capacitances are pre-charged to the offset instead of a conventional reset. During the complementary phase $\overline{\Phi_R}$ the binary weighted elements are connected to V_{Ref} or ground under the control of the bits, b_i. Thanks to the pre-charge to the offset the output voltage becomes

$$V_{Out} = -\frac{\sum_0^{n-1} b_i 2^i C_U}{2^n C_U}, \qquad (3.22)$$

assuming that the offset does not change in the period from Φ_R to $\overline{\Phi_R}$.

The pre-charge to the offset requires that the op-amp is used in unity gain configuration, thus relying on the slew-rate and bandwidth of the op-amp. Additionally, a feedback factor equal to 1 during the offset pre-charge and $1/2$ during the conversion makes the compensation of the circuit complicated. For high-speed applications when the offset is not important the reset of the array is done with a grounded switch while the reset of the feedback capacitor is obtained by simply shorting its terminals.

Observe that because of the reset phase the output is valid only during $\overline{\Phi_R}$ while during the other phase the output of the op-amp is zero or the offset.

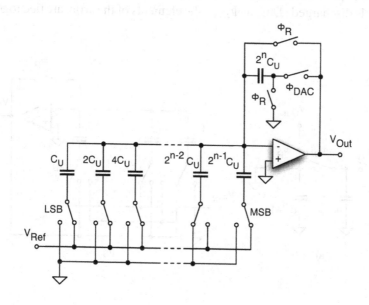

Figure 3.29. Capacitive MDAC with offset cancellation.

The specifications of the reconstruction filter become more difficult unless a track-and-hold maintains the value of the output during the reset periods.

3.4.3   "Flip Around" MDAC

The *MDAC* studied in the previous Subsection uses $2 \cdot 2^n - 1$ unity elements, twice as many as is needed in the other studied configurations. The "flip around" architecture shown in Fig. 3.30 avoids this increase while preserving the benefit of the use of the virtual ground. The basic idea is to charge k unity capacitances to V_{Ref} and to connect them in parallel to $2^n - k$ discharged capacitances (Fig. 3.30 (a)). The charge on the charged capacitances is shared with the discharged ones and leads to the required attenuation of V_{Ref}

$$V_{Out} = V_{Ref} \frac{kC_U}{2^n C_U}. \tag{3.23}$$

The circuit of Fig. 3.30 (a) works only at the concept level as it is sensitive to the top-plate parasitic capacitances that must be accounted for the charge sharing. Moreover, the circuit does not buffer the generated voltage.

The scheme of Fig. 3.30 (b) obtains the buffer functions together with a parasitic insensitivity. During the reset phase, Φ_R, the bottom plates of a fraction of the binary weighted array is charged to V_{Ref}, the rest is discharged to ground. All the top plates are connected to ground thus ensuring that the top-plate parasitic is discharged. During Φ_{DAC} the elements of the array are tied together

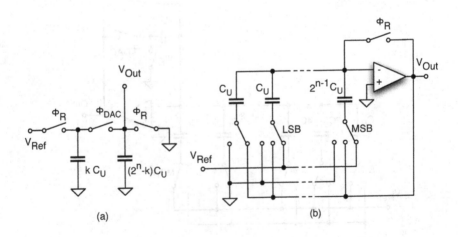

Figure 3.30.   Flip around DAC. (a) Basic idea. (b) Parasitic insensitive circuit.

by the right switch and connected to the output of the Op-Amp. The charge sharing is determined by the charge on the top plates while the low impedance of the Op-Amp provides the charge possibly required by the bottom plate parasitic capacitances.

The offset and the input referred noise of the op-amp is added to the voltage given by the charge sharing. The circuit can be made offset insen-

> **Keep in Mind**
>
> Any capacitor-based archi-tecture requires a reset phase to make sure that the array is initially discharged. Since the output voltage is not valid dur-ing the reset, a track and hold sustains the output.

sitive if the top plate of the array is pre-charged to the offset. This is possible if during Φ_R the op-amp is in the unity gain configuration and replaces the grounded switch for charging or discharging the top plates of the array.

3.4.4 Hybrid Capacitive-Resistive DACs

The resistive *DAC* and the capacitive *DAC* can be combined together to give hybrid segmented solutions. Since the capacitive based architecture requires a buffer to preserve the output voltages the resistive based *DAC* is used for

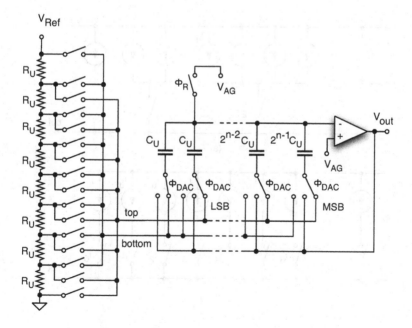

Figure 3.31. Hybrid architecture made by a Kelvin divider followed by a flip-around MDAC.

the coarse conversion while the capacitive *DAC* provides the fine conversion. Fig. 3.31 shows a hybrid architecture with a 3-bit Kelvin divider followed by a capacitive flip-around *MDAC*. During the reset phase the switches select one segment of the resistive divider used for charging the selected fraction of the array to V_{MSB+1}-V_{AG} and the remaining part, including the extra unity element, to V_{MSB}-V_{AG}. The phase Φ_{DAC} performs the normal control of a flip around *MDAC* giving

$$V_{out} = V_{MSB} + \Delta_{MSB} \frac{k}{2^n} \tag{3.24}$$

where k refers to the *LSB* driving the capacitive *MDAC*.

3.5   CURRENT SOURCE BASED ARCHITECTURES

This category includes current switching architectures with binary weighted, unary current sources or segmented methods which use different selection strategies for obtaining optimum linearity.

3.5.1   Basic Operation

An *n-bit* current steering *DAC* switches k out of $2^n - 1$ unity current generators toward the output node under the control of the *n-bit* digital input,

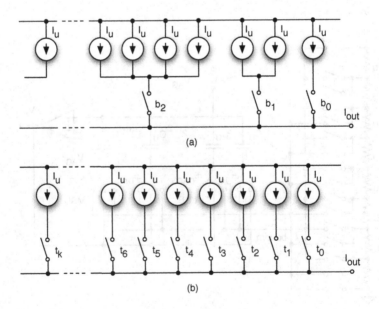

(a)

(b)

Figure 3.32.   (a) Binary weighted control. (b) Unary weighted control.

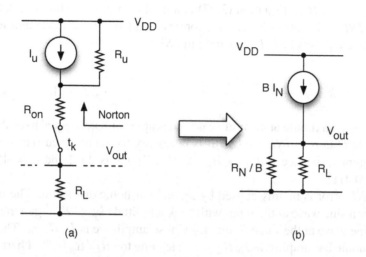

Figure 3.33. (a) Simple model of a current steering *DAC* cell. (b) Equivalent total circuit.

$b_{n-1}, \cdots, b_2, b_1, b_0$. With a binary weighted control 2^i unity elements are connected in parallel and switched together by the *i-th* bit. With a unary control it is necessary to transform the binary input into a thermometric signal to switch the unity current generators one by one. Fig. 3.32 shows the two possible arrangements. The output current is

$$I_{out} = I_u \left[b_0 + 2b_1 + 2^2 b_2 + \cdots + 2^k b_k + \cdot + 2^{n-1} b_{n-1} \right]. \qquad (3.25)$$

Real current sources and real switches do not realize the ideal functions used in Fig. 3.32. It is therefore necessary to account for at least the shunt resistance R_u across each current source and the on-resistance R_{on} of the switch, as shown by the equivalent circuit of a single cell in Fig. 3.33 (a). The use of the Norton equivalent circuit obtains a current source, I_N and a grounded resistance R_N

$$I_N = \frac{I_u R_u + V_{DD}}{R_u + R_{on}}; \quad R_N = R_u + R_{on}. \qquad (3.26)$$

Assume that $k = b_0 + 2b_1 + 2^2 b_2 + \cdots + 2^{n-1} b_{n-1}$ is the number of cells that are switched-on; suppose also that the current is injected into a resistance R_L. The global equivalent circuit becomes the one of Fig. 3.33 (b) where the Norton equivalent current sources are summed up and the k equal R_N resistors are connected in parallel. The output voltage becomes

$$V_{out} = k I_N \frac{R_L \cdot R_N / k}{R_L + R_N / k} = I_N R_L \frac{k}{1 + \alpha k} \qquad (3.27)$$

where $\alpha = R_L/R_N$. Equation (3.27) is a nonlinear relationship that deteriorates the *INL* and causes distortion. Moreover, equation (3.27) indicates a gain error. The endpoints fit *INL* measured in *LSB* is

$$INL(k) = \frac{k\left[1 + \alpha(2^n - 1)\right]}{1 + \alpha k} - k; \quad k = 0, \cdots, 2^n - 1. \tag{3.28}$$

This has a maximum at the mid-scale and is approximately equal to $\alpha \cdot 2^{2n-2}$. In order to obtain an *INL* <1 *LSB* it is necessary to use unity current sources whose shunt resistance is $R_u > R_L \cdot 2^{2n-2}$. If R_L is $25\,\Omega$ and $n = 12\text{-}bit$, $R_u > 100\,M\Omega$.

The *INL* error is mainly caused by second harmonic distortion. The use of (3.27) for a sine wave at the input with peak amplitude $k_p = 2^{n-1}$ generates an output sine wave at the same frequency whose amplitude is $I_N R_L k_p$. The second harmonic has amplitude $I_N R_L \alpha k_p^2/2$ leading to $(R_L/R_u) \cdot 2^n/4$ harmonic distortion. Therefore, with $R_u = 100\,M\Omega$ and R_L is $25\,\Omega$ the *THD* of a 12-*bit* DAC becomes $-72\,dB$, a low figure but not enough for some applications.

Since differential circuits eliminate even order errors the use of differential current steering *DAC*s almost avoids the limit caused by the finite current source shunt resistance.

Be Aware

The output resistance of the unity current source causes second order harmonic distortion. The use of differential architecture relaxes the requirements on the unity current source.

Note that the use of a load resistance (typically the termination resistance of a coaxial cable) is only used for very high speed. For medium speed it is possible to use an operational amplifier that receives current through the virtual ground. The voltage at the output of the current steering *DAC* is constant and the distortion effect vanishes.

Example 3.5

Simulate the effect of a finite output resistance in the unity current sources and a non-zero on-resistance of the switches. Estimate the INL and the harmonic distortion for a 12-bit DAC with $R_u = 200M\Omega$, $R_L = 25\Omega$ and $R_{on} = 100\Omega$. Calculate the distortion for a differential implementation.

Solution

The use of the approximate formulas (3.28) for INL and harmonic distortion predicts the following values: INL = 0.522, HD = −78 dB. The computer simulations are obtained by using the file Ex3_5.

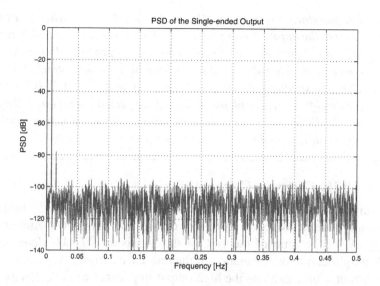

Figure 3.34. Spectrum of the single ended output.

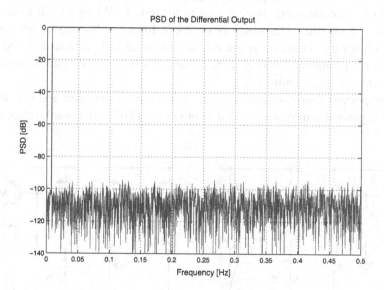

Figure 3.35. Spectrum of the differential output.

The input signal is a sine wave transformed into a 12-bit discrete amplitude signal with frequency such as to accommodate an integer number of sine waves in the sequence used to calculate the FFT.

The simulation results give INL = 0.539, HD = −78 dB, values very close to the approximate estimation. The simulation of the differential implementation uses two complementary input signals whose DAC conversions are subtracted to give the differential output. Fig. 3.34 gives the spectrum for the single ended output; Fig. 3.35 is the result for the differential implementation. As expected the second harmonic at −78 dB below the maximum amplitude sine wave almost vanishes for the differential case.

3.5.2   Unity Current Generator

The simple or cascode current mirrors shown in Fig. 3.36 are the basis for making unity current sources with *BJT* or *CMOS* transistors. The emitter area ratio Q_1/Q_2 or the ratio of the *W/L* of the transistors M_1-M_2 determine a possible amplification or attenuation of the reference current.

The current source exploits the high output impedance at the collector of a bipolar transistor or at the drain of an *MOS* transistor. The output resistance of the simple *MOS* current mirror is $r_{ds,2}$ while the one of the cascode is $r_{ds,2}\, g_{m,4}\, r_{ds,4}$. Since the gain $g_{m,4}\, r_{ds,4}$ is in the order of one hundred (or more for bipolar elements) having $r_{ds,2}$ in the ten of $k\Omega$ range leads to output resistance of several $M\Omega$, which is sufficient for a single-ended circuit with medium linearity. However if the output cannot be fully differential and the linearity requirements are very demanding a double cascode can be used to implement the current mirror.

An important design decision concerns the choice between a single main reference generator mirrored by all the cells of the *DAC* or many sub-references

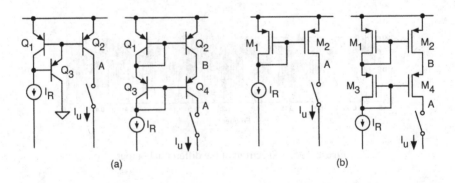

Figure 3.36.   (a) Simple and cascode BJT current mirror. (b) Simple and cascode MOS current mirror.

that are mirrored locally to serve fractions of the *DAC* cells. In the former case the distributed signal is a voltage while the latter solution distributes currents that are brought near the mirroring group for local use but require extra circuitry for duplicating the main reference and obtaining the required number of accurate replicas.

Keep Note

It is worth using current sources with high output resistance to secure linearity but in many situations it is necessary to ensure the best trade-off speed accuracy.

When many bipolar cells mirror the same reference the total base current can be a large fraction of the reference thus making it necessary to use a circuit that compensates for the base current as it is done by Q_3 in the simple bipolar current mirror.

For *MOS* circuits the gate voltage of all the mirroring elements are equal as the *dc* gate current is zero, but the source voltages can be different because of the parasitic drop along the V_{DD} connections (or the V_{SS} connections for the complementary case). The relative variation of the overdrive voltage diminishes if $V_{gs} - V_{th}$ is large; therefore, a large overdrive leads to smaller errors. Actually, what matters is the difference between the errors. Using a tree of matched metal lines for the supply connections makes the voltage drops along the V_{DD} connections equal and equalizes the unity currents.

Another possible source of error is the systematic or random mismatch of the fabrication parameters that control the saturation current of a MOS transistor

$$I_D = \beta \left(V_{gs} - V_{th} \right)^2 \tag{3.29}$$

where $\beta = \mu C_{ox}(W/L)$; μ is the surface carrier mobility, and W and L are the width and length.

Assume that β is the average value for the mirror and that $\Delta\beta$ is the mismatch. Also, V_{th} is the average of the thresholds and ΔV_{th} is the mismatch. Neglecting second order terms the currents in M_1 and M_2 are approximated by

$$I_1 = \overline{I} \left(1 + \frac{\Delta\beta}{\beta} + \frac{2\Delta V_{th}}{V_{gs} - V_{th}} \right) \tag{3.30}$$

$$I_2 = \overline{I} \left(1 - \frac{\Delta\beta}{\beta} - \frac{2\Delta V_{th}}{V_{gs} - V_{th}} \right) \tag{3.31}$$

where $\overline{I}$ is the average current. The above equations show that the relative error on β directly affects the current error while ΔV_{th} is divided by $(V_{gs} - V_{th})/2$. Therefore, similar to the V_{DD} drop error the use of a large overdrive attenuates the limit of the threshold voltage mismatch.

The variation of the fabrication parameters causes an error in the *MOS* saturation current due to a correlated and a random term, the variance of the random part being

$$\frac{\Delta I^2}{I^2} = \frac{\Delta \beta^2}{\beta^2} + \frac{4 \Delta V_{th}^2}{(V_{gs} - V_{th})^2}. \tag{3.32}$$

The variance of β and V_{th} depends on the process and are modeled by the relationships

$$\frac{\Delta \beta^2}{\beta^2} = \frac{A_\beta^2}{WL}; \quad \Delta V_{th}^2 = \frac{A_{VT}^2}{WL}, \tag{3.33}$$

where the parameters A_β, and A_{VT} are normally provided by the foundry.

The above study shows that for a given process the accuracy of the current mirror improves by increasing the gate area of the used transistor; namely, $\Delta I / I$ halves if the gate area increases by a factor of 4.

Example 3.6

The $\Delta I / I$ that obtains a given required yield in a 12-bit current steering DAC is 0.3%. Estimate the transistor sizes for a 0.4 V overdrive. The value of μC_{ox} is 39 $\mu A/V^2$ and the current in the unity element is 4.88 μA. Plot the error $\Delta I / I$ as a function of the area whose normalized value changes from 1/10 to 10. Use the following accuracy parameters: $A_{VT} = 2\,mV \cdot \mu$, $A_\beta = 0.3\% \cdot \mu$.

Solution

The use of equations (3.32) and (3.33) leads to

$$(WL)_{min} = \left(A_\beta^2 + \frac{4 \cdot A_{VT}^2}{(V_{gs} - V_{th})^2} \right) \Big/ \frac{\Delta I^2}{I^2} \tag{3.34}$$

giving $WL = 12.11 \mu^2$. Since $\beta = I_U / V_{ov}^2 = 30.5\,\mu A/V^2$, $W/L = 0.8$. The obtained aspect ratio is small because the unity current is small and the used overdrive is large. Using the area and aspect ratio it results $W = 3.1\mu$ and $L = 3.9\mu$.

Equations (3.32) and (3.33) show that the error $\Delta I / I$ is inversely proportional to the square root of WL. Accordingly, an increase or a reduction by a factor 10 of WL improves or worsens the error by $\sqrt{10}$. Fig. 3.37 shows the result. The use of an aggressive scaling factor reduces the silicon area and the parasitic capacitances but the yield diminishes significantly.

For the solution of this example use the file Ex3_6.

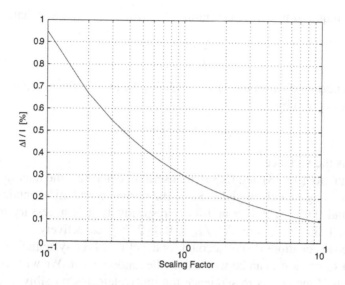

Figure 3.37. $\Delta I/I$ versus the scaling factor.

3.5.3 Random Mismatch with Unary Selection

The effect of the random and systematic error, $\Delta I = \Delta I_r + \Delta I_s$, of the unity current sources on the DAC linearity depends on the selection method used. With the unary approach the thermometric signals select the first k elements of a given sequence and if the code increases by one then an additional element is selected. The endpoint-fit error for k unity current sources selected is

$$\Delta I_{out}(k) = \sum_1^k \Delta I_{r,j} - k\overline{\Delta I_r} + \sum_1^k \Delta I_{s,j} - k\overline{\Delta I_s} \qquad (3.35)$$

where $\overline{\Delta I_r}$ and $\overline{\Delta I_s}$ are average errors used to cancel the possible gain error and the subscript j denotes the generic j-th unity current source.

The terms of the first sum are uncorrelated while the terms of the second can be fully or partially correlated depending on the selection scheme. Consider only the first two terms whose variance is given by

$$\Delta I_{out,r}^2(k) = k \cdot \Delta I_r^2 - \frac{k^2}{2^n - 1}\Delta I_r^2 \qquad (3.36)$$

and has a maximum at the mid-range $k = 2^{n-1}$

$$\Delta I_{out,r,max}^2 \simeq 2^{n-2}\Delta I_r^2. \qquad (3.37)$$

If it is required that the maximum *INL* error must be lower than half *LSB* then

$$\frac{\Delta I_r}{I_u} < 2^{-n/2} \qquad (3.38)$$

With a normal distribution of $x = \Delta I_r / I_u$ the probability of having an error equal to x is given by

$$p(x) = \frac{1}{\sigma\sqrt{2\pi}} e^{\frac{-x^2}{2\sigma^2}} \qquad (3.39)$$

where σ is the variance.

The equations (3.38) and (3.39) are useful to determine the maximum error $\Delta I_r / I_u$ leading to a given yield. Since the normal distribution obtains a yield of 0.99 and 0.999 at $2.57\,\sigma$ and $3.3\,\sigma$ respectively it is necessary to ensure $\Delta I_r / I_u < 0.39 \cdot 2^{-n/2}$ and $\Delta I_r / I_u < 0.3 \cdot 2^{-n/2}$ respectively.

The above equations do not account for the effect of the systematic mismatch which, in some cases, can be worse than the random term. We will see shortly that one used method is to sequence the unity elements possibly affected by systematic errors using a pseudo-random path thus transforming the systematic error into a pseudo-random error. Even with pseudo-random errors instead than real random errors the above study is valid if the selection sequence makes the distribution of $\Delta I_s / I_u$ normal.

3.5.4   Current Sources Selection

The unity current sources are often arranged in a two-dimensional array whose optimum shape is a square with $2^{n/2}$ lines and columns if the number of bits n is even. The simplest thermometric selection is sequential by lines and columns starting from one corner of the array. Fig. 3.38 shows the block diagram of a possible 8-*bit DAC* with 70 selected cells (shadowed) corresponding to the input code (01000110). A possible grading error in the x and y directions causes *INL*. If the gradients are γ_x and γ_y and the cell spacing in the x and y directions are Δx and Δy respectively then the current of the cell in the i,j position is

$$I_u(i,j) = \overline{I_u}(1 + i \cdot \gamma_x \Delta_x)(1 + j \cdot \gamma_y \Delta_y) \qquad (3.40)$$

where $\overline{I_u}$ is the nominal unity current.

The sequential selection of Fig. 3.38 gives an error that changes linearly with i. The resulting *INL* is a sequence of parabolic arcs with curvature due to the x gradient sitting on another arc with curvature due to the y gradient. Using a more complex selection technique reduces the *INL*. An example is a symmetrical selection of lines and columns around the centroid of the array. In this case the linear grading is compensated by every pair of unity cells thus obtaining a lower

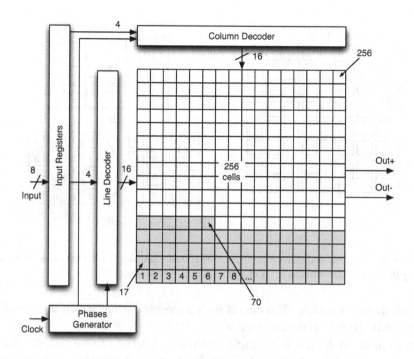

Figure 3.38. 8-bit DAC with thermometric selection of current cells.

INL. The method is normally adequate for lines or columns up to 8 elements for which the sequence of the selected elements is 1, 8, 2, 7, 3, 6, 4, 5.

The method depicted in Fig. 3.39 (a) uses a shuffling of lines and columns. The goal is to randomize the mismatches and to keep the accumulated error low. The first cell $(1, 1)$ is in the middle of the array. The second cell $(1, 2)$ is in the bottom left corner. The

Notice

The unary selection ensures monotonicity and enables flexibility but requires one control signal per element. This makes the unary selection unpractical for 8-bit or more.

third $(1, 3)$ is in the top right corner and so forth. The average of the currents of the second and the third cell is almost equal to the one of the first. In addition to shuffling elements many architectures use dummy cells around the array to ensure the same boundary layout of the actual cells.

Fig. 3.39 (b) limits the error by using multiple replicas of the reference: the array is divided into four sectors that locally mirror the reference for each cell of the sector. The small distance of the sector cells from the local reference limits

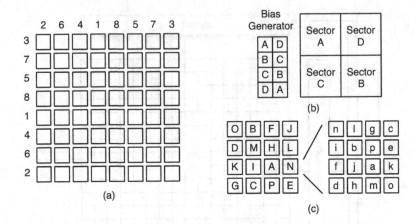

Figure 3.39. (a) Line and column shuffling. (b) Use of multiple current references for minimizing the threshold mismatch. (c) Random walk selection of unity cells.

the threshold mismatch. The circuit uses a master current which is carefully mirrored in the bias generator section.

In addition to the use of multiple references, instead of shuffling the lines and columns of the sector, Fig. 3.39 (c) divides the sectors into quadrants that are sequenced randomly. The cells inside each quadrant are also selected using a random walk.

Since the *INL* of the *DAC* is the accumulation of errors a thermometric selection with the random walk of Fig. 3.39 (c) transforms the correlated error into pseudo-noise allowing the equation derived for random errors to be applied to the gradient errors. The switching sequence used in Fig. 3.39 (c) from A to P for the 16 quadrants and from a to p inside a quadrant is one of the possible solutions for obtaining a random walk. It has been optimized to compensate for the quadratic like residual errors. The Q^2 *Random Walk* method has given excellent linearity for 12-*bit* resolution with a reasonable silicon area.

We have studied that the resistance of the metal lines from the V_{DD} connections give rise to a systematic error. The scrambling techniques and the Q^2 *Random Walk* method mitigates this limit as they obtain a scrambling or a random walk of the drop voltage on the V_{DD} connections as well.

3.5.5 Current Switching and Segmentation

A binary-weighted *DAC* combines 2^{k-1} unity current sources in parallel and uses the *k-th* bit to switch the entire parallel connection *on* or *off*. This approach has the advantage that virtually no decoding logic is required. In contrast, when using the unary method each unity current source requires a logic control signal generated by the binary-to-thermometer decoder. The complexity of the logic increases exponentially with the number of bits and requires a significant silicon area even for few bits.

The large area required for the decoder is the only limit of the unary method which is otherwise superior to the binary weighted approach in most aspects. Unary selection gives better switching performances as the magnitude of a glitch is proportional to the

Unary of binary weighted?

The significant advantage of the binary-weighted selection is a small digital section. The larger digital logic of unary selection obtains minimum glitches distortion, good *DNL* and *INL*, and has intrinsic monotonicity.

number of switches that are actually switching. In contrast, when using binary selection the number of elements that switch is not proportional to the change of the input code; namely, at the mid-scale transition all the switches are exercised; and all but one switch at a quarter or at three quarters of the full scale. Going across these points by just one *LSB* causes a large glitch. In comparison, the thermometric approach switches a number of elements that is proportional to the amplitude of the step: for small steps the glitch is small, for large steps the glitch is large. The result is that the nonlinearity is minimally affected.

Even worse than the error caused by glitches, the binary weighted method has very poor *DNL* at critical points. The worst case is again at the mid-code. Consider the effect of the random error for the mid-point transition. The ΔI of the MSB generator is $\sqrt{2^{n-1}}\Delta I_r$; the error of the remaining part of the array is almost the same, $(\sqrt{2^{n-1}} - 1)\Delta I_r$. Therefore, the maximum possible *DNL* is

$$DNL_{max} = \pm 2\sqrt{2^{n-1}}\Delta I_r/I_u. \qquad (3.41)$$

Also, a large *INL* occurs on the same points caused by the step variations in the *DNL*. In contrast the unary method gives the same maximum *DNL* equal to $2\Delta I_r/I_u$ for any code transition.

Example 3.7

Estimate the worst-case DNL versus the bin for a 12-bit binary-weighted current steering DAC. The error $\Delta I_r/I_u$ is 0.5%.

Solution

The DNL depends on the number of unity current sources that switch on and off at the code transition. The file Ex3_7 provides the Matlab description for solving the example. The functions dec2bin and bin2dec obtain the conversion of an input signal into its binary representation or back to the decimal value. The function num2str

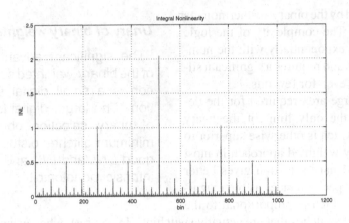

Figure 3.40. Maximum DNL in a binary-weighted DAC with 0.5% random error.

> *transforms the vector of bits that change at the code transition into a*
> *string.*
> *Fig. 3.40 gives the result.   Observe that the worst DNL is at the*
> *mid-code and, as expected, it is very close to* $2 \cdot 64 \cdot 0.005 = 0.452$
> *LSB. Other critical transitions are at a quarter and three quarters of*
> *the full scale. The DNL is* $1/\sqrt{2}$ *the value at the mid-scale. Half a*
> *way the DNL is halved and so forth.*

Another limit of the binary-weighted approach is that it does not ensure monotonicity: at critical transitions the current of generators that are switched off can differ from the ones that are switched on by more than one *LSB*.

According to the above considerations the unary selection is the preferred solution as the binary weighted is good only for $4 - 5$ bits. In order to limit the decoder area a strategy frequently used is to combine the unary and the binary weighted methods using segmentation. A relatively small number, n_L of *LSB* is converted with a binary-weighted selection while the remaining $(n - n_L)$ *MSB* bits use a unary selection. The unity current of the *MSB* is $2^{n_L} I_U$. When the *DAC* uses a large number of bits the remainder $(n - n_L)$ is divided into two parts for a two-step segmentation. Thus, the n bits are given by three terms: the n_L LSBs the n_I intermediate bits and the $n_M = n - n_L - n_I$ MSBs. The weight of the *ISB* is 2^{n_L} and that of the *MSBs* is $2^{n_L+n_I}$.

The current of the three *DACs* as shown by Fig. 3.41 are added together by simply connecting the three outputs together. The controls of the current sources of the intermediate *DAC* are 2^{n_I} signals while the ones of the *MSB DAC* are 2^{n_M} signals, both of which are thermometric.

The area of a segmented architecture depends on the area of the unity current sources and the area of the circuitry necessary to generate and distribute the control signals. The maximum allowed *DNL* determines the value of the gate area *WL* of the *MOS* transistor used to generate I_U in the binary weighted *LSB DAC*. Using (3.33) and (3.41) gives

$$WL > 2^{n_L+1} \left(A_\beta^2 + \frac{4 \cdot A_{VT}^2}{(V_{gs} - V_{th})^2} \right) \bigg/ DNL_{max}^2 \qquad (3.42)$$

showing that for a given DNL_{max} the area of the unity current generator must increase as 2^{n_L+1} increases. It can be expressed as $A_U = A_u \cdot 2^{n_L}$ (A_u is the area of the unity current source with unary selection, $n_L = 0$). It can be assumed that the area of the logic circuitry necessary to generate and distribute a single thermometric code increases with the number of n_M bits and can be

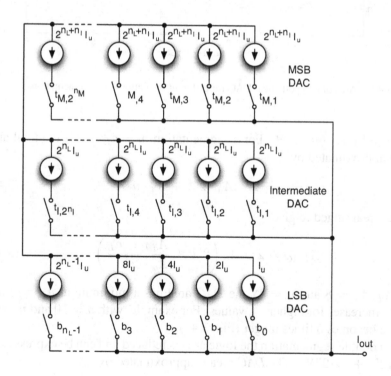

Figure 3.41.   Conceptual schematic of a current steering segmented *DAC*.

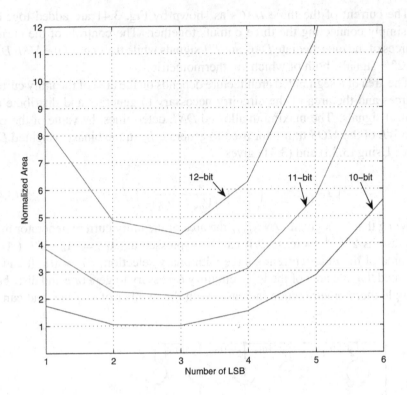

Figure 3.42. Area of a hypothetical DAC for different resolutions and segmentations.

expressed by $A_d \cdot n_M 2^{n_M}$. For a segmentation $n = n_L + n_M$ the total area can be approximated by

$$A_{DAC} = 2^n \cdot A_u \cdot 2^{n_L} + A_d \cdot n_M 2^{n_M} \qquad (3.43)$$

which is rearranged to give

$$A_{DAC} = 2^n \cdot A_u \left(2^{n_L} + \frac{A_d}{A_u} \frac{n - n_L}{2^{n_L}} \right). \qquad (3.44)$$

If $A_d/A_u = 8$ and $n = 12$ the *DAC* area has a minimum at $n_L = 3$ and sharply increases for higher n_L values. For example, with $n = 10$ and $n_L = 6$ the area becomes 5 times bigger (Fig. 3.42).

For a double segmentation the logic area is reduced and can be expressed by $A_d(n_I 2^{n_I} + n_M 2^{n_M})$. The *DAC* area is approximated by

$$A_{DAC} = 2^n \cdot A_u \left[2^{n_L} + \frac{A_d}{A_u 2^n} \left(n_I 2^{n_I} + n_M 2^{n_M} \right) \right]. \qquad (3.45)$$

Thanks to the reduced area for the logic, the total area for the same parameters as above: $A_d/A_u = 8$ and $n = 12$, becomes 2.5 times smaller with $n_L = 2$ and $n_I = n_M = 5$.

3.5.6 Switching of Current Sources

The switching method used in the Figures 3.36 and 3.41 is only conceptual. When the current path opens the transistor of the current source falls into the triode region and the voltage of node A (and the voltage of B for the cascode current mirror) of Fig. 3.36 goes quickly to V_{DD}. When it is switched on the output current must discharge the node A (and B) to its previous value and this produces a transient that slows down the circuit operation. In addition, the non-linearity associated with the charging and discharging transients is a possible source of harmonic distortion.

Using the right switching strategy avoids pushing the transistors into the triode region when the current generator is not being used and keeps the transistors in saturation. To obtain this the current source is not switched off but is routed towards a dummy connection. Fig. 3.43 (a) shows the schematic of a generic *k-th* cascode current mirror that sends the current to the output when phase $\Phi_{s,k}$ is low, and routes the current to the dummy load when $\Phi_{d,k}$ is low.

The complementary logic signals $\Phi_{d,k}$ and $\Phi_{s,k}$ are obtained by the clock and are the input digital code for driving the switching transistors $M_{k,1}$ and $M_{k,2}$. A simple way to obtain the complementary phases is to use an inverter as illustrated in Fig. 3.43 (b). However, the method is not satisfactory as the delay of the inverter makes the switching-on and the switching-off of the phase

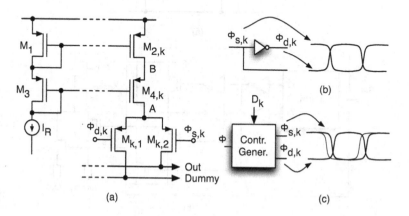

(a)   (b)   (c)

Figure 3.43. (a) Switching scheme for the k-th current mirror. (b) and (c) Possible control phases.

$\Phi_{d,k}$ delayed with respect to the $\Phi_{s,k}$ transition. Therefore, the two transistors are both *off* or both *on* for short periods of time.

When the switching transistors are both *on* the current can flow on both the output connections. As long as the overlapping time is small the situation is beneficial since the voltage of the above mentioned node A is still controlled. In addition, it will also smooth the transition between the voltages in the two switching conditions. In contrast, if both transistors are *off* the current does not find any path to ground and will quickly charge node A towards V_{DD}.

Fig. 3.43 (c) shows an optimum phase scheme. The selection phase, Φ_k (which is synchronous with the clock) drives a block whose output generates two complementary phases. They are a delayed version of the selection signal with a downward crossing point: the crossing is such that the *p-MOS* switches

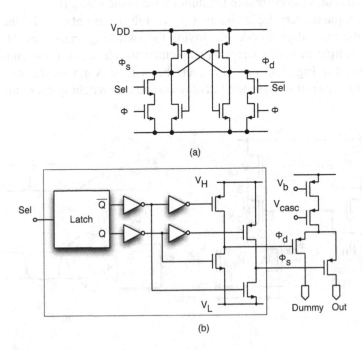

(a)

(b)

Figure 3.44.   (a) Ratioed logic latch used to generate the switching control. (b) A more complex and flexible solution.

are both *on* for a short while. This ensures that the current path to ground never goes *off* and the transistors of the current source always remain in saturation.

Fig. 3.44 (a) shows a simple and effective circuit implementation of the optimum phase scheme. It is ratioed logic latch with only eight transistors. When Φ goes *on* the output node controlled by the high selection discharges to ground. The other output goes to V_{DD} after the inverter delay. Consequently the crossing point of Φ_s and Φ_d occurs at a low voltage.

The solution of Fig. 3.44 (b) requires more hardware but enables flexibility in the timing control. A pair of inverters squares the outputs of the latch. A second pair of inverters determines the delay used to control the actual driver of the switches. The swing of phases is given by the voltages V_H and V_L whose value is suitable choosen for ensuring the off state with the minimum charging and discharging of the gates of the differential switch. Indeed, the voltage swing of the phases must be large enough to firmly open and close the switches but not too large because the charge injection due to the parasitic coupling between gate and output node can be problematic. This is the reason why the circuit of Fig. 3.44 (b) provides the option of limiting the swing of the phases.

3.6 OTHER ARCHITECTURES

The *DAC* conversion methods described in the previous sections are the most frequently used. However, there are many other possible algorithms and architectures suitable for more specific needs. We will not go into the details of the many solutions proposed, but we will mention three of them: the single ramp/dual ramp, the algorithmic and the duty-cycle converter. The first two have been studied for low-cost digital audio before the advent of integrated $\Sigma\Delta$ *DACs*. Both build the required analog voltage and when the result is ready a *S&H* samples the result and makes it available for the time required to build a new analog sample. The duty cycle converter generates pulses with full-scale amplitude that are then averaged by a low-pass filter.

The ramp converter belongs to the category of converters that use an intermediate quantity to obtain the result, as it conceptually obtains the conversion as the cascade or a digital-to-time followed by a time-to-amplitude conversion. This method, also studied in basic courses is similar to the ramp analog-to-digital converter that uses the single ramp or the dual-ramp algorithm.

Be Advised

The data conversion field is very prolific of algorithms and methods. This cause difficulty for the user in understanding (and knowing) the benefits and limits to select the most appropriate solution for a given application and technology.

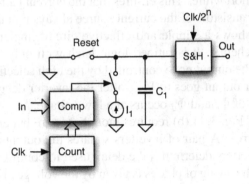

Figure 3.45. Single ramp DAC.

Fig. 3.45 shows the schematic of a single-ramp converter simply consisting of a counter, a digital comparator, a current source (I_1), a capacitor (C_1), and two switches. The output voltage is replicated by a *S&H*. At the beginning of each conversion cycle both switches are closed, and the counter is reset to zero. The voltage across the capacitor is initially held at 0 V through the reset switch. Then the current flows into the capacitor causing a ramp that is stopped when the counter reaches the input data. The obtained voltage, assuming an input code k, is

$$V_{out} = \frac{kI_1}{C_1 f_s}. \tag{3.46}$$

Since for n-bits the counter needs 2^n clock periods to reach the full scale, the sampling of the voltage across the capacitance is at a rate equal of $f_s/2^n$.

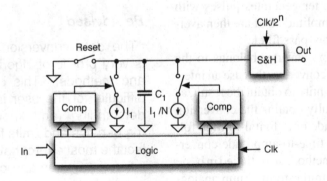

Figure 3.46. Dual-ramp DAC.

The method is simple but is able to achieve 14-*bit* of linearity without trimming. However, the clock frequency for obtaining 14-*bit* and audio band is unpractical (20 *kHz* signal band needs a clock at 720 *MHz*). The problem can be solved by using two current sources (Fig. 3.46) with the second attenuated by N (normally N equals 32 or 64). The *MSB* controls the timing of the larger current generators and the *LSB* corresponding to N controls the small generator. The number of clock periods required to reach the full scale diminishes by N enabling an equal reduction of the clock frequency.

The algorithmic approach also builds the analog output in successive steps but minimizes the required number of clock periods by using weights that are generated with successive steps of a given algorithm. If, for example the algorithm used is the multiplication by two the first weight corresponds to the *LSB*, the second to the next bit and so forth. Therefore, it is only necessary to select the weights that correspond to digital input bits equal to 1. The method is attractive but the main problem is to design an exact analog multiplier by 2. The accuracy that can be obtained enables resolutions in the 10–12-*bit* range.

Another possible algorithmic solution that trades accuracy with the number of clock periods required to perform the conversion is accumulation. The ramp converter obtains the output voltage using a linear ramp, that is actually the accumulation of unity elements. The algorithmic converter based on the accumulation algorithm uses a staircase generator as basic block whose output is equal to the input step multiplied by the number of steps. The conceptual scheme of operation is shown in Fig. 3.47 using two staircase generators, one cascaded to the other. Two logic signals are used to start and stop the staircases.

Assume that the input signal k can be written as

$$k = p_1 \cdot x + p_2 \cdot y; \quad p_1 < p_2 \tag{3.47}$$

requiring x times the weight p_1 and y times the weight p_2 to obtain k. In order to obtain the output the first ramp generator runs for p_1 clock periods and makes

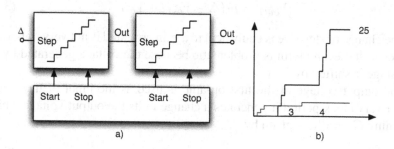

Figure 3.47. Conceptual scheme describing the operation of an algorithmic converter and its possible output waveform.

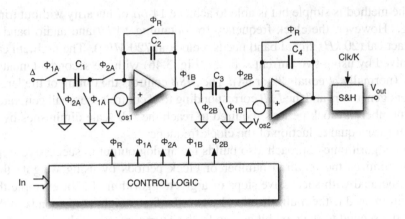

Figure 3.48. Possible circuit implementation of the algorithmic converter.

the weight Δp_1 available. Then, the second ramp uses this weight as its input, and runs for x clock periods to obtain $\Delta p_1 x$. At this point the second ramp generator is stopped and the first ramp generator is run for $p_2 - p_1$ clock periods to increase its output to Δp_2. This is the weight required for the second part of the splitting of k generated in y clock periods. Assuming that there are no waiting periods the conversion is completed in $p_2 + x + y$ time slots.

A possible circuit implementation is shown in Fig. 3.48 that uses two switched capacitor integrators with phases controlled by a logic for realizing the staircase generators. After a reset phase (that zeros the two outputs every active clock period) the capacitance C_1 injects a charge into C_2 which is proportional to Δ. Since the operation is non-inverting, then if $C_1 = C_2$ the output steps by Δ. In reality a possible offset alters this result as the capacitance C_1 remains charged to V_{os1}. Therefore, the output voltage after p_1 clock periods is

$$V_{out1} = p_1(\Delta - V_{os1}) + V_{os1}. \qquad (3.48)$$

The change of step size is equivalent to a gain error and is not very important. However the second term is problematic because the voltage generated by the first stage is shifted by V_{os1}.

The output voltage of the first op-amp is used as the input of the second one. Every clock period it generates a change in its the output voltage which, assuming $C_3 = C_4$, is given by

$$\Delta V_{out2} = p_1(\Delta - V_{os1}) + V_{os1} - V_{os2} \qquad (3.49)$$

showing that, actually, a possible problem comes from $V_{os1} - V_{os2}$ as this term shows up every time the second op-amp accumulates the output of the first

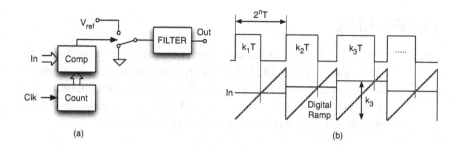

Figure 3.49. (a) Duty-cycle DAC. (b) signal waveforms.

op-amp. As a result, the error is amplified by $x + y$ and if $x + y$ changes in a non-linear manner with the code harmonic distortion results. The solution to the problem is to have

$$x + y = const \tag{3.50}$$

which will give rise to an ineffective global offset term in the output signal.

The condition (3.50) is a constrain on relationship (3.47) but it can be satisfied. An integrated implementation showed that the method can obtain 15-*bit* with a reasonable *SNDR* (74 *dB*).

The duty-cycle converter is a slow converter that transforms the input signal into a sequence of pulses whose duty-cycle is a digital fraction of a $2^n T$ period. The possible architecture, based on the scheme of Fig. 3.49 is made by a counter, a digital comparator and two switches that give rise at the output to 0 or V_{ref}. The pulse starts with the counter and finishes when the digital ramp at the output of the counter reaches the input signal. Therefore, the duration of the pulse is kT if the digital input is k. The spectrum of the output is mirrored at $\pm m f_N = 1/(2^n T)$ and attenuated by a *sinc*. The low-pass filter removes the high frequency terms and obtains the result.

This method is suitable for slow and medium resolution applications. It is a first example of an oversampled architecture but its effectiveness is much less than other solutions that will be studied shortly.

PROBLEMS

3.1 Calculate the area of the waveform generated by the response of a unity gain buffer with finite slew-rate and bandwith as a function of the input step amplitude. Use the approximate equations (3.7) with $SR = 10^8 V/s$, $\tau = 1\,ns$ and $T = 5\,ns$. Assume that the maximum input code change corresponds to $0.8\,V$.

3.2 Consider the resisitive DAC shown in Fig. 3.10 and study the effect of the error in the value of the current generators I_L. Namely, study the effect when only one current generator is correct and when both are larger than the expected value by a given percentage. Study also the effect of a non-infinite resistance in both current generators.

3.3 Repeat the Example 3.1 but assume that the error on the resistivity increases with the square of the distance from the middle of the resistive divider. The value at the two ends is 1.3 times the value in the middle.

3.4 Estimate the harmonic distortion caused by a 2%/resistor in the x direction and a 0 gradient in the y direction of the unity resistance value. The DAC is a Kelvin divider that uses an array of 32×32 unity resistances interconnected in a serpentine fashion. Estimate the result with a full scale and half scale sine wave. Explain the result with an approximate analysis.

3.5 Modify the file provided for the solution of Example 3.2 to obtain a statistical analysis. It should perform 100 consecutive simulations and draw the histogram of the SFDR with a full scale input sine wave.

3.6 A linear array of 1024 unity resistances is affected by a 0.1% gradient per element. Determine the number of calibration points needed to give 80 dB SFDR. Note that the voltages at the calibrated points will go to their ideal values.

3.7 Use the file Ex3_3 for studying a voltage-mode ladder DAC. Determine the effect of a 2% error in just one of the arms or the riser resistors. Identify the element which affects the INL the most.

3.8 Simulate using a circuit level simulator (Spice or equivalent) the operation of a 6-bit R–$2R$ architecture with MOS transistors used to replace the resistors. Match the different sections of the R–$2R$ ladder by connecting an increasing number of basic R–$2R$ cells in parallel such that the currents in every cell match. Estimate the linearity of the obtained static input-output response.

3.9 Modify the file provided for the solution of Example 3.4 for a statistical analysis such that it performs 100 simulations and obtains a histogram of the absolute value of the peak INL.

3.10 A 12-bit capacitive divider DAC uses two attenuators, one after 4 bits, and the other after 8 bits. Determine the value of the two attenuators and write a computer program that plots the input output characteristics with an error ϵ_1 and ϵ_2 affecting the attenuator capacitors.

3.11 Determine the z-transfer function from the positive input to the output of the op-amp in the circuit of Fig. 3.29 and plot the result as a function of z with $|z| = 1$.

3.12 Assume that the unity capacitance of the array of Fig. 3.30 (a) has a top plate parasitic equal to $0.01C_U(1 + 0.001V_C)$ where V_C is the voltage across the capacitance. Determine by computer simulation the transfer characteristics of an 8-bit DAC with $V_{Ref} = 2V$ and the harmonic distortion with a full scale sine wave. Assume that the output voltage is measured with an ideal voltage buffer.

3.13 The flip around DAC of Fig. 3.30 (b) compensates for the op-amp offset by pre-charging the capacitor array to the offset during Φ_R via the unity gain connection. Assume that during the reset phase the top plate of the capacitance array is instead connected to ground. Estimate the effect of the offset of the op-amp on the input-output transfer characteristics. The DAC is controlled by 10-bit.

3.14 Estimate the effect of the offset of the op-amp used in the hybrid DAC of Fig. 3.31. The resistive divider comprises of 32 unity resistance while the flip around capacitive MDAC converts 5-bit. Assuming the offset is equal to 3.4 LSB derive the input-output transfer characteristics.

3.15   Consider a current steering DAC and calculate the effect of a non-linear output resistance $R_u = \overline{R_u}(1 - \alpha V_{out})$. The non-linearity is possibly caused by the output voltage which pushes the current source transistor towards the linear region. Determine, with $R_{on} = 0$, the equation equivalent to (3.28) giving the INL as a function of the input code.

3.16   Repeat Example 3.5 but use 14 bit, $R_L = 25\Omega$, $R_{on} = 0$, and $R_u = 100M\Omega$. Find the required R_u that obtains a SFDR= 90 dB. Verify the improvement with a fully differential architecture.

3.17   Assume that the current selection of a current steering DAC is done according to the line and column shuffling of Fig. 3.39 (a). The gradient in the unity current sources is by 1% per line or column. Estimate the INL of the DAC using computer simulations. Study the effect of other shuffling techniques.

3.18   Use equations (3.44) and (3.45) to estimate the current steering DAC area for $A_d/A_u = 20$ and n=14. Perform the calculations for the simple and double segmentation schemes, and for both cases determine the optimum design.

REFERENCES

Books and Monographs

D. A. Johns and K. Martin: *Analog Integrated Circuits Design*. John Wiley and Sons, New York, NY, 1997.

F. Maloberti: *Analog Design for CMOS VLSI Systems*. Kluwer Academic Press, Boston, Dordrecht, London, 2001.

R. van de Plassche: *CMOS Integrated Analog-to-Digital and Digital-to-Analog Converters*. Kluwer Academic Press, Boston, Dordrecht, London, 2003.

D. H. Hoeschelle: *Analog-to-Digital and Digital-to-Analog Convertsion Techniques*. John Wiley and Sons, New York, NY, 1994.

Journals and Conference Proceedings

General Issues

M. J. M. Pelgrom, A. C. J. Duinmaijer, and A. P. G. Welbers: *Matching properties of MOS transistors*, IEEE Journal of Solid-State Circuits, vol. 24, 1290–1297, 1989.

J. Huang: *Resistor termination in D/A and A/D converters*, IEEE Journal of Solid-State Circuits, vol. 15, pp. 1084–1087, December 1980.

Y. Chiu, B. Nikolic, and P. R. Gray: *Scaling of analog-to-digital converters into ultra-deep-submicron CMOS*, IEEE Custom Integrated Circuits Conference, pp. 375–382, 2005.

J. Shyu, G. C. Temes, and K. Yao: *Random errors in MOS capacitors*, IEEE Journal of Solid-State Circuits, vol. SC-17, pp. 1070–1076, 1982.

Resistor Based DACs

D. J. Dooley: *A complete monolithic 10-b D/A converter*, IEEE Journal of Solid-State Circuits, vol. 8, pp. 404–408, December 1973.

J. A. Schoeff: *An inherently monotonic 12 bit DAC*, IEEE Journal of Solid-State Circuits, vol. 14, pp. 904–911, 1979.

R. J. van de Plassche and D. Goedhart: *A monolithic 14-bit D/A converter*, IEEE Journal of Solid-State Circuits, vol. 14, pp. 552–556, 1979.

J. R. Naylor: *A complete high-speed voltage output 16-bit monolithic DAC*, IEEE Journal of Solid-State Circuits, vol. 18, pp. 729–735, 1983.

K. Maio, S. I. Hayashi, M. Hotta, T. Watanabe, S. Ueda, and N. Yokozawa: *A 500-MHz 8-bit D/A converter*, IEEE Journal of Solid-State Circuits, vol. 20, no. 6, pp. 1133–1137, 1985.

M. Pelgrom: *A 50MHz 10-bit CMOS digital-to-analog converter with 75 Ω buffer*, IEEE International Solid-State Circuits Conference, vol. XXXIII, pp. 200–201, 1990.

S. Brigati, G. Caiulo, F. Maloberti, and G. Torelli: *Active compensation of parasitc capacitances in a 10 bit 50 MHz CMOS D/A converter*, in IEEE Custom Integrated Circuit Conference, pp. 719–722, 1994.

M. P. Kennedy: *On the robustness of R-2R ladder DAC's*, IEEE Transactions on Circuits and Systems, vol. 47, pp. 109–116, Feb. 2000.

Lei Wang, Y. Fukatsu, and K. Watanabe: *Characterization of current-mode CMOS R-2R ladder digital-to-analog converters*, IEEE Transactions on Instrumentation and Measurement, vol. 50, pp. 1781–1786, 2001

Capacitor Based DACs

R. E. Suarez, P. R. Gray, and D. A. Hodges: *All-MOS charge redistribution analog-to-digital conversion techniques - Part II*, IEEE Journal of Solid-State Circuits, vol. SC-10, pp. 379–385, 1975.

P. R. Gray, D. A. Hodges, D. A. Hodges, Y. P. Tsividis, and J. Chacko, Jr.: *Companded Pulse-Code Modulation Voice Codec Using Monolithic Weighted Capacitor Arrays* , IEEE Journal of Solid-State Circuits, vol. SC-10, pp. 497–499, 1975.

J. F. Albarrán and D. A. Hodges: *A charge-transfer multiplying digital-to-analog converter*, IEEE Journal of Solid-State Circuits, vol. 11, pp. 772–779, 1976.

Y. S. Yee, L. M. Terman, and L. G. Heller: *A Two-Stage Weighted Capacitor Network for D/A-A/D Conversion*, IEEE Journal of Solid-State Circuits, vol. 14, pp. 778 - 781, 1979.

G. Manganaro, S. Kwak, and A. R. Bugeja: *A dual 10-b 200-MSPS pipelined D/A converter with DLL-based clock synthesizer*, IEEE Journal of Solid-State Circuits, vol. SC-39, pp. 1829–1838, 2004.

Current Based DACs

T. Miki, Y. Nakamura, M. Nakaya, S. Asai, Y. Akasaka, and Y. Horiba: *An 80-MHz 8-bit CMOS D/A converter*, IEEE Journal of Solid-State Circuits, vol. 21, pp. 983–988, 1986

A. Cremonesi, F. Maloberti, and G. Polito: *A 100-MHz CMOS DAC for video-graphic systems*, Journal of Solid-State Circuits, vol. 24, pp. 635–639, 1989.

J. Bastos, A. M. Marques, M. S. J. Steyaert, and W. Sansen: *A 12-bit intrinsic accuracy high-speed CMOS DAC*," IEEE Journal of Solid-State Circuits, vol. 33, pp. 1959–1969, 1998.

G. A. M. van Der Plas, J. Vandenbussche, W. Sansen, M. S. J. Steyaert, and G. G. E. Gielen: *A 14-bit intrinsic accuracy q^2 random walk CMOS DAC*, IEEE Journal of Solid-State Circuits, vol. 34, pp. 1708–1718, 1999.

A. R. Bugeja, B. Song, P. L. Rakers, and S. F. Gillig: *A 14-b, 100-MS/s CMOS DAC designed for spectral performance*, IEEE Journal of Solid-State Circuits, vol. 34, pp. 1719–1732, 1999.

A. R. Bugeja and B. Song: *A self-trimming 14-b 100-MS/s CMOS DAC*, IEEE Journal of Solid-State Circuits, vol. 35, pp. 1841–1852, 2000.

K. O'Sullivan, C. Gorman, M. Hennessy, and V. Callaghan: *A 12-bit 320-MSample/s current-steering CMOS D/A converter in 0.44 mm^2*, IEEE Journal of Solid-State Circuits, vol. SC-39, pp. 1064– 1072, 2004.

T. Ueno, T. Yamaji, and T. Itakura: *A 1.2-V, 12-bit, 200MSample/s current-steering D/A converter in 90-nm CMOS*, 2005 IEEE Custom Integrated Circuits Conference, pp. 747–750, 2005

M. Choe, K. Baek, and M. Teshome: *A 1.6-GS/s 12-bit return-to-zero GaAs RF DAC for multiple nyquist operation*, IEEE Journal of Solid-State Circuits, vol. SC-40, pp. 2456–2468, 2005.

Other DACs

W. D. Mack, M. Horowitz, and R. A. Blauschild: *A 14 bit dual-ramp DAC for digital-audio systems*, IEEE Journal of Solid-State Circuits, vol. 17, pp. 1118–1126, 1982.

M. Pelgrom, M. Rooda: *An Algorithmic 15-bit CMOS Digital-to-Analog Converter*, IEEE Journal of Solid-State Circuits, vol. 23, pp. 1402–1405, 1988.

Chapter 4

NYQUIST RATE ANALOG *TO* DIGITAL CONVERTERS

This chapter deals with the architecture, features and limits of Nyquist-rate analog/digital converters. We shall start with a full-flash architecture capable of obtaining the conversion in only one clock period. Following this, we shall study the two-step solution whose algorithm requires at least two clock periods. Next, we shall discuss the folding and the interpolation methods. The interleaved technique permits the designer to take advantage of the cooperative action of many converters working in parallel. We shall consider the benefits and the limit of this approach. Then, we shall analyze the successive-approximation algorithm before studying a widely used sequential scheme: the pipeline architecture. Finally, we shall consider some techniques that are useful for special needs.

4.1   INTRODUCTION

Depending on the bandwidth of the input signal, Nyquist-rate data converters can require either one or multiple clock cycles to implement the conversion algorithm. Since small signal bandwidths enable long conversion periods the algorithm can use high frequency clocks with the sampling period extended over many clock cycles; in contrast, for large bandwidths it is necessary to maximize the time allowed for the circuit operation and so the number of clock periods required for implementing the algorithm must be at a minimum.

It is useful to evaluate, even approximately, the maximum frequency of operation of high-speed *ADC*s. For this we start from the speed of the technology or, better, the technology unity gain frequency, f_{Tech}, whose value determines the maximum obtainable unity gain bandwidth (f_T) of an op-amp or an *OTA*. f_T is lower than f_{Tech} by a factor α, which is at least 2-4, but ultimately, depends on the desired accuracy of the converter.

Since *A/D* conversion is sampled data system, it is necessary to provide enough time for settling analog signals, thus requiring a suitable margin, γ, between the op-amps f_T and the clock frequency.

In order to estimate γ suppose that the input V_{in} is a step at $t = 0$. A single pole band-limitation gives rise to an output $V_{out}(t)$ given by

$$V_{out}(t) = V_{in}(1 - e^{-t/\tau}) \tag{4.1}$$

where the time constant τ is

$$\tau = \frac{1}{2\pi\beta f_T} \tag{4.2}$$

and where β is the feedback factor of the op-amp (or *OTA*) feedback network.

Since an *n-bit ADC* needs an accuracy better than $2^{-(n+1)}$, the settling time must be $t_{sett} > \tau \cdot (n+1)ln(2)$. Therefore, recalling that the time allowed for settling is half-clock period, it results

$$f_{CK} < \frac{\pi\beta f_T}{(n+1)ln(2)} \tag{4.3}$$

$$\gamma = \frac{f_T}{f_{CK}} > \frac{(n+1)ln(2)}{\pi\beta}. \tag{4.4}$$

The foreseen order of the anti-aliasing filter sets a given margin λ, which is the ratio between sampling rate and signal band. Moreover, since the conversion algorithm can use multiple clock periods (say k), the conversion rate is therefore given by $f_{CK}/(\lambda k)$.

Example 4.1

A 10-bit ADC uses a 1.6 GHz technology. The ADC algorithm requires two clock periods to complete the conversion. Assume $\alpha = 2$ and $\beta = 0.5$. The anti-aliasing specification uses one octave in the transition-band. Calculate the maximum band of the input signal.

Solution

The OTA f_T is 800 MHz, with $\beta = 0.5$ the time constant is $\tau = 0.398\, ns$. The settling of the OTA requires $t_{sett} = 6.93 \cdot 0.398\, ns =$

2.7 ns to reach half LSB error with a full swing input. The resulting γ is 4.85 and to maximum allowed clock frequency is

$$f_{CK} = \frac{1}{2 \cdot t_{sett}} = \frac{f_T}{\gamma} = 164.8 \, MHz.$$

Since the conversion requires two clock periods the sampling frequency becomes $f_s = 82.4 \, MHz$. The transition band of the anti-aliasing filter is one octave: f_B from f_B to $f_s - f_B$. The signal band becomes

$$\frac{f_s - f_B}{f_B} = 2 \rightarrow \lambda = 3; \quad f_B = \frac{f_s}{3} = 27.5 \, MHz.$$

The obtained low signal band depends on conservative numbers used. More relaxed accuracies enable higher signal bands.

When the converter uses only one clock cycle the architecture is named full-flash. Other architectures such as the two-step flash or the use of folding methods require two (or three) clock cycles, while several clock cycles are used by the successive approximation, the algorithmic technique and other slow methods.

The conversion algorithms that employ multiple clock cycles are normally more accurate, use less power and need less chip area their few (or just one) clock period counterparts. Furthermore, the throughput of slow converters increases by using the interleaved method which utilizes many converters in parallel or by the pipeline architecture which operates by cascading a number of stages. Although these schemes allow faster sampling-rates they provide output after a latency time, that can give rise to instability when the converter is used in a control system loop.

4.2 TIMING ACCURACY

The error due to the sampling jitter, determined in Chapter 1, is

$$\delta V_{in} = \delta T_{ji} \cdot \frac{dV_{in}}{dt}. \tag{4.5}$$

Knowing this, since the performance of an *n-bit* converter requires a timing error less than $1/2 \, LSB = V_{fs}/(2^n + 1)$, a 12-*bit* ADC with 1 *V* peak sine wave at 20 *MHz* requires a timing error below 1 *psec*.

The above statement highlights the need for great care in phase generation and distribution for resolutions higher than 10–12 *bit* and input frequencies of several ten of *MHz*. Furthermore, the inaccuracy of the sampling time is not only caused by the clock jitter but also depends on the finite rise/fall time of the

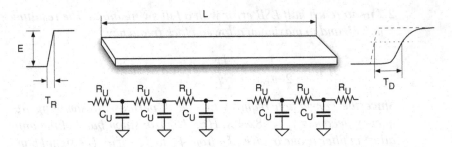

Figure 4.1. Distributed RC model of a metal interconnection carying a clock signal.

clock, the propagation delay, and other non linear effects which cause signal dependent clock delay.

With a high-speed clock the mismatch impedance at the input pin causes reflections and clock timing errors. To avoid this precise applications will input, instead of a square wave, an input sine wave with carefully matched impedances. The sine wave is then amplified and squared inside the chip to obtain a main clock whose jitter will be very low provided that the squaring circuit uses low noise amplifiers for detecting the zero crossings.

Other timing errors occur as a result of the propagation of digital signals along metal interconnections which causes dispersion and delay. Fig. 4.1 shows a distributed *RC* that models the interconnection. The time response to a step with finite slope signal $e_i(t)$

$$e_i(t) = \frac{E}{T_R}t \quad for \;\; 0 \leq t \leq T_R$$
$$e_i(t) = E \quad\quad for \;\; t > T_R \quad\quad\quad (4.6)$$

is given by

$$e_{out}(t) = V_r(t) - V_r(t - T_R) \quad\quad\quad (4.7)$$
$$V_r(t) = \frac{E}{T_R}\left\{(t + \frac{\tau}{2})erf\left[\sqrt{\frac{\tau}{4t}}\right] - \sqrt{\frac{\tau t}{\pi}}e^{\frac{-\tau}{4t}}\right\} \quad\quad (4.8)$$

where V_r is an intermediate function and $\tau = R_u C_u L^2$. The delay is roughly estimated by

$$T_D = \frac{\tau}{4} = \frac{R_u C_u L^2}{4} \quad\quad\quad (4.9)$$

where R_u and C_u are the resistance and capacitance per unity length. Observe that if the width increases then the resistance per unity length decreases but the capacitance per unity length increases by the same extent. Therefore, the delay is not affect by changing the width of the interconnections. For a typical submicron technology the product $R_u C_u$ is in the order of some $10^{-17} s/m^2$ leading to delays in the order of 1 ps over only a few hundreds of micron of interconnection.

Example 4.2

Use equations (4.7) and (4.8) to estimate the clock waveform after 150 μm and 300 μm of metal interconnections. $R_u = 0.04$ Ω/μ and $C_u = 2.5 \cdot 10^{-16} F/\mu m$. The rise time of the clock is 50 fsec. Estimate the clock slope at half of the input step amplitude.

Solution

The file Ex4_2 is the basis for the computer simulation. Fig. 4.2 gives the waveforms. The use of equation (4.9) obtains the value of τ that equals 0.225 psec and 0.9 psec for 150 μm and 300 μm metal interconnections respectively. These values are close to the simulated results at half input amplitude. The two slopes are 0.98 $V_{FS}/psec$ and 0.28 $V_{FS}/psec$ respectively showing that a noise of 40 mV_{FS}

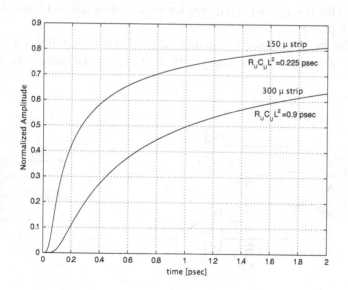

Figure 4.2. Clock waveforem after a 150 μm and 300 μm metal line.

> *in the crossing threshold would cause a jitter as low as 39 fsec and*
> *12 fsec respectively.*

4.2.1   Metastability error

A metastability error occurs when the output of a comparator is undefined. Typically a sampled-data comparator is realized using a pre-amplifier and a latch (Fig. 4.3). During one phase the input signal is pre-amplified and during the latch phase a regenerative circuit fixes the logic level. If the input differential voltage, $V_{in,d}$, is not large enough the comparator output may be undefined at the end of the latch phase giving an error in the output code and possibly causing a code bubble error in the thermometric output of some converter architectures.

The differential latch of Fig. 4.3 is the positive loop of two transconductors whose regenerative time constant, τ_L, is

$$\tau_L \simeq \frac{C_p}{g_m}. \tag{4.10}$$

The metastability error probability can be approximated by

$$P_E = \frac{V_0}{V_{in} A_0} e^{-t_r/\tau_L} \tag{4.11}$$

where V_0 is the voltage swing required for valid logic levels and t_r is the period of the latch phase.

Typically the latch period t_r equals $1/(2f_s)$, so the probability of a metastability error increases exponentially with the sampling frequency and at high

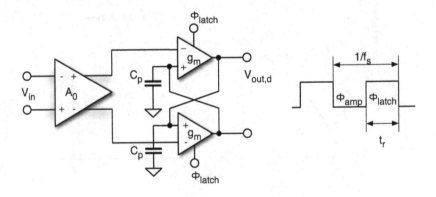

Figure 4.3.   Typical comparator used in data converters.

frequencies becomes equal to 1 (since more than 1 is not a valid result, if (4.11) gives more than 1 the result means $P_E = 1$).

At reasonable operating frequencies the exponential is a small value; moreover, we see that the probability of error is inversely proportional to the differential input amplitude, V_{in}. Since the comparator must ensure an almost certain output for V_{in} larger than half *LSB*, it is required to have P_E less than a given $P_{E,max}$ for $V_{in} = V_{FS}/2^{n+1}$, leading to the condition

$$f_s \cdot ln \left[\frac{V_0 2^{n+1}}{P_{E,max} V_{FS} A_0} \right] < \frac{1}{2\tau_L}. \qquad (4.12)$$

If, for example, $V_{FS} \simeq V_0$, $n = 8$, $\tau_L = 2 \cdot 10^{-10}$ and $A_0 = 10^3$ the sampling frequency that ensures an error rate equal to $P_{E,max} = 10^{-4}$ is equal to $f_s = 1/(17 \cdot \tau_L) = 293\,MHz$.

4.3   FULL-FLASH CONVERTERS

An analog-to-digital converter must identify the quantization interval that contains the input signal. A direct way to achieve this operation (Fig. 4.4) is to compare the input signal with all the transition points between adjacent

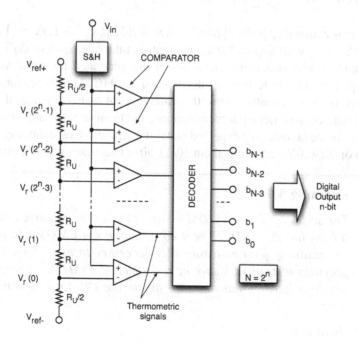

Figure 4.4.   Basic block diagram of the full-flash converter.

quantization intervals. The result of these comparisons highlights the limit at which the input is larger than one of the thresholds giving information that can then be transformed into the digital code. This "brute force" (but effective) approach leads to the full-flash architecture. The name of the method comes from the fact that all the comparators operate in parallel and procure quickly (like a flash) the result in just one clock period.

An *n-bit* quantizer identifies 2^n bins with $2^n - 1$ transition points; therefore, the architecture shown in Fig. 4.4 requires $2^n - 1$ reference voltages and $2^n - 1$ comparators which output logic 1 up to a given level and logic 0 above. A *ROM* decoder (or equivalent circuit) can then translate this thermometric representation into an *n-bit* digital output.

4.3.1   Reference Voltages

The simplest way to generate the reference voltages is to use a resistive Kelvin divider connected between the positive and the negative references, V_{ref+}, V_{ref-}. Some implementations like the one shown in Fig. 4.4 use $R_U/2$ as ending elements giving rise to an *i-th* reference voltage equal to

$$V_r(i) = V_{ref-} \frac{i - 1/2}{2^n - 1}(V_{ref+} - Vref-); \quad i = 1, 2^n - 1. \qquad (4.13)$$

The quantization step is the dynamic range divided by $2^n - 1$, $\Delta = (V_{ref+} - V_{ref-})/(2^n - 1)$ with first and last quantization intervals equal to $\Delta/2$.

The random and systematic errors affect the generated reference voltages as it happens when the Kelvin divider is used as *DAC*. Therefore, for a flash type converter, it is essential to use the same kind of resistive material, to use matched contacts and metal interconnections, and to ensure the same resistance orientation in the layout. The expected matching with modern technologies is in the order of 0.1-0.05% enabling about 10-11 *bits* of accuracy without trimming.

Example 4.3

The series of $(2^6 - 1)$ 50 Ω resistors makes the resistive divider of a 6-bit full-flash ADC. The used technology is CMOS 0.18 μm but for ensuring good matching it is necessary to use 2.5μm wide of polysilicon strips with specific resistance of 25 $\Omega/\square$. Assume a 300 ppm/μm linear gradient and estimate the INL for different layout styles.

Solution

Fig. 4.5 (a) shows the layout of a resistive divider made by a single straight sequence. The first and the last element are one square of polysilicon (25 Ω) while all the other elements are two squares. The

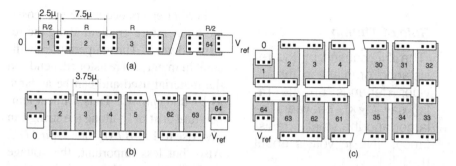

Figure 4.5. Resistive divided layed out using different layout strategies.

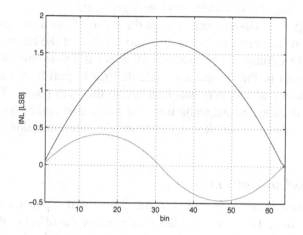

Figure 4.6. INL for the layout b) and c) of the previous figure.

resistance pitch is 7.5 µm leading to a total lenght of 480 µm. Assuming that the resistance at the beginning of the string equals the nominal value, the resistance at the end of the string becomes 1.14 the nominal value. The use of the file Ex4_3 for studying the problem by computer simulations obtains about 1.7 LSB INL.

Fig. 4.5 (b) shows a more compact arrangement of same unity resistances whose pitch is reduced to 3.5 µm. Fig. 4.5 (c) folds the single straight line of Fig. 4.5 (b) making equal the resistor values at the beginning and the end of the string. The reduced 3.5 µm pitch diminishes the INL to 0.5 LSB while the folding leads to the s-type INL shown in Fig. 4.6 which maximum is 0.45 LSB. The compact layout with folding gives 0.15 LSB INL.

> **Rule of Thumb**
>
> The INL caused by the resistance gradient in the divider depends on the ratio, say (1+k), between the larger and the smaller resistance: INL in volt is about $0.18 \cdot k\, V_{FS}$.

Further errors occur due to a possible temperature gradient along the resistive divider. The temperature coefficient in integrated resistors depends on the material used and can be as large as $10,000\, ppm°C$. Thus, a linear temperature drift of $7°$ along the string can give rise to 1 *LSB INL* for a *7-bit ADC*. Also, but less important, the voltage coefficient gives rise to a change of the resistance values along the string. Since the resistance changes almost proportionally to the drop voltage across the divider the effect is equivalent to a linear gradient.

Another important design parameter is the value of the unity resistance to be used in the references generator. The various taps of the resisitive divider are connected to a comparator which can often act as a time-variant load. The divider must react to the comparator variable load to pull back the tap voltages to an accuracy of less than half *LSB* before the latch phase. The type of comparator, and the speed and resolution of the converter and its power budget are the elements that, via simulation, determine the optimum value of the unity resistance.

4.3.2 Offset of Comparators

Since comparator offset is added to the differential input it modifies the threshold transition. Therefore, in a flash architecture the offset of the *i-th* and the *(i-1)-th* comparators alter the *i-th* quantization interval Δ which becomes

$$\Delta_i = V_{thr,i} - V_{thr,i-1} = \Delta - V_{os,i} + V_{os.i-1}. \qquad (4.14)$$

In order to ensure no missing codes or monotonicity for a given yield the maximum value of the offset must be lower than $1/2LSB$ divided by the number of sigma required to obtain that yield. Thus, for example, an *8-bit* flash with $1\,V$ full scale requires a comparator offset lower than $0.6\,mV$ to ensure a 99.9% yield, as the corresponding sigma with a normal distribution of errors is $\sigma = 3.3$.

Since the offset is mainly caused by the pre-amplifier of the comparator it is necessary to properly design this first stage and to optimize the layout for having a minimum threshold, transconductance parameter μC_{ox} and aspect ratio (W/L) mismatches in the input differential pair and active loads. The threshold error ΔV_{Th} depends on the gate area and is estimated by

$$\Delta V_{th} = \frac{A_{VT}}{\sqrt{WL}}. \qquad (4.15)$$

If the length of the *MOS* transistors are close to their minimum then the μC_{ox} and the $\Delta W/W$ mismatches are negligible and much lower than the relative length mismatch $\Delta L/L$. This error multiplied by I_D/g_m (current and transconductance of the input pair) determines the corresponding contribution to the input referred offset. For *MOS* in saturation $I_D/g_m = (V_{GS} - V_{Th})/2$. Since offset sources caused by threshold and $\Delta L/L$ mismatch are uncorrelated their quadratical superposition gives

$$V_{os,MOS} = \sqrt{\frac{A_{VT}^2}{WL} + \left[\frac{V_{GS} - V_{Th}}{2}\right]^2 \frac{\Delta L^2}{L^2}}. \tag{4.16}$$

For a typical $0.18 \, \mu m$ *CMOS* process $A_{VT} \approx 1mV/\mu m$, to give a first term lower than $0.6/\sqrt{2} \, mV$ requires $WL > 5.5 \, \mu^2$. The second term is the same if $(\Delta L/L) < 2 \cdot 0.42/(V_{GS} - V_{Th})$ requiring $(\Delta L/L) < 1\%$ for $120 \, mV$ overdrive.

The bipolar counterpart of (4.16) superposes quadratically the V_{BE} mismatch and the error accounting for the emitter area mismatch

$$V_{os,BJT} = \sqrt{\Delta V_{BE}^2 + \left(\frac{kT}{q}\right)^2 \frac{\Delta A^2}{A^2}}. \tag{4.17}$$

The first term is small as it depends on the emitter current mismatch ΔI_E that hands $\Delta V_{BE} = V_{BE} \log\left[(I_E + \Delta I_E)/I_E\right]$. The second term is also small when compared to the *MOS* counterpart because the area mismatch is smaller than the length mismatch; moreover, the multiplying factor is $kT/q = V_T = 26 \, mV$ instead of half the overdrive. Consequently, the offset of *BJT* comparators is relatively smaller that the offset of the *CMOS* counterparts.

About the Offset

The offset of CMOS circuits is a few mV. The offset of bipolar circuits is a fraction of mV. A good design can obtain a minimum systematic offset. A good layout compensates for the random offset caused by fabrication inaccuracies.

Example 4.4

Estimate the offset of a CMOS and a bipolar comparator. In the MOS implementation the aspect ratio of the input transistors of the preamplifier is W/L = 5 μm/0.18 μm. The transconductance g_m is 0.8 mA/V for a bias current equal to 80 μA. The 0.18 μ CMOS technology has $A_{VT} = 2 \, mV/\mu m$. The length mismatch is $\Delta L/L = 1.8\%$.

For the bipolar preamplifier use $I_E = 0.8\,mA$ with $\Delta I_E/I_E = 0.5\%$.
The area mismatch is 0.2%.

Solution

The use of $(V_{GS} - V_{Th})/2 = I_D/g_m$ in equation (4.16) determines the one-sigma MOS offset

$$V_{os,MOS} = \sqrt{\frac{4 \cdot 10^{-6}}{0.9} + \left(\frac{0.08}{0.8}\right)^2} \cdot 0.018^2 = 2.77\,mV$$

with almost equal contributions from threshold and length mismatch: 2.1 mV and 1.8 mV.
Since the current in bipolar transistors changes exponentially with $V_{BE} : I_E = I_s e^{V_{BE}/V_T}$ the error $\Delta I_E/I_E = 1\%$ causes a ΔV_{BE} given by

$$\Delta V_{BE} \simeq V_T \cdot \Delta I_E/I_E = V_T \cdot 0.005 = 0.13\,mV$$

therefore the one-sigma offset of the bipolar comparator is

$$\Delta V_{os,BJT} = 26 \cdot 10^{-3}\sqrt{(0.005)^2 + (0.002)^2} = 0.14\,mV$$

that is 20 times lower than the offset of the MOS.

4.3.3   Offset Auto-zeroing

The typical offset of *CMOS* circuits is in the range of some *mV*, too large a value for ensuring a good yield where, for example, having a 99% yield requires an one-sigma offset of less than $0.5/2.57 \simeq 0.2\,LSB$ (2.57σ of the offset must be less than $1/2\,LSB$) and to have a 99.9% yield requires $0.5/3.3 \simeq 0.15LSB$ (3.3σ of the offset must be less than $1/2\,LSB$). It is therefore necessary to reduce the offset down to fractions of *mV* using only simple and inexpensive methods as a flash converter already uses a fairly large number of comparators. The commonly used technique is the auto-zero which is a two phase sampled-data method: one phase is used for measuring and storing the offset and the other phase is for the offset cancellation.

The schematic of Fig. 4.7 shows the circuit implementation of the auto-zero technique. During phase Φ_{az} the amplifier is in the unity gain configuration and makes the offset available at the inverting input. The capacitor C_{OS} is charged to the offset minus the input voltage V_1. During the complementary phase, $\overline{\Phi}_{az}$, the open loop configuration amplifies the input by A_0 which, thanks to the voltage shift of the capacitor C_{os} by $V_{os}(\Phi_{az}) - V_1$, gives rise to a differential input equal to

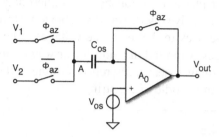

Figure 4.7. Auto-zero technique.

$$V_{d,in} = V_+ - V_- = V_{os}(\bar{\Phi}_{az}) - V_2 + [V_1 - V_{os}(\Phi_{az})]. \qquad (4.18)$$

Provided that the offset does not change during the auto-zero period then the circuit achieves a perfect offset cancellation. This is usually the case as the offset is generally considered to be a *dc* signal which only changes due to environmental fluctuations or aging. However, the input referred generator also affects the noise, including the $1/f$ term: the auto-zero technique is partially effective in canceling the $1/f$ noise but increases the white part of the noise as the lack of correlation between two successive samples doubles the white term.

Another problem that is generated by the offset cancellation circuit is clock feed-through. A fraction of the charge of the *MOS* conductive channel is injected into the auto-zero capacitor. For the scheme of Fig. 4.7 the critical opening is through the switch connected in feedback. Compensations techniques, studied in the next Chapter, reduce this charge injection and can obtain equivalent offsets equal to fractions of *mV*.

Typically, the flash *MOS* converters adapt the comparator offset cancellation scheme of Fig. 4.7 as follows: one input terminal connects to a tap of the resistive divider and the other to the input of the flash: terminals V_1 and V_2 become the differential inputs of comparators used in Fig. 4.4. Notice that the difference between input and reference is not calculated by a differential pair but comes from subtracting the two different voltages at the node A of Fig. 4.7 during the two operation phases.

Since the parasitic capacitance of the left plate of C_{os} (node A, Fig. 4.7) is charged to V_1 during one phase and

Does the Auto-zero helps?

The offset of a CMOS comparator can be reduced from few *mV* to fractions of *mV* as clock-feedthrough cancellation schemes do not fully cancels the limit. For an effective cancellation use fully differential architectures.

to V_2 during the other phase, it establishes a sampled-data parasitic coupling across the inputs: the resistive divider must be able to restore all of the nodes A from the input voltage back to the reference voltages. The required charge to and from the references is limited in a non-linear way by the resistive divider, as restoring the tap voltages near a reference terminal is faster than restoring the voltage of taps in the middle of the string.

Example 4.5

An 8-bit full-flash converter uses a resistive divider made by 256 unity 25 Ω resistors connected across 0 and 1 V. The parasitic capacitance of the switching nodes is 6.25 fF. Simulate with Spice the voltage recovery of the tap nodes after the conversion of 0.5625 V.

Solution

The equivalent network modeling the resistive divider is an RC ladder made by 256 cells (Fig. 4.8 (a). Since the network is complex and it is not really necessary to determine the waveforms in fine details we simplify the circuit by grouping together a number of cells (16) and approximate the group with an equivalent RC network as shown in Fig. 4.8 (b).
The Spice simulation of the simplified circuit, made by the file Spice4_5, gives rise to the responses shown in Fig. 4.9. The figure refers to the bins #80, #144 and #240 whose expected voltages are 0.3125, 0.5625 and 0.9325 respectively. Observe that the voltage of the bin #144, that is is the same of the one of the previous conversion, does not

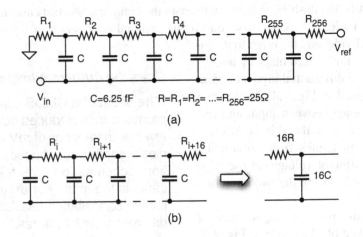

Figure 4.8. Equivalent circuit of the resistive divider (a) and its macro-cell simplification (b).

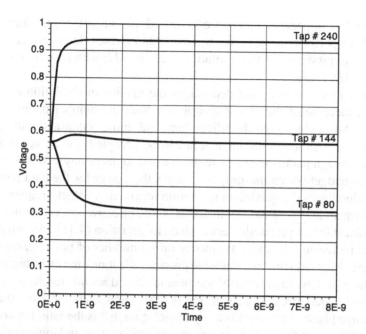

Figure 4.9. Transient response of the voltage at different taps.

remain constant because the resistive string must provide the charge required by other nodes.

Immediately after the switches close the voltage of all the taps is 0.5625V, followed by transients that achieve 1/2 LSB error after 5.7 ns, 3.9 ns and 4.3 ns for the bin #80, #144 and #240 respectively. The tap #240 is close to the low impedance V_{ref} but since it must swing more than the bin #144 it takes a bit more for settling.

If the used clock enables, for example, 5 ns for setting the references the tap #80 is not able to reach the 1/2 LSB accuracy, giving rise to a possible code error if the next sample is near 0.3125 V.

4.3.4   Practical Limits

The speed and resolution of full flash converters depend on a number of issues that establish a practical limit on the use of the architecture. The first issue concerns the very small unity resistances that the Kelvin divider must use for high-resolution and high-speed. In turn, a low resistance loading the divider requires a reference voltage with a very low output impedance from *dc* to the sampling frequency. For this there are two possible approaches: using an external reference with a solid on-chip filter capable of dumping any ringing caused by the bonding inductance, or using an on chip reference made by a band-gap and a very low impedance buffer. Both methods are suitable for medium speed and medium performances; however, precise and stable references become the critical design issue when the resolution exceeds 8-*bit* with conversion speeds in the 100 *MS/s* range.

Another practical limit that determines the maximum resolution is the exponential increase of the circuit complexity with the number of bits: every additional bit doubles both the silicon area and, more importantly, the power consumption. Although the large area can be accepted, the power is a very important design parameter because the specifications of any system establish the power budget which, in turn, determines the power for the data converter whose value limits the maximum resolution for the users clock frequency.

The flash architecture uses comparators whose effectiveness can be measured by the metastability probability error given by equation (4.11). Assuming P_E constant, increasing the clock frequency or the number of bits requires either increasing A_0 or reducing τ_L. Since the pre-amplification time in the high-speed applications is low, the output of the preamplifier does not reach its *dc* level but instead experiences a transient that is terminated when the latch starts its regenerative phase. The result is a "dynamic" gain that is the ratio between the voltage obtained at the end of the pre-amplification phase and the input. Since the pace of the transient is $C_p/g_{m,A}$ (the ratio between the parasitic capacitance and the transconductance of the preamplifier), a preamplification period equal to $1/(2f_{ck})$ gives a "dynamic" gain equal to

$$A_0 = \frac{g_{m,A}}{2f_{ck}C_p}. \qquad (4.19)$$

Since increasing the number of bits requires more gain and the gain is inversely proportional to the frequency, multiple bits and high speed requires large $g_{m,A}/C_p$ ratios. The transconductance of *MOS* transistors in saturation is proportional to the square root of the current; therefore, for doubling speed, doubling the number of levels or for adding an additional bit it is necessary to use four times as much power in the comparator preamplifier.

In order to have a quantitative idea of the required power, consider a 7-*bit* 500 *MHz* flash that needs preamplifiers with dynamic gain $A_d = 20$. Assume that the trade-off between offset and aspect ratio of the input pair gives rise to an overdrive equal to 200 *mV* and $C_p = 0.4\,pF$. Since the transconductance is also expressed by $g_m = 2I_D/V_{ov}$, the current of the input pair is

$$I_P = 2I_D = 2V_{ov}f_{ck}C_p = 80\,\mu A \qquad (4.20)$$

which gives rise to a current consumption of 10 *mA* just for the preamplifiers.

Another important limit is the capacitance load on the input sample-and-hold caused by the parasitic of a single comparator multiplied by the number of comparators. Indeed, the parasitic of a comparator caused by the input transistor is typically very low but the total effect, for many bits, becomes significant. If the capacitance of a single comparator is, for example, $10\ fF$ the capacitive load of the *S&H* caused by an *8-bit* flash is $2.5\ pF$, a large value if the clock frequency is many hundred *MHz*.

Use of Full-Flash

Full-flash converters are the optimum for very high-speed but the resolution cannot be very high as a number of limits make the implementation unpractical.

Another critical issue concerns the current that the *S&H* must deliver for charging or discharging the comparators' capacitances. The charge on the parasitics after sampling the references is $2^n C_p V_{ref}/2$. A full scale input voltage drains an equal amount of charge from the *S&H* which must be provided in a fraction α of the measure period. The resulting current pulse depends on the speed of the *S&H* but its peak value is

$$I_{S\&H,peak} > \frac{f_s 2^n C_p \Delta V_{in,max}}{2\alpha}. \tag{4.21}$$

For $\Delta V_{in,max} = 1/, V$, $2^n C_p = 2.5\ pF$, $\alpha = 0.1$ and $f_s = 500\ MHz$ the current $I_{S\&H,peak}$ becomes larger than $16\ mA$.

The limits discussed above are such that with present technologies it is impractical to design an *8-bit* full-flash with conversion speeds higher than 500 *MS/s* or a *6-bit* flash operating at more than 2 *GS/s*.

4.4 SUB-RANGING AND TWO-STEP CONVERTERS

When the resolution is higher than *8-bit* then instead of a full-flash it can be more convenient to use a sub-ranging or a two-step algorithm which both secure a better speed-accuracy trade-off. The sub-ranging or the two-step implementation require two (or three) clock periods to complete the conversion but they use a smaller number of comparators thus benefitting silicon area, power consumption and parasitic capacitance loading on the *S&H*.

Fig. 4.10 shows the basic scheme of a sub-ranging or a two-step architecture. It uses a sample-and-hold at the input to drive an *M-bit* flash-converter which estimates the *MSBs* (coarse conversion). The *DAC* then converts the *M-bits* back to an analog signal which is subtracted from the held input to give the coarse quantization error (also called the residue). Next, the residue is converted into digital by a second *N-bits* flash which yields the *LSB* (fine conversion). The digital logic combines coarse and fine results to obtain the $n = (M+N)\text{-}bit$ output.

The use of the gain stage distinguishes between the sub-ranging and the two-step architecture as the sub-ranging scheme does not use any amplification. The gain of the amplifier used in the two-step method increases the amplitude of the residue for a better *LSB* estimation. Moreover, if the gain equals 2^M the dynamic range of the amplified residue equals that of the input making it

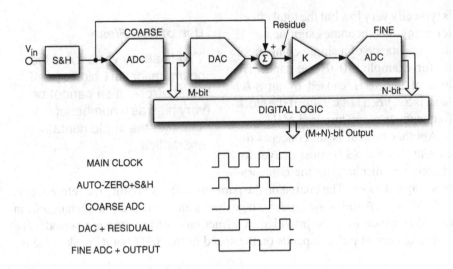

Figure 4.10. Block diagram of sub-ranging (K=1) and two-step architectures (K>1).

possible to share the reference voltages between the coarse and the fine flash converters.

Fig. 4.10 shows the timing control of the two-step (or sub-ranging) architecture requiring four logic signals to be derived from the main clock: auto-zero and *S&H*, coarse *ADC* conversion, *DAC* plus residue generation, and fine conversion plus output. Assuming that half clock period is enough to accomplish each function or group of functions, the algorithm needs two clock periods. In some cases the converter uses three clock periods when it is necessary to use a specific time-slot for one of the combined functions or to assign two time-slots to a critical step.

It is evident that the number of comparators needed for a two-step or a sub-ranging is much lower than the number needed for a full-flash architecture. For example, for 8-*bits* $M = N = 4$ the scheme uses $2(2^4 - 1) = 30$ comparators, which is 8 times less than the $(2^8 - 1) = 255$ comparators required by the 8-*bit* full-flash counterpart. The spared area and power are much more than what is required to design the DAC and residue generator; moreover, the *S&H* is only loaded by 2^M comparators.

The obvious disadvantage is the reduced conversion-rate as it is necessary to use two or three clock periods to complete the conversion. Nevertheless, since the speed of the *S&H* is the bottleneck of medium resolution full-flashes, the clock of the sub-ranging scheme can end up being higher than the one of the full-flash as the reduced parasitic capacitance enables faster *S&H*.

4.4.1   Accuracy Requirements

The accuracy of each individual block used in the architecture of Fig. 4.10 determines the overall accuracy of the converter. In particular the gain error, the offset errors and the input referred noise of the *S&H* establish equivalent limitations on the *ADC* as it is the first block of the conversion chain. Regarding the noise we know that the *kT/C* limit (studied in Chapter 1) caused by the sampling should be lower than the quantization noise. However, since for medium resolution the quantization step is 1 *mV* or more, the capacitance that gives rise to a $\sqrt{kT/C}$ noise voltage smaller that 1/2 *LSB* is not very large. For example, 0.5 *pF* causes noise equal to 90 μV. Therefore, with resolutions up to 10-*bits* the input capacitance is not a problem for the *S&H* design.

Since the residue, determined by coarse *ADC*, *DAC* and gain factor *K*, is

$$V_{res}(V_{in}) = K\left[V_{in} - V_{DAC}(i)\right] \qquad (4.22)$$
$$for: \ V_{Coarse}(i-1) \le V_{in} < V_{Coarse}(i);$$

ideal *ADC* and *DAC* give rise to a residue that is a perfect sawtoothed non linear function of the input with amplitude confined between 0 and $V_{FS} \cdot K/2^M$. However, limitations of the *ADC* and *DAC* cause errors on the break points and amplitude of the sawtooth.

A positive error $\epsilon_{ADC}(i)$ in the *i-th* transition threshold of the *ADC* moves the break point onward and leads to residue which is larger than $V_{FS}\cdot K/2^M$ causing an error at the break point of $K \cdot \epsilon_{ADC}(i)$. Assuming an ideal *DAC* response the downward shift is exact and the error immediately after the break becomes zero. This kind of situation occurs for the transfer characteristic of Fig. 4.11 (a) around the 000 → 001 transition where the corresponding threshold is

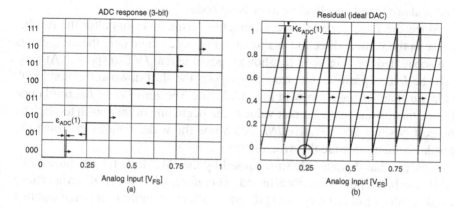

Figure 4.11.   (a) Response of a real coarse ADC (3=bit). (b) Residue with ideal DAC.

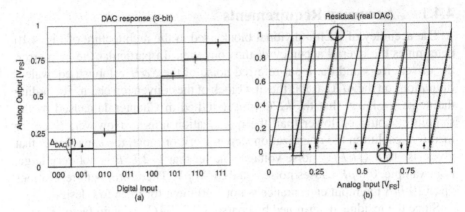

Figure 4.12.   (a) Response of a real DAC (3=bit). (b) Residue with real DAC and ideal ADC.

slightly higher than 0.125 ($V_{FS} = 1$): the first break occurs after the expected value and the residue amplitude continues increasing for a short while after $V_{in} = 0.125\,V_{FS}$ (Fig. 4.11 (b), $K = 2^M$).

Similarly, a negative error in the threshold transition moves the break point backward; the downward shift associated with the change of the code controlling the *DAC* brings the residue to a negative value as happens at the second break point of Fig. 4.12. Then, the error goes to zero at the next ideal break point: $V_{in} = 0.25\,V_{FS}$, Fig. 4.11.

A residue that exceeds the 0 and 1 interval is out of the *LSB* flash range and establishes an *LSB* code of all zero or all ones which remains unchanged until the input re-enters the $0 - 1$ boundaries. On the other hand if the residue does not reach the 0 or the 1 limit then the *LSB* flash does not switch from zero to full scale or vice-versa leading to missing codes.

Since the *DAC* generates the subtractive terms its errors alter the residue curve in the vertical direction as shown in Fig. 4.12 where only the *DAC* errors are accounted for. Fig. 4.12 (a) shows a possible real *DAC* response. At code 001 the *INL* is positive (up arrow); therefore, the shift down at the first break is larger than 1 (down arrow in Fig. 4.12 (b)). The zero *INL* at the third code brings the residue to the correct value at the beginning of the fourth tooth. The biggest positive and negative *INL* determine the wider regions in which the residue remains outside the dynamic range.

Observe that the possible shift caused by the *DAC* lasts for an entire tooth while the break points are not affected. Therefore, while the *ADC* errors cause local regions of inaccuracy the *DAC* errors affect an entire *LSB* range and this makes the requirement of *DAC* accuracy more demanding than that of the *ADC*.

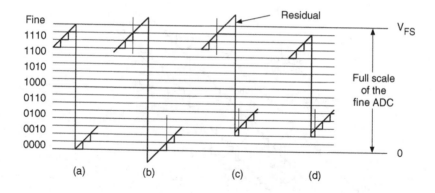

Figure 4.13. Residue plots around a break point and the LSB quantization.

The local error caused by the *ADC* is easily detectable with extra thresholds placed outside the ±1 region. Once the error is identified its cancellation is possible by using suitable correction techniques. Another possibility is to limit the range of the residual well inside the nominal intrerval so that possible mismatches does not bring the residual value outside the dynamic range of the fine converter. This is actually what is done in pipeline data con-

Notice

The errors of the ADC used in a two-step flash can be easily corrected because they are localized around the break points. The errors of the DAC are more critical as their effect extend over entire fine conversion intervals.

verters (studied shortly) that cancel the *ADC* errors by using digital correction.

The residue signal is quantized by the fine flash to determine the *LSBs*. The possible errors affecting the residue give rise to the four possible situations illustrated in Fig. 4.13 for a 4-*bit* fine conversion. The case a) is ideal as the residue drops from V_{FS} to 0 after the *MSB* transition: the fine *ADC* generates the code 1111 before the transition and the code 0000 after. In the case b) the residue exceeds the limits on both sides making the output of the fine *ADC* stacked at the 1111 and 0000 codes for inputs outside the 0–1 range. The case c) shows what happens when the residue does not reach 0. The fine quantizer generates after the 0010 code the break skipping the codes 0000 and 0001. The case d) shows that if the residue does not reach either 1 or 0 then there are missing codes before and after the threshold transition.

Fig. 4.14 shows the input-output transfer curve for three different cases: ideal response (left curve), transfer characteristics with real *ADC* and ideal *DAC* (middle curve), and response with both *ADC* and *DAC* real (right curve).

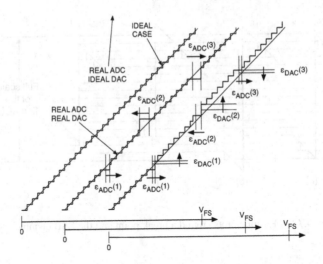

Figure 4.14.   Static response of the two-step flash for ideal case and real *ADC* or *DAC*.

It can be noted that errors in the middle curve only differ from the ideal response around the *MSB* transitions and that 1 or 2 *LSB* from the *MSB* transitions the response correctly matches the interpolating line. The right curve, due to the *INL* of the *DAC* is above and below the interpolating line.

Example 4.6

Study with computer simulations a two-step 4+4 bit converter. Model the transfer characterisitcs of ADCs e DAC with a random DNL and include distortion terms up to the fifth order. Use an input full-scale sine wave to determine the output spectrum and the equivalent number of bits. Study the effect on the output spectrum of the ADC and DAC non-idealities.

Solution

The file Ex_4.6.m enables the study of both full-flash and two-step architecture with a flag that permits the user's choice. The Matlab file uses the functions statchar.m to generate the static characteristics: it divides the analog range (firstbin - lastbin) into nbin-1 intervals and estimates an ideal response. Then, it adds offset and a random term representing the DNL. The correlated part of the DNL is accounted for with the function dist.m that alters the static response without modifying the end points values. The function uses a polynomial approximation up to the fifth order. The input signal is a linear ramp

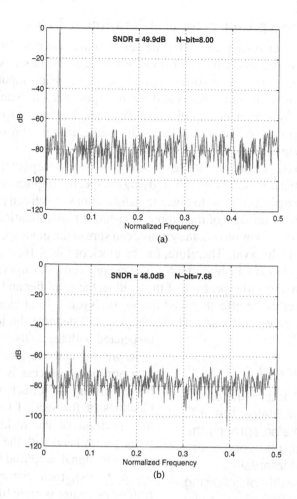

Figure 4.15. Output spectra for ideal (a) and real ADC and DAC responses (b).

(to plot the transfer characterisitc) or a sine wave. Even in this case a flag enables the selection. Fig. 4.15 show the output spectrum for $f_s/f_{in} = 35.31$. The top plot depicts the ideal case. The SNR is 49.9 dB corresponding to 8-bit. The bottom spectrum accounts for a random DNL which maximum is 0.3 LSB and a 0.01 third order distortion coefficient. Equal non-idealities affect ADCs and DAC. The spectrum tones evident in the figure gives rise to 48 dB SFDR corresponding to a 0.5 bit loss. The use of the provided files permits a more extensive study of limitations and features of the full-flash and the two-step flash that the reader can do autonomously.

4.4.2   Two-step Converter as a Non-linear Process

The residue of a two-step converter (i.e., the quantization error of the coarse *ADC*) can be viewed as a non-linear transformation of the input as shown by the block of Fig. 4.16. Its response is non-univocal: many inputs give the same outputs with a linear in pieces and equal slopes characteristic. After the non-linear transformation an amplifier provides a possible interstage gain. The equations on the right side of Fig. 4.16 are the mathematical expression of this non-linear response.

The diagram is not just a repeated explanation of the two-step (or the subranging) algorithm but is a useful view for studying the spectral implications of using non-linear systems. In contrast to linear transformations which can preserve the band-limited characteristics of their input, non-linear transformations generate extra tones which, in sampled-data systems, can spread the input spectrum over the entire Nyquist Interval. Therefore, for the block of Fig. 4.16, the generated residue voltage can occupy the entire Nyquist range, even if the input signal does not. To preserve the wide spectrum of the residue, the amplifier and the second flash must therefore be effective well above the Nyquist limit since possible attenuations or phase shifts around the Nyquist will degrade the information associated with the *LSBs*.

> **Remember that**
>
> Any non-linear transformation that facilitates the A/D conversion also spreads the input spectrum over the entire Nyquist interval.
> The bandwidth of following circuits must be significantly larger than f_N!

Representing the residue generation as a non-linear process is also useful for estimating the effect of the non-idealities of the second flash. Since the spectrum of the residue is only weakly correlated with the input (provided the signal amplitude is at least a few *MSBs*), then distortion in the residue generator is more likely to produce white noise than to produce tones at multiples of the input frequency. Therefore, the linearity of the residue

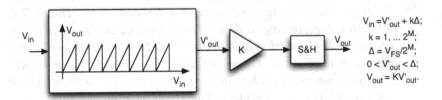

$$V_{in} = V'_{out} + k\Delta;$$
$$k = 1, \dots 2^M;$$
$$\Delta = V_{FS}/2^M;$$
$$0 < V'_{out} < \Delta;$$
$$V_{out} = KV'_{out}.$$

Figure 4.16.   Non-linear block suitable to generate the residue voltage.

generator is not typically critical for the *SNDR* or the *SFDR* which mainly result from the non-idealities of the first *ADC* and the *DAC*.

4.5 FOLDING AND INTERPOLATION

The previous sub-section models the residue generator as a non-linear transformation of the input: the dynamic range is divided into a given number of *MSB* parts with linear input-output relationships inside each sector. Another suitable (and equivalent) non-linear transformation is the folding shown in Fig. 4.17 that splits the input range into a number of sectors (4 in Fig. 4.17 (a) and 8 in Fig. 4.17 (b)) to obtain a linear response inside each sector with alternate positive and negative equal slopes. The non-linear responses can be viewed as the multiple folding of a ramp (two times to generate four sectors, three times to generate the non-linear responses of Fig. 4.17 (b)). This is why the method is named folding.

A single folding bends the input around $1/2\,V_{FS}$ and gives rise to two sectors (1-*bit*) with peak amplitude $1/2\,V_{FS}$. Folding two times leads to four sectors (2-*bit*) with peak amplitude $1/4\,V_{FS}$. Folding three times corresponds to three bits whose peak value becomes $1/8\,V_{FS}$ and so forth. Since multiple folding reduces the output range, the number of intervals which will be required to quantize the folded signal diminishes accordingly. For example, after an *M-bit* folder it will only be necessary to use $2^{n-M} - 1$ comparators to complete the *n-bit* conversion. Obviously, it is necessary to know which segment the input is in to determine the *MSBs*, which are then combined with the *LSBs* given by the folded signal quantization.

Fig. 4.18 is a conceptual block diagram of the folding converter. The *M-bit* folder produces two signals: the analog folded output and the *M-bit* code which identifies which segment the input is in. The gain stage possibly augments the dynamic range to become V_{FS}. The *N-bit ADC* then determines the *LSBs* which the digital logic combines with the *MSBs* to give an overall output of $n = (N + M)$ bits.

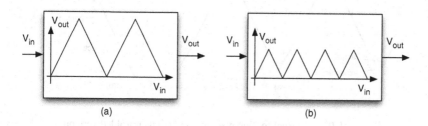

Figure 4.17. Non-linear blocks that obtain input folding.

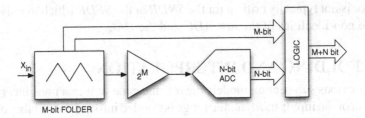

Figure 4.18.   Basic architecture of a folding converter.

The input regions around the folding points are critical since it is necessary to have sharp changes in the input-output response slope. This feature can not be properly obtained by real circuits using either bipolar or *MOS* transistors, as their responses are always somewhat rounded. Moreover, operating in different segments establishes different delays because it is necessary to charge or discharge the parasitic capacitances of the switching elements for procuring the transition between sectors.

In addition to the above limits it is necessary to account for finite bandwidth and slew-rate as the folding circuit is normally used for high conversion rates and medium-high resolutions.

4.5.1   Double Folding

Fig. 4.19 shows the input-output characteristic of a possible real folder and its corresponding unfolded plot. The folder linearity is good in the regions midway between the folding points and becomes bad as the input approaches the segment borders. Since the *N-bit* quantization of a folded response corresponds to the *(N+M)-bit* quantization of an unfolded curve, the unfolded response can be

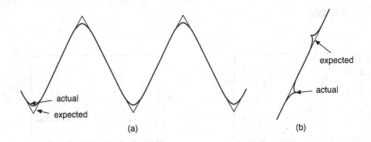

Figure 4.19.   (a) Real folding response and (b) its unfolded version.

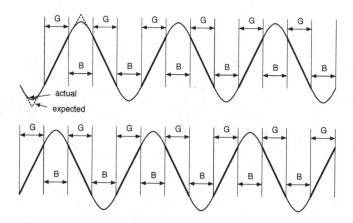

Figure 4.20. The use of double folding avoids non-linear regions.

used to determine *INL* by taking the difference between actual and expected curves and measuring the result using the *n-bit* quantization step.

The actual curve is always rising; therefore, monotonicity is ensured; however, the rounding near the transition points often give rise to an *INL* which can make the method impractical.

The solution to this problem is to use two folders which permits the designer to discard the bad regions and utilize only the good ones. One transfer characteristic is shifted with respect to the other by a quarter of the folding period thus ensuring that one or other folder will always be in its linear region for all input amplitudes (Fig. 4.20). Only the linear regions need to be quantized by the *LSB* flashes whose dynamic range is sometimes partially overlapped thus making the number of comparators equal to, or slightly higher than 2^N.

Finally, the logic that combines the *MSB*s and *LSB*s must take into account the sign of the slope in the used segment and decide which folder provides the best linear response.

4.5.2 Interpolation

An interpolator generates an electrical value that is intermediate between two other electrical quantities by using, for voltage inputs: resistive or capacitive dividers and, for current inputs: schemes based on current mirrors.

The simple circuit of Fig. 4.21 (a) is a resistive divider whose output voltage V_{inter} is given by

$$V_{inter} = \frac{V_1 R_2 + V_2 R_1}{R_1 + R_2}. \tag{4.23}$$

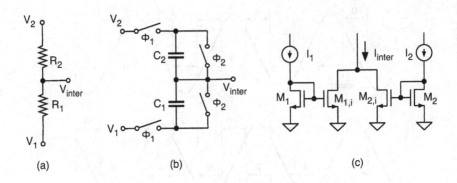

Figure 4.21.   Simple circuits that achieve interpolation.

If the two resistors are equal then interpolation is median; different resistors give rise to a fractional result. Moreover, the use of multiple resistances achieves a multiple interpolation as is used in the Kelvin divider *DAC*.

The circuit of Fig. 4.21 (b) obtains interpolation with two capacitors C_1 and C_2. They are discharged during phase Φ_1 and generate the output voltage during the complementary phase, Φ_2. The interpolated voltage is

$$V_{inter}(\phi_2) = \frac{V_1 C_1 + V_2 C_2}{C_1 + C_2}. \qquad (4.24)$$

The discharging phase, necessary for removing possible residue charges affecting the initial conditions, and the interpolation phase denote a sampled data operation leading to a valid output voltage only during half of the sampling interval. Obviously, the use of the interpolated voltage must avoid charge leakage and show a minimum parasitic capacitance on the output node that, for a more accurate calculation, should be accounted for in equation (4.24).

The use of current mirrors enables current interpolation as shown in Fig. 4.21 c). Mirrored replicas of the currents I_1 and I_2 are weighted by using the aspect ratios between transistors M_1, $M_{1,i}$, M_2, and $M_{2,i}$. If the desired interpolation factor is α, then

$$\frac{(W/L)_{1,i}}{(W/L)_1} = \alpha; \quad and \quad \frac{(W/L)_{2,i}}{(W/L)_2} = 1 - \alpha \qquad (4.25)$$

making the interpolated current

$$I_{inter} = \alpha \cdot I_1 + (1 - \alpha) \cdot I_2. \qquad (4.26)$$

The accuracy of the interpolation factors depend on the matching of the components used in the circuits. Therefore, the use of symmetric or common centroid layouts for resistors or capacitors gives, for present technologies, accuracies in the order of 0.1%; slightly worse values are obtained for well laid out *MOS* current interpolators whose overdrive voltage must be also fairly large with respect to the expected threshold mismatch.

4.5.3 Use of Interpolation in Flash Converters

The design of flash converters can take advantage of the interpolation technique if the comparator is implemented by a preamplifier followed by a latch. The use of interpolation, as shown in Fig. 4.22, reduces the number of preamplifiers by generating the median of adjacent pre-amplifier outputs. This interpolated voltage is then used by intermediate latches. The method works if the outputs saturate for an input that is equal to or greater than the immediate upper threshold or equal to or lower than the immediate lower threshold. The overlapped non-saturated regions interpolated by the resistors, as shown in Fig. 4.22 (b), determine crossings that are midway between the preamplifiers zero crossings. Moreover, equal slopes at the zero crossings equalize the speed and the metastability error of the latches. Only far from the zero crossing does the slope of the interpolated curve diminish. However this is not a problem since at these points the differential signal is large enough to properly control the latch.

The number of preamplifiers (and reference voltages) diminishes by a factor of two; therefore, the capacitive load on the *S&H* diminishes, thus enabling

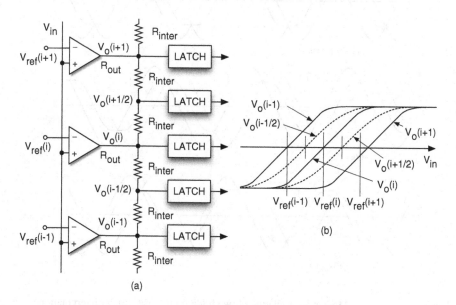

Figure 4.22. a) Use of interpolation in flash converters. b) Outputs and interpolated responses.

an easier *S&H* design and, in turn, allowing either less power consumption or higher speed. Moreover, the reduced number of reference voltages reduces the charge pumping effect that, as studied previously, causes a settling limit.

If the comparator architecture uses the cascade of two gain stages for boosting the speed it is possible to use interpolation twice; after both the first and second gain stages, thus obtaining a further reduction of the *S&H* parasitic load.

The method can also be extended to multiple interpolation where a series of four or eight equal resistors connected between neighboring pre-amplifiers provide multiple inputs for latches. The interpolating networks can be single ended or fully differential depending on the type of output of the preamplifier and the type of signal required by the input of the latches.

4.5.4 Use of Interpolation in Folding Architectures

The interpolation technique can also be used in a folding architecture by allowing multiple interpolators to take the place of the fine flash converter. Fig. 4.23 (a) shows the interpolation of two folded responses, V_{F1} and V_{F2}, shifted by half a segment. The shape of the interpolation curve is more rounded than the generating signals (actually, it is almost flat over a given range).

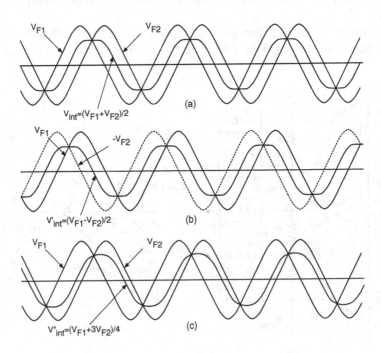

Figure 4.23. (a), (b) Median interpolation of two folding response. (c) 3/4 interpolation.

However, what is important is to have the zero crossings midway between the two zero crossings of the generating signals. The interpolation of V_{F1} and V_{F2} produces one set of the mid zero crossings while interpolating V_{F1} and $-V_{F2}$ gives rise to a second set of mid zero crossings, as shown in Fig. 4.23 (b). The result is a doubled number of zero crossings whose occurrences can be conveniently detected by additional comparators.

> **Observe**
>
> Interpolation can avoid the use of the flash converter in a folding scheme by procuring $2^N - 1$ zero-crossings among zero-crossings points of the interpolating curves.

Since uneven resistances or capacitors change the interpolation factor, it is possible, as shown in Fig. 4.23 (c), to obtain a shifted zero crossing. If a resistive interpolator used $R_1 = 3R_2$ the zero crossings is at $1/4$ distance from the zero crossing of V_{F2}. Using $R_1 = 7R_2$ yields a crossing distance of $1/8$ and so forth. Therefore, multiple interpolations can give rise to a sufficient number of zero crossing to obtain the *LSB* conversion. The use of interpolation requires using resistors as for a flash converter but the use of parallel divider for obtaining interpolation reduces the cross talk between *LSB* channels and possibly increases the speed.

4.5.5 Interpolation for Improving Linearity

The resistances of the interpolation network load the preamplifiers they are connected to. Therefore, if the preamplifier output resistance is not much smaller than the load, then possible drops alter the generated voltages; the error does not affect the overall operation but works in a beneficial manner. To be precise, the voltage error is dependent on the generated voltage and, for zero output, goes to zero. Consequently, the load of the interpolating network does not affect the zero crossing which is what really matters in the converter.

Beyond this, an interpolating network and a bank of preamplifiers with finite output resistances will average the offset of the preamplifiers and, in turn, improve linearity.

The simple circuit of Fig. 4.24 (a) models the preamplifiers as a real voltage generator with output resistance R_{out}. In general, we can state that the reciprocal influence of nodes is negligible for large distance between nodes and depends on the ratio between R_{out} and R_{int}. If the ratio is very small the interpolating network is almost completely ineffective. Moreover, the interpolating network causes a current flowing from or into the *i-th* preamplifier node given by

$$I_{P,i} = \frac{V_{I,i+1} + V_{I,i-1} - 2V_{I,i}}{2R_{int}} \tag{4.27}$$

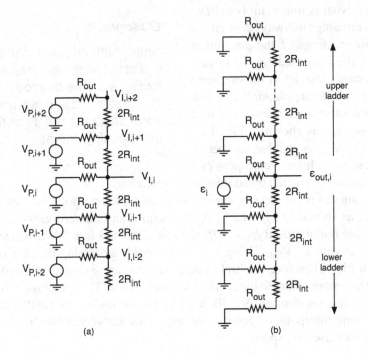

Figure 4.24.   Models for studying the interpolating network.

showing that if the *i-th* interpolated voltage is the average of its neighbors then its current is zero. Therefore, since ideal output voltages are not affected by the interpolation network, we need only consider the effect of the errors.

Assume that an offset ϵ_i changes the output of the *i-th* preamplifier from $V_{P,i}$ to $V_{P,i} + \epsilon_i$. The error at the output $V_{I,i}$ is attenuated by the resistive division between R_{out} and the parallel connection of the upper ($R_{U,i}$) and lower ($R_{L,i}$) networks, at the *i-th* node: $R_{T,i} = R_{U,i}R_{L,i}/(R_{U,i} + R_{L,i})$. Therefore, the error after interpolation becomes

$$\epsilon_{I,i} = \epsilon_i \frac{1}{1 + R_{out}R_{T,i}} = \epsilon_i T_{i,i} \qquad (4.28)$$

where $T_{i,i}$ is the attenuation of the *i-th* offset at the *i-th* interpolated output.

Using the scheme of Fig. 4.24 (b) it is also possible to estimate the error induced by ϵ_i at other taps of the network. The result a set of attenuation factors $T_{i,j}$ which values diminish as $|i - j|$ increases.

For a single integrated circuit the effect of the offsets must be superposed linearly but for performing a statistical study (as required for determining the yield) the effects must be superposed quadratically as the offsets of different preamplifiers realized with different runs are uncorrelated. For both cases it results

$$\epsilon_{tot,i} = \sum_{j=1}^{2^n-1} \epsilon_j \cdot T_{i,j}; \quad \sigma_{I,os}^2 = \sigma_{os}^2 \sum_{j=1}^{2^n-1} T_{i,j}^2. \tag{4.29}$$

Remind that in practical cases the factor $T_{i,j}$ is approximately equal to zero for $|i - j| > 4$, then the averaging effect of the interpolating network benefits short distance error thus mainly reducing the *DNL*.

Example 4.7

Estimate with Spice simulations the improvement of the DNL due to the resistive interpolation smoothing the response of 63 preamplifiers. The interpolating resistance and the preamplifier output resistances are equals. Consider only 5+5 neighbors and assume that the static offset is a random variable. Since the offsets are static errors their contributions should be added linearly for estimating the DNL.

Solution

This solution uses the Spice description of the scheme of Fig. 4.24 limited to 11 cells. Also, it assumes a preamplifiers response that is linear around zero and saturates if the input signal exceeds $\pm 5\Delta$. The preamplifier output is plotted in Fig. 4.25. The same figure shows that interpolation spreads out a little the response but, as expected preserves the zero crossing. For example, an input equal to 2Δ gives rise to an output of $0.85 \, V_{FS}$ without interpolation and $0.706 \, V_{FS}$ with interpolation. The simulation that obtains the curves of Fig. 4.25 applies equally spaced pre-amp voltages and measures the

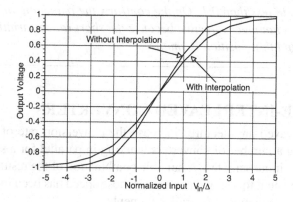

Figure 4.25. Possible output voltages of preamplifiers without and with interpolation.

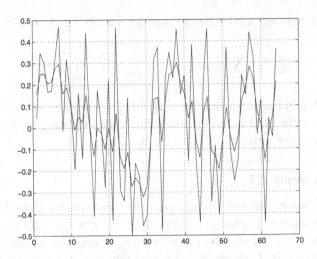

Figure 4.26.   Smoothing action of the interpolation on a possible DNL.

interpolated results. For measuring the attenuation factors only one pre-amp generator set to 1. The voltage on the neighbor nodes gives: $T_{i,i} = 0.44$; $T_{i,i-1} = T_{i,i+1} = 0.17$; $T_{i,i-2} = T_{i,i+2} = 0.06$; $T_{i,i-3} = T_{i,i+3} = 0.02$; $T_{i,i-4} = T_{i,i+4} = 0.01$. Observe that the last two attenuation factors are so small to make negligible the contributions of the ± 3 and ± 4 neighbours.

The obtained results are than used as coefficients of a spatial filter (file Ex4_7) applied to a random signal modeling the comparators' offset. Fig. 4.26 shows the input and output of such a spatial filter. As expected the effect is on short distance variations of the offset. For example, the error of the comparator #22 goes down from 0.46 arbitrary units to less than 0.1. On the contrary, the reduction for comparator #38 is not much effective because the short range variation of errors around that position is small.

4.6   TIME-INTERLEAVED CONVERTERS

Time-interleaved architectures increase the conversion rate of a data converter by using a number of converters working in parallel for a simultaneous quantization of input samples. A suitable combination of the results makes the operation equivalent to a single converter whose speed has been increased by a factor equal to the number of parallel elements.

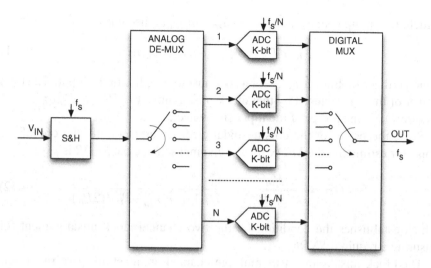

Figure 4.27. Time-interleaved architecture.

The architecture shown in Fig. 4.27 uses an input sample-and-hold running at its full speed $f_s = 1/T_s$ to acquire the samples to be converted. Then, an analog selector delivers the input samples to the N parallel *ADC*s whose conversion frequency is f_s/N. Finally, the digital multiplexer sequentially selects the output of each channel to obtain the full speed *ADC* code.

An alternative solution that avoids the demanding specification associated with the full speed *S&H* employs one *S&H* in each path. However, this requires careful generation and distribution of the control phases as misalignment degrades the dynamic performances.

Other important limitations are the offset and the gain mismatch between channels. These error sources, which do not occur in high-speed single converter applications become significant for interleaved architectures as they are transformed into dynamic errors by the system operation.

4.6.1 Accuracy requirements

A clock misalignment in the sampling time causes an error similar to clock jitter. However, since the misalignment is a fixed delay, the error occurs with an NT_s periodicity. If the clock misalignment between the *K-th* channel and the first channel is δ_K, the error caused is

$$\epsilon_{ck,K}(nT) = \delta_K \frac{dV_{in}}{dt}\bigg|_{nT} ; \quad n = i \cdot N + K \qquad (4.30)$$

which, for an input sine wave $V_{in} = A_{in}sin(\omega_{in}t)$, becomes

$$\epsilon_{ck,K}(nT) = \delta_K A_{in}\omega_{in}cos(\omega_{in}nT) \tag{4.31}$$

which is the sampling, at f_s/N, of a co-sinusoidal replica of the input. The total power of the error caused by that clock misalignment is $P_{\epsilon_{ck}} = P_{in}\delta_K^2\omega_{in}^2/N$ (where P_{in} is the power of the input sine wave).

Since the power of the co-sinusoidal error is summed with the one of the noise for estimating the $SFDR$ for a 2-channel scheme the $SFDR$ is

$$SNDR = \frac{P_{in}}{P_n + P_{\epsilon_{ck}}} = \frac{P_{in}}{P_n}\frac{1}{1 + \delta_2^2\omega_{in}^2 \cdot P_{in}/(2P_n)} \tag{4.32}$$

which establishes the condition on the two channel clock misalignment for ensuring a required $SNDR$.

The clock misalignment of multiple channel architectures give rise to errors whose effects must be superposed linearly after accounting for the delay between each channels and one used as reference. Moreover, because of the down-sampling, the power $P_{\epsilon_{ck}}$ gives rise to images located at $kf_s/(2N)\pm f_{in}$.

Let us consider now the offset beginning with the observation that identical offsets in all the ADCs only causes an overall offset, however any mismatch in the offsets causes tones. For example, an offset influencing a single channel appears at the output every N clock periods giving rise to a pulse with amplitude equal to the offset and with duration $1/f_s$. The leads to tones at f_s/N and its multiples. More generally, mismatched offsets in various channels lead to a repetitive pattern with period N/f_s that, again, causes tones at f_s/N and its multiples. Since an interleaved architecture is normally used for the conversion of a wide-band signal, the signal band is often a large fraction of the Nyquist interval; therefore, even tones at relatively large fractions of the sampling frequency are likely to fall into the signal band.

The allowable offset mismatch is established by the $SFDR$ specifications. If, in a worst case, the interleaved architecture experiences an alternate sequence of positive and negative offset V_{os}, then the effect is an additive square wave whose highest tone has an amplitude of $4V_{os}/\pi$. Therefore, with a full range input sine wave, the $SFDR$ is

$$SFDR \simeq 20 \cdot log\frac{\pi V_{FS}}{8V_{os}}. \tag{4.33}$$

If the $SFDR$ requirement for a maximum input equals the SNR, then the offset mismatch must be $V_{os} < \pi/8 \cdot \sqrt{8/12} \cdot \Delta \rightarrow 0.32\,LSB$, which is a difficult condition to achieve, especially when a high yield must be ensured because the offset variance must equal the estimated offset must be divided by the number of σ required to obtain the yield. For example, with 10-*bit* and $1\,V_{FS}$ a 99% yield would require an offset variance as low as $\sigma_{off} < 0.19\,mV$ as required to satisfy the condition necessary to obtain a 99% yield: $2.57\sigma_{off} < 0.5\,LSB$.

Similar to offset errors, identical gain errors are not a problem as they cause an equal gain error on the overall architecture. However, the mismatch between the channel gains causes tones due to the multiplication of the input signal with the periodic sequence of the channel gains. The worst case occurs for gains alternating be-

> **Keep Note!**
>
> Offset and gain mismatch are often insurmountable limits that prevent using high-resolution time-interleaved converters unless trimming or calibration techniques are utilized.

tween $(1 + \epsilon_G)$ and $(1 - \epsilon_G)$ which gives rise to an error equal to the multiplication of the input signal with a square wave of amplitude $2\epsilon_G$ at $f_s/2$ or its sub-multiples.

The largest spur tone occurs at $f_s/2 \pm f_{in}$ (or $f_s/4 \pm f_{in}$, or ...) and has amplitude

$$A_{spur} = \frac{4}{\pi}\epsilon_G \cdot A_{in} \tag{4.34}$$

where A_{in} and f_{in} are the amplitude and frequency of the input sine wave.

Equation (4.34) gives rise to the *SFDR*

$$SFDR \simeq 20 \cdot log\frac{\pi}{4\epsilon_G} \tag{4.35}$$

which, for example, with a gain error of only 0.1% causes $-58\,dB\,SFDR$.

Example 4.8

A four channel time-interleaved uses 10-bit, 60 MS/s, 1 V_{FS} ADCs. The required SFDR is 70 dB with -6 dB_{FS} input sine wave at 120 MHz. Determine the requested offset and gain mismatch. What is the maximum clock misalignment that gives rise to a SNDR larger than 59 dB with a full scale input?

Solution

Either offset and gain mismatch, or clock misalignment generate spurs at different frequencies. If the spurs are not piled up, the parameters that achieve the target SFDR are estimated separately.

The approximate equations studied in this subsection leads to the following conditions

$$V_{os} = 10^{-3.5}\pi V_{FS}/8 = 0.12\,mV$$

$$\epsilon_G = 10^{-3.5}\pi/4 = 2.5 \cdot 10^{-4}$$

that are difficult to obtain especially for the offset and the clock control. Notice that the number of bits is not used in calculations.

The use of equation (4.32) shows that for having an SFDR 3 dB less than the SNR it is necessary to have a clock misalignment that verifies the condition

$$\delta_{mis}^2 \omega_{in}^2 \cdot P_{in} = NP_Q$$

yielding

$$\delta_{mis} = \frac{\sqrt{4 \cdot 8}}{\sqrt{12} \cdot 2\pi \cdot 120 \cdot 10^6 2^{10}} = 2.11ps.$$

The result is not particularly critical but for higher resolution or higher input frequencies the issue becomes problematic and would require a special care in the clock distribution.

4.7   SUCCESSIVE APPROXIMATION CONVERTER

The successive approximation algorithm performs the *A/D* conversion over multiple clock periods by exploiting the knowledge of previously determined bits to determine the next significant bit. The method aims to reduce the circuit complexity and power consumption using a low conversion rate by allowing one clock period per bit (plus one for the input sampling).

For a given dynamic range $0 - V_{FS}$ the *MSB* distinguishes between input signals that are below or above the limit $V_{FS}/2$. Therefore, comparing the sampled input with $V_{FS}/2$ obtains the first bit as illustrated by the timing scheme

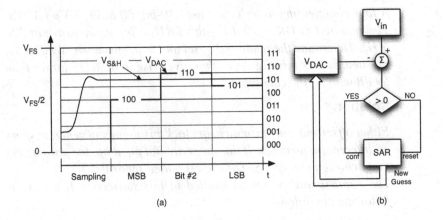

Figure 4.28.   Timing (a) and flow diagram (b) of the successive approximation technique.

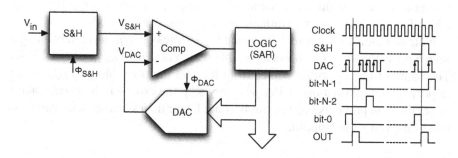

Figure 4.29. Basic circuit diagram of the successive approximation algorithm.

of Fig. 4.28 (a). The knowledge of the *MSB* restricts the search for the next bit to either the upper or lower half of the $0 - V_{FS}$ interval. Consequently, the threshold for determining the second bit is either $V_{FS}/4$ or (as it is for the case of the figure) $3V_{FS}/4$. After this, a new threshold is chosen and the next bit can be estimated. The timing diagram of Fig. 4.28 describes the operation for three bits but, obviously, the search can continue for additional clock cycles to determine more bits. The voltages used for the comparisons are generated by a *DAC* under the control of a logic system known as the successive approximation register (*SAR*) as shown in Fig. 4.28 (b). Notice that the input common mode range of the comparator must equal the dynamic range of the converter.

The method uses one clock period for the *S&H* and one clock period for the determination of every bit thus requiring $(n + 1)$ clock intervals for an *n-bit* conversion. Sometimes, if the *S&H* settling period is significantly longer than the time required for each comparison, then it can be convenient to use two clock periods for the sampling and one per every bit totalling $(n + 2)$ clock intervals for an *n-bit* conversion.

Fig. 4.29 shows a typical block diagram of a successive approximation converter. The *S&H* samples the input during the first clock period and holds it for *N* successive clock intervals. The digital logic controls the *DAC* according to the successive approximation algorithm whose flow diagram is shown in Fig. 4.28 (b): initially the *SAR* sets the *MSB* to 1 as a prediction of the *MSB* value. If the comparator confirms the predicted value then the value is retained, otherwise the *MSB* is set to zero. On the next clock period the *SAR* makes another prediction by setting the value of the next bit to 1. Again the comparator confirms if this assumption was correct and, after confirmation, the algorithm proceeds in the same way predicting each successive bit until all *n-bits* have been determined.

At the start of the next conversion, while the *S&H* is sampling the next input, the *SAR* provides the *n-bit* output and resets the registers. Fig. 4.29 shows the timing sequence for this conversion: the voltage V_{DAC} changes at the rising edge of Φ_{DAC} and remains available for the entire clock period.

Note that the *SAR's* control is such that V_{DAC} tracks $V_{S\&H}$ thus establishing a search path. Fig. 4.30 (a) shows a possible search path for $V_{S\&H} = 0.364\,V_{FS}$. The name of the algorithm comes from the fact that the voltage V_{DAC} is an improving approximation of $V_{S\&H}$: every step the error can be occasionally larger than the previous one but surely is not larger than successive divisions of two of the full scale amplitude.

4.7.1   Errors and Error Correction

An error in the bit estimation modifies the search path in such a way that the error propagates along all the successive steps. Assume, as shown in Fig. 4.30, that V_{DAC} changes from a level well below $V_{S\&H}$ to a level just above (fourth clock period). It may happen that the recovery of the comparator from overdrive is not fast enough and eventually the comparator determines a logic 1 instead of a logic 0. The next V_{DAC} voltage brings the search path in a wrong direction and following the search path results in a final code of 01110000 instead of 01101110, an error of 2 *LSB*. Since this kind of error typically occurs at the beginning of the conversion cycle when large steps in the search path lead to a large comparator overdrive, then possible error correction methods must expand the search range near the end for accommodating the initial inaccuracy. The account for this expansion, the conversion will require extra clock cycles to complete the algorithm.

The accuracy of the successive approximation converter obviously depends on the accuracy of the *S&H*, the comparator and the *DAC*. The designer must

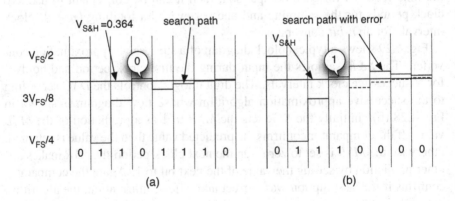

Figure 4.30.   a) Correct search path. b) Search path with error at the fourth clock period.

study the single effects and their superposition that, in general, causes a random term and harmonic distortion on the conversion results that influence the *DNL* and the *INL*.

Example 4.9

Simulate the static behavior of an 8-bit successive approximation converter. Use a slow input ramp and estimate INL and DNL. Model the non-ideality of the DAC using a second and a third harmonic distortion term.

Solution

Consider a slow ramp starting from 0 and achieving the full scale in $k\,2^8$ sampling periods. The converter acquires $k\,2^8$ samples producing, for an ideal response, a flat histogram with k samples per bin. Since a bin of the histogram containing less than k samples denotes a quantization interval smaller than 1 LSB and a content of more than k sample means a Δ exceeding 1 LSB, the histogram obtains the DNL and the INL plots with 1/k LSB accuracy.

The file Ex4_9.m uses the m-file SuccAppr.m for implementing the successive approximation algorithm of an 8-bit SAR. The slow ramp is such that k = 50 leading to an accuracy of the DNL estimation equal to 2% LSB. The simulation uses an input ramp with added a

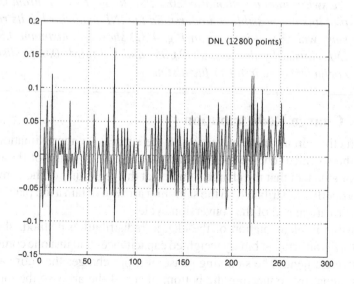

Figure 4.31. DNL with an average of 50 samples per bin (accuracy 2%).

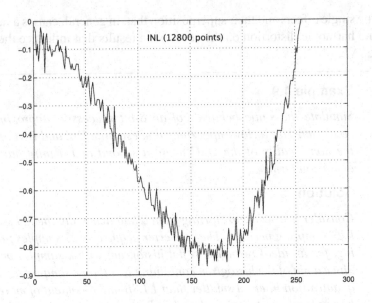

Figure 4.32. INL plot with second and third harmonic distortion terms.

small noise (0.001 variance) for modeling the behavior of the source
generator noise and the time-sampling jitter. The result of Fig. 4.31
shows fluctuations of the DNL that averages the plot thus reducing
the swing from a systematic ±0.2 LSB to ±0.15 LSB. From the LSB
plot it is not possible to understand the INL behavior but its running
sum, whose plot is given in Fig. 4.32, shows a maximum 0.85 LSB
INL caused by the second (0.005) and the third (0.005) distortion
coefficients use in the m-file dist.m.

4.7.2 Charge Redistribution

An effective circuit implementation of the successive approximation algorithm is the so called charge redistribution scheme. The name of the method
comes from the fact that the charge sampled at the beginning of the conversion
cycle is properly redistributed on the sampling array to obtain a top plate voltage
close to zero at the end of the conversion cycle.

A possible implementation of the charge redistribution method, shown in
Fig. 4.33, uses an array of binary weighted capacitances and just one comparator
as an active element. The sampling phase, Φ_s, pre-charges the entire array to
the input signal by connecting the bottom plates of the array to the input and

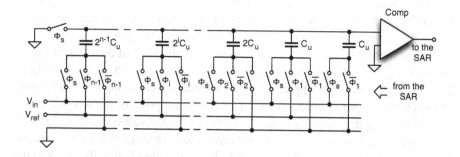

Figure 4.33. Charge redistribution implementation.

the top plates to ground. The charge on the entire array is

$$C_{Tot} = 2^n C_u V_{in}. \tag{4.36}$$

After the sampling phase the *SAR* begins the conversion (clock period 1) by first connecting the bottom plate of the biggest capacitance ($2^{n-1}C_u$) to V_{ref} and the remaining part of the array to ground. The superposition principle determines the voltage on the top plate, applied to the comparator, to be equal to

$$V_{comp}(1) = \frac{V_{ref}}{2} - V_{in}. \tag{4.37}$$

Since this voltage is the difference between the *MSB* voltage and the input, it is only necessary to compare it to ground. The comparator result determines the *MSB* and enables the *SAR* to establish the conditions for the next bit calculation. If the *MSB* is 1 the connection of $2^{n-1}C_u$ to V_{ref} is confirmed and the capacitance $2^{n-1}C_u$ is tentatively connected to V_{ref} during the *2nd* period.

Depending on the value of the already determined *MSB* the new top plate voltage becomes

$$V_{comp}(2) = \frac{3V_{ref}}{4} - V_{in} \quad or \quad V_{comp}(2) = \frac{V_{ref}}{4} - V_{in} \tag{4.38}$$

for *MSB* =1 or *MSB* = 0 respectively. This voltage is then used to determine the next bit and the algorithm continues until all the *n-bit* are generated.

Observe that parasitic capacitances affect the top plate voltage. Actually, the total parasitic C_p attenuates the generated voltages by a factor of α

$$\alpha = \frac{C_u 2^n}{C_u 2^n + C_p}. \tag{4.39}$$

However, the attenuation factor is not a significant limit as it only reduces the comparator input but does not changes its sign, which is the information relevant for determining the bit. This feature is a consequence of the pre-charging to zero of the top plate during the sampling phase. That voltage is zero during the sampling and is almost zero at th end of the conversion cycle.

The advantage of the charge redistribution method is that the input common mode range of the comparator is brought to zero without using op-amps or *OTAs*. Furthermore, only the comparator and the dynamic charging and discharging of the capacitive array determine the power consumption of the scheme.

Variations of the scheme of Fig. 4.33 can use the auto-zero of the comparator to compensate for offset errors. In this case the gain stage is connected in the unity gain configuration and used to pre-charge the top plate of the array to the offset during the sampling phase. Another modification is the use of capacitive attenuation for limiting the capacitive spread in the binary weighted array. For this, one or more series capacitors are used to divide the array into sections that are all connected to ground during the sampling phase.

4.8   PIPELINE CONVERTERS

A pipelining data converter uses a cascade of individual stages which each perform one of the elementary functions required by a sequential algorithm. Essentially, the pipeline unwinds over space what should be done over time by a sequential scheme.

The simplest sequential method is the two-step algorithm that uses two clock periods, one for converting the *MSBs* and the other for the *LSBs*. The pipelined version of the two-step obtains *MSBs* and *LSBs* in one single clock period by rearranging the timing control given in Fig. 4.10: the first stage samples and determines the *MSBs* of an input sample while the second stage calculates the residue and *LSBs* of the previous sample. At the same time digital logic assembles the bits and provides the digital output of the sample which entered two clock periods before. Notice that the two-step method can be expanded to a multi-step algorithm and implemented as a pipeline architecture.

Another sequential algorithm which can be realized in a pipeline is successive approximation. The pipeline obtains one bit per stage instead that one bit per clock period. Each stage of the pipeline generates two outputs; the required bit, and the difference between the input and the internal *DAC*, the residual. The accuracy of the analog signals must comply with the number of bits to be determined from that stage onward.

The pipeline can also generate multiple bits per stage and, if this is the case, each stage requires a multi-bit *ADC* to obtain the digital output and a multi-bit *DAC* to generate the input to the next stage. The total resolution of the pipeline architecture is given by the sum of the bits at each stage. Note, that the number of bits at each stage can either equal or differ from one another depending on design trade-offs.

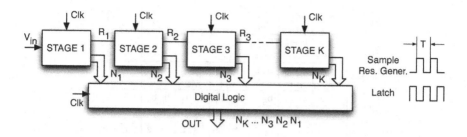

Figure 4.34. Pipeline architecture.

Fig. 4.34 shows the conceptual block diagram of a pipeline architecture with K stages. The timing scheme assumes that the clock has a 50% duty cycle and that one clock phase is used for sampling and the other phase is used for latching the comparators. After latching, during the next sampling phase, each stage generates an analog output to be sampled by the next stage of the pipeline. The first stage generates N_1 bits, the second stage determines N_2 bits and so forth.

> **Keep in Mind**
>
> The pipeline architecture generates the converted output with a latency time that increases with the number of stages.
> Be aware of the latency time when using the converter in a feedback loop!

Therefore the entire pipeline gives rise to $N_1 + N_2 + \cdots + N_K$ bits. The digital logic combines the bits from each stage and generates the output words at a rate of f_s albeit with a delay of $(K+1)$ clock periods (one for the sampling of the input an one each stage). The latency time, caused by this delay, is a consequence of the pipeline operation. This is a minor limit and will not cause problems for most applications unless they use the converter in a feedback loop. The operation of the digital logic is simple as it is necessary to properly delay the bits of the cells and combine them side by side. We shall see shortly that with digital correction the operation of the digital logic is a bit more complicated.

As an example of timing operation Fig. 4.35 illustrates the sequential control of a 10-*bit*, 5-*stages*, 2-*bit* per stage pipeline. Assuming an analog input value is sampled on the *n-th* clock period. On the *(n+1)-th* clock period the bits b_9 and b_8 are generated. On the next clock period (the *(n+2)-th* clock interval), the circuit generates the bits b_7 and b_6. This continues until the *(n+5)-th* interval during which the bits b_1 and b_0 are determined. Finally, the *(n+6)-th* is used by the digital logic to combine the bits and make the result available.

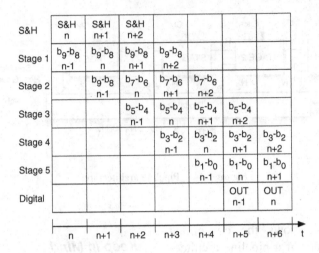

S&H	S&H n	S&H n+1	S&H n+2				
Stage 1	b_9-b_8 n-1	b_9-b_8 n	b_9-b_8 n+1	b_9-b_8 n+2			
Stage 2		b_9-b_8 n-1	b_7-b_6 n	b_7-b_6 n+1	b_7-b_6 n+2		
Stage 3			b_5-b_4 n-1	b_5-b_4 n	b_5-b_4 n+1	b_5-b_4 n+2	
Stage 4				b_3-b_2 n-1	b_3-b_2 n	b_3-b_2 n+1	b_3-b_2 n+2
Stage 5					b_1-b_0 n-1	b_1-b_0 n	b_1-b_0 n+1
Digital						OUT n-1	OUT n
	n	n+1	n+2	n+3	n+4	n+5	n+6 t

Figure 4.35. Timing control of a 2-bit per stage 10-bit pipeline.

The block diagram of a generic pipeline stage is shown in Fig. 4.36. The *ADC* generates *j-bit* while the *DAC* converts the result into analog using the same number bits. However, as we will see shortly, architectures which use digital correction have a *DAC* resolution which is lower than the resolution of the *ADC*. The subtraction of the *D/A* converter output from V_{in} gives the quantization error of V_{in} which after amplification determines the new residue voltage

$$V_{res}(j) = \{V_{res}(j-1) - V_{DAC}(b_j)\}K_j. \tag{4.40}$$

The dynamic range of the residue equals that of the input if, for an n_j bit *DAC* the gain is 2^{n_j}. This condition is frequently used because it allows the same reference voltages to be used for all stages.

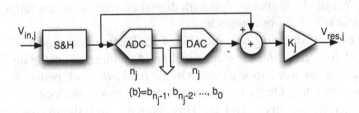

Figure 4.36. Block diagram of a pipeline stage.

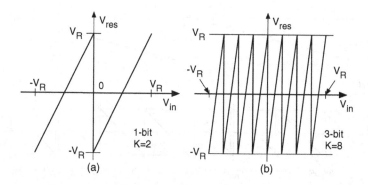

Figure 4.37. Residue generator transfer characteristic. One bit (a) and three bit stage (b).

Fig. 4.37 (a) shows the ideal input-output transfer characterisitcs of the residue generator with input range $-V_R + V_R$, 1-*bit ADC* and *DAC* and $K = 2$. With a negative input the *DAC* subtracts $-V_R/2$ making the initial point after the multiplication by 2 equal to $-V_R$. At $-V_R/2$ the residue crosses 0 and ramps up to $+V_R$ just before the input zero crossing which is the point at which the *DAC* changes from $-V_R/2$ to $+V_R/2$ bringing the residue down to $-V_R/2$.

For 3-*bit* per stage and $K = 8$ the residue plot shown in Fig. 4.37 b) has 7 breaks at the 7 transitions between the 8 quantization intervals of the *DAC*. Since the amplitude of the quantization error is $V_R/4$ the multiplication by 8 makes the dynamic range of the residue equal to the input, $\pm V_R$.

4.8.1   Accuracy Requirements

Recall that the number of remaining bits to be estimated decreases throughout the pipeline; therefore, the accuracy requirements and the design difficulties are greater for the first few stages of the pipeline than for the last. The most demanding requirements are those on the input *S&H*, as the non-idealities of the first block lead to offset, gain error and non-linearity for the entire pipeline.

The non-idealities of *ADC, DAC* and interstage amplifier used in the pipeline stages cause limitations similar to the ones studied for the two-step scheme. Threshold errors within the *ADC* cause the residue amplitude to be either greater or less than full-scale at the break points, as shown in Fig. 4.38. This figure shows the signal generated in with a real *ADC* when used with an ideal *DAC* and interstage amplifier. The responses of both the 1-*bit* and 3-*bit* residue highlight localized anomalies which, at this point, are not errors since the residue generator can still correctly provides the difference between the analog input and the quantized signal generated by the *DAC*. An error will not occur until the *ADC* of the next stage, as it will not be able to properly convert residue

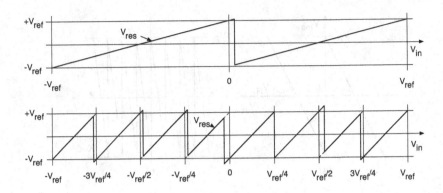

Figure 4.38.   Residue response with real ADC and ideal DAC. Top: 1-bit, bottom: 3-bit.

signals outside of the $\pm V_{ref}$ range of operation. This observation is the basis of the correction strategy used by the digital correction method studied in the following sub-section.

Previously, it was observed for the two-steps that *DAC* errors modify the residue over an entire *LSB*s segment, thus affecting the *INL* of the entire architecture. Therefore, even for the pipeline the accuracy of any *DAC*, referred to the input of the entire architecture, must be better than the required *INL* and lower than 1-*LSB* to ensure monotonic response. Since the input referred residue of the *k-bit*, is divided by 2^k, the accuracy requirement for the *DAC* used for generating that residue is relaxed by the same factor. Therefore, after a few stages the *DAC* linearity becomes minor concern.

Since an interstage gain error, $G = 2^{n_j}(1+\delta G)$, either increases or decreases the slope of the residue, it causes a residue error that is zero in the middle and maximum at the endings of the segments. At the break points the error inverts its sign giving rise to a step change equal to

$$\Delta V = \delta G \cdot 2V_{ref} \qquad (4.41)$$

which, referred to the input, must be less than the desired *INL* (again, less than 1 *LSB* for ensuring monotonicity).

4.8.2   Digital Correction

Errors caused by the non-idealities in the *ADC* can be cancelled out by using a digital correction technique. The method is described here for 1-*bit* per stage but can be extended to any multi-bit architectures. It was previously observed that the dynamic range of the residue can exceed the expected limits but that this is not necessarily a direct flaw. The error occurs when the next stage can not properly convert an out-of-range signal.

One solution is to reduce the inter-stage gain so that the out-of-range conditions are avoided. The method can be problematic because the used coding scheme is binary and an attenuation by a factor different from $1/2$ is difficult to account for. Another method is to add additional levels in the *ADC* of the stage. The addition of these redundant levels avoids, on one hand, generating out-of-range residuals and, on

> **Remind**
>
> Digital correction compensates for ADC error only. The errors caused by the DAC or residue generator are not reduced (nor increased) by digital correction!

the other hand, provides information to the digital domain (hence the name digital correction) and assuming this information is properly used, can fully compensate for the *ADC* error.

Consider a 1-*bit DAC* which usually requires a 1-threshold *ADC*. Obviously, since the quantization is at its minimum, then in order to ensure sufficient redundancy for digital correction, it is necessary to use at least a 2-threshold *ADC*. Since a 1-*bit ADC* uses one threshold while a 2-*bit ADC* employs 3 levels the use of 2 thresholds is normally referred to as 1.5-*bit* conversion.

A 1.5 *bit* per stage pipeline with digital correction uses thresholds nominally symmetrical around zero. The input range is divided into three regions: one below the lower threshold, one across zero, and the other above the upper threshold as shown in Fig. 4.39 (a). If the separation of $V_{th,L}$ and $V_{th,H}$ is sufficiently large then inputs below the lower threshold are said to be "certainly negative" and inputs above the upper threshold are said to be "certainly positive". Uncertainty arises in the middle region as inputs near the zero crossing could lead to a residue that exceeds the limits.

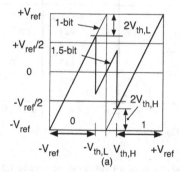
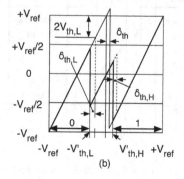

Figure 4.39. Residue response with 1-bit per stage and 1.5 bit per stage for digital correction.

The residue generator adds $V_{ref}/2$ to the input when the *ADC* provides a "certain" zero and subtracts $V_{ref}/2$ when the output is a "certain" one but does not perform any operation in the uncertain region. Therefore, in the uncertain region the residue is the simple amplification by 2 of the input, as shown in Fig. 4.39 (a) with correct thresholds or in Fig. 4.39 b) with incorrect thresholds. The residue at $V_{th,L}$ reaches $V_{ref} - 2V_{th,L}$ and immediately after becomes $2V_{th,L}$; similarly at the second threshold, $V_{th,H}$, the residue swings from $2V_{th,H}$ to $-V_{ref} + V_{th,H}$. Possible errors in the thresholds, change the values of the residue at the break points but remain within the $\pm V_{ref}$ limits if the absolute value of $\delta_{th,L}$ or $\delta_{th,H}$ is lower than $V_{th,L}$ or $V_{th,H}$ respectively. For the case shown in Fig. 4.39 (b) the input range for which the result is a "certain" zero diminishes while the input range that surely provides a logic one increases.

The information provided by the 1.5-*bit* converter is a flag in an *LSB* position that is set to 1 for showing uncertain region. Therefore, 00 and 10 ensure a certain zero or a certain one, while 01 shows that the bit determination is postponed and depends on the output of successive stages. The code 11 is not used since the output is 1.5-*bit*.

Consider an input is in the uncertain region close to $V_{th,H}$; the output will 01 giving a provisional 0 but setting the uncertain flag to 1. Since the residue equals the multiplication by 2 of the input, the result is close to $2V_{th,H}$, which is outside the region of uncertainty and larger than $V_{th,H}$. The next stage will certainly determine a "certain" 1 (10) which combined with the 01 of the previous stage gives the expected 10. Similarly, if the input is in the uncertain region but close to $-V_{ref}$ then the 1.5-*bit* conversion again gives rise to 01. This time however, the certain zero of the next stage (10) confirms the previous zero while also setting the current stage bit to 1. If the input is in the uncertain region and is

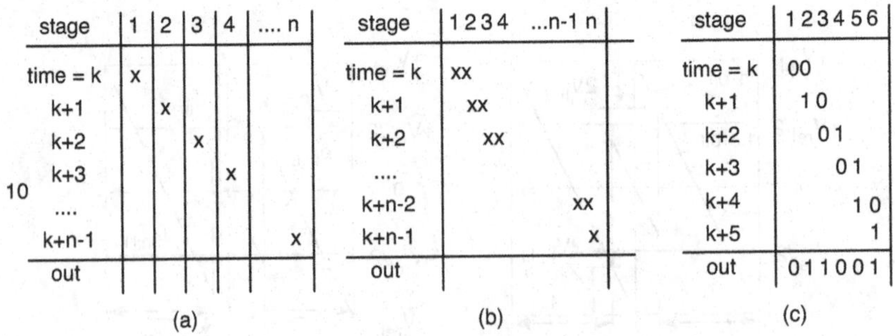

Figure 4.40.   Digital combination of the bits at the output of a 1-bit per stage (a) and 1.5 bit per stage pipeline (b). (c) Numerical example for 6-stages.

very close to zero, the interstage amplification may not be enough to to pull the signal out of the uncertain region and the bit determination is further postponed.

The digital correction logic sums the outputs while accounting for both the delay and consequent weight of each of the interstage gains. Since the gain of 1.5-*bit* is 2 the bits are shifted by one position making the weight of the uncertain flag equal to the *MSB* of the next stage. Fig. 4.40 shows the operation of the digital logic for 1-*bit* and 1.5-*bit* per stage. Fig. 4.40 (a) shows that for the 1-*bit* architecture, the digital logic need only account for the interstage delay and combines the bits side by side at the output of the cells. The 1.5-*bit* per stage needs adders to account for the possible 01 outputs; therefore, as shown in Fig. 4.40 (c), the 01 output of the third stage becomes 1 when it is corrected by the 10 of the next two stages, while the 10 of the fourth and the fifth stages are fixed to zero. The last stage does not use an extra threshold because a possible uncertainty can not be confirmed by successive comparisons.

Example 4.10

Study with behavioral simulations the ADC non-ideality of a 1.5-bit per stage pipeline converter. Plot the residue response at the output of the first three stages. Study the effect of a small error on the thresholds of first and second ADC and explain the obtained results.

Solution

The model file Ex4_10 and the m-file Ex4_10_launch enable the Simulink simulation of a pipeline with three stages but the description can be easily extended to any number of stages. The m-file provides the five parameters used for modeling the pipeline stage: the two threshold of the ADC, the two DAC levels and the interstage gain. The values, for an ideal stage, are $V_{DAC,L} = -0.5$, $V_{DAC,H} = 0.5$ and the Gain is 2. The ADC thresholds can have any value under the condition to be less than half scale. The used nominal values are $V_{th,L} = -0.25$ and $V_{th,H} = 0.25$.

The simulation with ideal parameters gives residue plots that depend on the chosen thresholds. Use, for example, $V_{th,L,1} = -0.16$ and $V_{th,H,1} = 0.18$ and $V_{th,L,2} = -0.26$ and and $V_{th,H,2} = 0.28$ to observe that the uncertainty regions modify the residue plot but the final digital result is not affected. Moreover, the use of an ADC with nominal thresholds in the third stage can give rise to a third residue that is not influenced by the previous thresholds errors.

These observations are confirmed by the plots of Fig. 4.41, Fig. 4.42 and Fig. 4.43. The first figure shows how the combination of threshold shifts can alter the residue and the ADC output. Since the first break

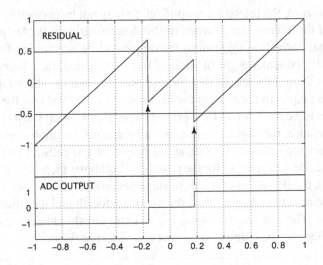

Figure 4.41. First stage residue ($V_{thL} = -0.16$; $V_{thH} = 0.18$) and ADC output.

points of the first stage, marked by arrows in Fig. 4.42, is – 0.16, the
first residue is – 0.32. It is out of the uncertainty region of the second stage and generates a (00) digital output. Only when the input
becomes – 0.13 the first residue crosses the lower second threshold
giving an uncertain (01) result. The third ADC output also accounts

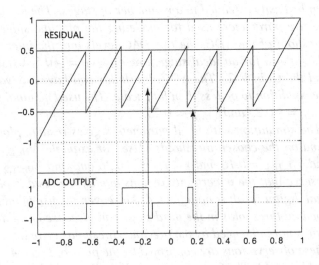

Figure 4.42. Second stage residue ($V_{thL} = -0.26$; $V_{thH} = 0.28$) and ADC output.

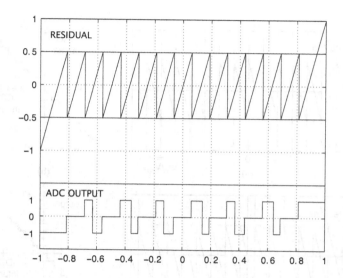

Figure 4.43. Third stage residue ($V_{thL} = -0.25$; $V_{thH} = 0.25$) and ADC output.

for the behavior of the second residue but in the used case, the error affecting the thresholds of the second stage are within given limits and such that the third residue is nominal being the third thresholds correct (Fig. 4.43).

Digital correction used in pipelines with multi-bits per stage is similar to the method described for the 1.5-*bit* per stage. The commonly used approach reduces the interstage gain by 2 thus halving the residue range as shown in Fig. 4.44. Possible out-of-range situations are detected by the *ADC* of the next stage whose digital result is properly combined with the outputs of the other stages to account for the reduced gain. Although reducing the interstage gain by 2 is much more than what typical mismatches would require, the factor 2 is an optimum choice because the signal processing is based on a binary radix. In some cases, the number of levels of the *ADC* are limited to the ones that are expected to be used.

Fig. 4.44 shows two possible residues both of which were generated by a 4-*bit* stage with interstage gain equal to 8. Even though both responses are adequate for a pipeline architecture the residue plots are different because the dynamic range of Fig. 4.44 (a) is divided into 16 equal segments while the plot of Fig. 4.44 (c) has 15 equal segments plus 2 half-segments.

The *DAC* outputs used to obtain the residues are shown in shown in Fig. 4.44 (b) and Fig. 4.44 (d) respectively. Since the latter characteristic uses 17 levels,

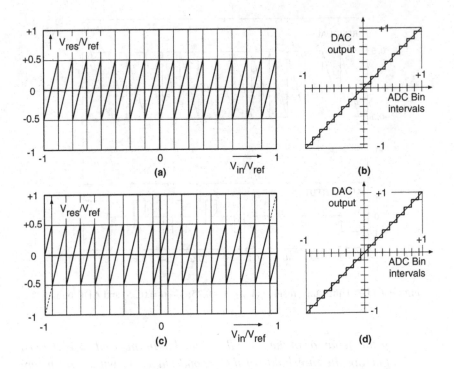

Figure 4.44. Two possible residue plots and the corresponding DAC outputs (4-bit per stage).

it requires the use of a 16 comparator flash to determine the 5-*bits* for the *DAC* control. However, since the first and the last segments of the residue are exercised by the first stage of the pipeline, the response can be modified as shown by the dashed line of Fig. 4.44 (c). The modified residue response only requires a 14 comparator flash whose output is 4-*bit*.

Since digital calibration methods require additional comparators then the power consumption increases. Moreover, the associated digital section is more complex and also consumes more power as it must perform, after the delay and the shift, the additions of the stages outputs.

4.8.3   Dynamic Performances

The dynamic performances of the pipeline converter depend on the slew-rate and bandwidth of the circuits used in both the *S&H* and the residue generator. Although a consistent study of the dynamic behavior requires extensive simulations at the transistor level, performing a behavioral analysis prior to the time-consuming transistor level study, can help in the definition of the active blocks specifications and in identifying the bottlenecks of the architecture.

The previous chapter studied an *S&H* with a real operational amplifier. Since the residue generator also uses an op-amp with real features, we can use a similar analysis to account for the finite gain-bandwidth, f_T, and the finite slew-rate, *SR*. A residue that, ideally, is a step with amplitude $\overline{V}_{out}$ becomes a ramp during the slewing period and turns into an exponential when the feedback takes control of the output voltage. The equations representing the transient are

$$
\begin{aligned}
V_{out}(t) &= SR \cdot t & t < t_{slew} \\
V_{out}(t) &= \overline{V}_{out} - \Delta V \cdot e^{-(t-t_{slew})/\tau} & t > t_{slew}
\end{aligned} \tag{4.42}
$$

$$
\Delta V = SR \cdot \tau; \qquad t_{slew} = \frac{\overline{V}_{out}}{SR} - \tau
$$

where $\tau = 1/(\beta 2\pi f_T)$ depends on the op-amp unity gain-bandwidth and the β feedback factor of the circuit used to generate the residue.

Since the *S&H* of the next stage of the pipeline samples the residue after $T_s/2$, the exponential settling gives rise to a non-linear error, $\Delta V \cdot e^{-(T_s/2-t_{slew})/\tau}$.

Observe that a small settling error is important in the first stages of the pipeline, but that the accuracy requirement decreases for later stages because the input referred effect is divided by all the preceding interstage gains. Therefore, the settling requirements can be relaxed along the pipeline making it is possible to use different speed specifications for op-amps used in different stages of the pipeline. This helps in reducing the design complexity and, also, reducing the total power consumption.

The above equations used in a behavioral simulator or in a circuit simulator running a behavioral model determine the op-amp specifications required to obtain a given *SNR* or *SNDR*.

Example 4.11

Employ the equations (4.42) to describe the speed limitation of the sample-and-hold and the residue generators used in a 100 Ms/s, 10-bit pipeline converter with $\pm$ 1 V reference made by 9 stages with 1.5-bit per stage and a final 1-bit stage. Study the effect of the op-amps' slew-rate and finite band-width on output spectrum, SNR and SFDR and determine the values of bandwidth and slew rate that ensure a SNR > 59 dB (more than 9.5 bit) and SFDR > 70 dB.

Solution

The pipeline model of Ex4_11 (Fig. 4.45) uses a modified version of the 1.5 bit stage of the previously studied Example by adding the delay between input and output and including the behavioral description of equation (4.42). The digital correction generates the quantized

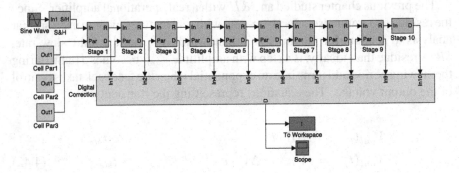

Figure 4.45. Behavioral block diagram of the pipeline.

output by accounting for the delay and the interstage gain. The file Ex4_11_launch.m enables the definition of static and dynamic parameters of S&H and pipeline stages. The specifications of the stages from 3 to 9 are equal; however, the user can possibly change the code and verify the limits caused by the last stages of the architecture.
The simulator makes possible a separate estimation of the limits of S&H, first, second stage and the rest of the pipeline leading to the results shown in Fig. 4.46 that plots the SNR and the SNDR as function

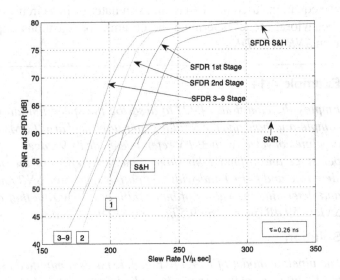

Figure 4.46. SNR and SFDR versus the slew-rate. The settling period is 5 ns.

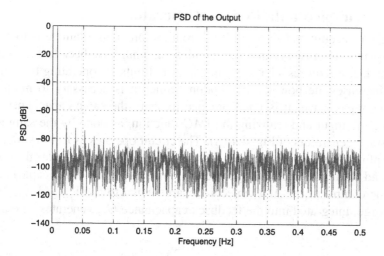

Figure 4.47. Spectrum of the digital output with the design parameters given in the Table.

of the slew-rate. The clock frequency is 100 MHz and τ is 0.26 ns. Since for an op-amp with feedback factor β the time constant is given by $\tau = 1 / (2\beta\pi f_T)$ the used value corresponds to $\beta f_T = 600MHz$. The results of Fig. 4.46 show that, for example, the SNR and the SFDR requirements are satisfied with a S&H which slew-rate is larger than 250 V/μs while maintaining ideal the rest of the circuit. The slew-rate requirements are less demanding when considering singularly first stage or second stage and is about 200 V/μs for the third stage only.

Obviously, when considering together the op-amp limits the performances must be better than what given in Fig. 4.46. Moreover, a less performing βf_T requires a higher slew-rate. A set of simulations with changed parameters secures an optimum design. For example, the parameters of the following Table gives rise to the spectrum of Fig. 4.47. The used values are a trade-off between slew-rate and βf_T ratio as increasing the slew-rates relaxes the op-amp f_T. The obtained results are SNR = 60.7 dB and SFDR = 70.9 dB.

Specifications	SR (V/μs)	βf_T (MHz)
S&H	300	600
First stage	270	500
Second stage	270	400
Stages from 3 to 9	250	350

4.8.4   Sampled-data Residue Generator

A design recommendation resulting from the previous example is that the residue generator must use an op-amp with a maximum feedback factor. A low feedback factor requires a large f_T and, in turn, limits the operating frequency and/or increases the power consumption. Since it is necessary to minimze the input capacitance, it is good practice to share the same capacitance for sampling the input and realizing the *DAC* function, as done by the sampled-data residue generators of Fig. 4.48 a). The circuit operates as follows: during the sampling phase Φ_S an array on N equal unity capacitors samples the input signal and during the residual generation phase Φ_R each unity capacitance under the control of the *ADC* injects a charge equal to $C_u(V_{in} - V_{DAC}(i))$. This charge, integrated into the feedback capacitance C_u, generates an output voltage

$$V_{res} = NV_{in} - \sum_1^N V_{DAC}(i) \qquad (4.43)$$

where the controls of $V_{DAC}(i)$ are the thermometric *ADC* output.

Since amplification by 2^n gives rise to a feedback factor equal to $\beta = 1/(2^n + 1)$ the unity gain frequency of the op-amp must increase exponentially with the number of bits of the stage. Therefore, more than 3-*bit* per stage is not suitable for high conversion rates.

The use of the scheme in Fig. 4.48 (a) with 1-*bit* gives rise to a 1/3 feedback factor (since the gain is 2 the input element is $2C_U$). A better result is obtained by the circuit of Fig. 4.48 (b) that is a residue generator with flip around of one of the two sampling capacitances. It improves the speed by obtaining

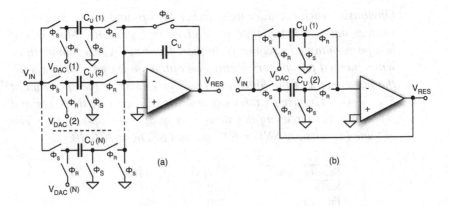

Figure 4.48.   (a) Sampled data n-bit residue generator ($N = 2^n$). (b) Flip around 1-bit residue generator.

$\beta = 1/2$ because input and feedback capacitance during phase Φ_R are equal. Both unity capacitances are used for sampling during Φ_S while while one of the two is flipped around during the residue phase. Since an already charged capacitance $C_u(2)$ receives the charge of the other $C_u(1)$, the voltage V_{DAC} is the addition of the two charges and the amplification by $-C_u(1)/C_u(2)$ of V_{DAC}. Nominally equal sampling capacitances obtain the by 2 amplification of the input; however, since the gain $-C_u(1)/C_u(2)$ is -1 it is necessary to double the references.

4.9   OTHER ARCHITECTURES

The Nyquist-rate architectures studied in the previous sections are only a fraction of the architectures currently in use. The discussed algorithms can be implemented with different solutions and also there are many other possible conceptual methods. A number of implementations have been popular in the past because they were suitable for a discrete implementation or appropriate for the technologies available at that time. Some of these methods eventually became obsolete and are not used anymore as they are not convenient for integrated technology. However, some of these methods may be re-used in the future to meet special requirements or implemented with new circuit solutions.

The reader must be aware of the above and as a complement to the architectures already studied this section considers three additional *ADC* techniques to highlight other methods that obtain specific features.

4.9.1   Cyclic (or Algorithmic) Converter

A 1-*bit* per stage pipeline determines the bits by converting the residue at the output of each stage. The cyclic (or algorithmic) method obtains the result by returning back to a sequential approach using just one cell over and over again to perform the operation of each of the pipeline stages. Therefore, an algorithmic converter requires $n + 1$ clock periods to determine *n-bit*.

The key operation in a pipeline stage is the multiplication by two required for obtaining the residue and the sampling of the result for establishing the next bit. The circuit of Fig. 4.49 uses two op-amps in a closed feedback loop to obtain this result (actually, since the circuit uses capacitances even *OTAs* are adequate). The conversion starts with the sampling phase Φ_S which is used to pre-charge one of the switched capacitor input structures to the input voltage. In addition, Φ_S also resets the unity capacitances in feedback around the op-amps and the capacitances C and $2C$. This makes the scheme offset insensitive as during the reset phase the capacitances are pre-charged to the offsets, or to the difference between the offsets.

The next period Φ_1 is used to inject the input signal into the loop. The non-inverting configuration and the unity gain are such that the input voltage

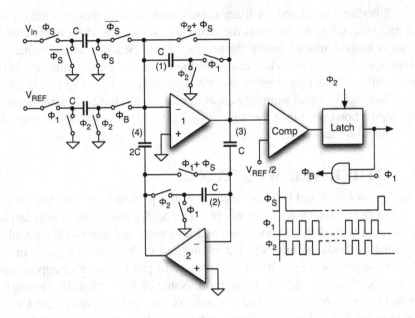

Figure 4.49. Circuit diagram of a cyclic converter.

is reproduced at the output of the first op-amp and stored onto the capacitance labeled (3). The comparator determines the *MSB* by comparing the output of op-amp 1 with $V_{REF}/2$. The second op-amp is another non-inverting amplifier that, after an extra half-clock period delay, pre-charges the capacitance labeled (4). Note that this capacitance is twice the unity value.

During the next Φ_1 the charge of capacitance labeled (4) generates a multiplication by two of the input signal. Meanwhile, the second input structure injects CV_{REF} into the inverting node if Φ_B is enabled. The voltage at the output of the *OTA* 1 becomes

$$V_{O1} = 2V_{in} - V_{REF} = 2\left(V_{in} - \frac{V_{REF}}{2}\right) \qquad (4.44)$$

which is the same as the residue of the first stage of a 1-*bit* per stage pipeline.

The next cycle operates in the same way to determine the next bit, and the process continues until all the bits have been determined.

4.9.2   Integrating Converter

Applications which can allow for very slow conversion-rates, such as digital multi-meters and panel meters, are appropriate for integrating architectures that provide high resolution, good noise performance and low power consumption.

The simplest form of an integrating *ADC* is the single-slope architecture that compares the integration of the input with a reference level and measures the time the integrator takes to reach this reference level. The method uses time as an intermediate quantity, and it corresponds to the cascade of an amplitude-to-time converter with a time-to-digital converter. The accuracy of the amplitude-to-time conver-

Key Features

Integrating converters are slow as they require a number of clock periods equal to or even twice the number of quantization steps. However, simple schemes easily obtain 14-bit or more.

sion requires a stable and accurate reference; moreover, the absolute accuracy and temperature coefficient depends on the absolute value of the time constant which, when determined by a resistor and a capacitor cannot be better than 0.1% even if external components are used. Furthermore, the temperature and voltage coefficient of the resistor and the capacitance directly affect the temperature coefficient and the linearity of the converter.

The dual-slope integrating architecture partially overcomes these limits by integrating the input voltage for a fixed number of clock periods, 2^n. It then "de-integrates" the reference voltage and counts the number of clock cycles k until it crosses zero. Since the peak amplitudes of rising and falling ramps are equal, we have

$$V_{in}\frac{2^n T_{ck}}{\tau} = V_{ref}\frac{k \cdot T_{ck}}{\tau} \tag{4.45}$$

Notice that k gives the result regardless of the value of τ, provided that τ is constant during the conversion cycle (the temperature effects are very slow relative to the conversion period time scale). Since equation (4.45) gives $V_{in} = k \cdot V_{ref}/2^n$ the value k is the digital conversion result.

Fig. 4.50 (a) shows the typical waveforms of the architecture. The conversion cycle begins with the resetting of the integrating capacitance and the offset auto-zero, and is followed by the integration of the input, lasting 2^n clock periods. Then logic signals are generated which reverse the integration direction, replace the input with the reference voltage and start a counter for measuring the "de-integration" period. When the integrator voltage reaches zero the counter stops and makes the circuit inactive until the 2^n-*th* periods countdown finishes.

Since the offset is an important limit it is necessary to use a very effective cancellation scheme like the one of Fig. 4.50 (b). During the auto-zero phase the loop made by the op-amp and the comparator's pre-amplifier is in unity gain configuration making the overall offset to be stored on capacitance C_{os} available. The integrating capacitance is pre-charged to the difference between the op-amp offset, $V_{os,amp}$, and the offset of the pre-amplifier, $V_{os,pre}$, procuring the cancellation of the residue term $V_{os,amp}/A_0$ due to the finite value of the

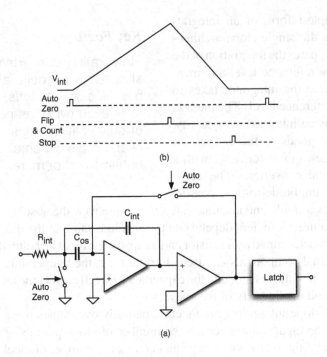

Figure 4.50.   Waveforms of a two-slope ADC (a) and (b) offset cancelled scheme.

op-amp gain, A_0. High resolution converters (like 4-*digit* or more) use external capacitances and resistor to ensure high linearity and to make effective the offset cancellation.

4.9.3   Voltage-to-Frequency Converter

The voltage-to-frequency converter (*VFC*) uses frequency as the intermediate quantity. It is the cascade of an amplitude-to-frequency converter and a frequency-to-digital converter. Basically, a *VFC* is an oscillator whose frequency, which is linearly proportional to the controlling voltage, is measured to provide the *A/D* conversion.

There are many ways to design voltage controlled oscillators but the key feature in *A/D* conversion is obtaining high linearity. A good solution is the current-steering multivibrator which, actually, generates a frequency under the control of a current and thus requires a voltage-to-current converter at its input. The current discharges a capacitor until a certain threshold is reached, after which the capacitor terminals are reversed and the cycle repeats itself as shown in Fig. 4.51 (a). The key features of this *VFC* are monotonicity, small area, low power consumption and linearity that can be as good as 14-*bits* with comparable

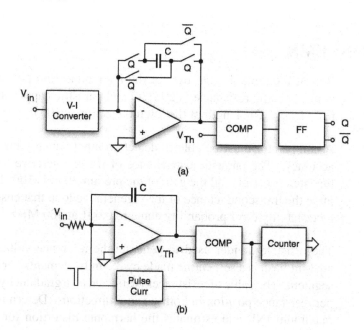

Figure 4.51. Block diagram of a typical voltage-to-frequency converter.

stability. The performance is limited by the threshold accuracy, and the linearity of both the *V-I* converter and the capacitor.

Another possible *VFC* implementation is the charge balance configuration of Fig. 4.51 (b) which integrates the input voltage over a capacitance; when the output reaches a given threshold a precise pulse of charge removes a fixed amount of charge from the capacitor, leading to a triangular waveform. This method is very demanding in its current pulse accuracy requirements but is quite accurate as it is capable of 16–18-*bit* linearity.

PROBLEMS

4.1   Repeat Example 4.2 but use for the interconnection polysilicon which specific resistance is $20\,\Omega/\square$ and parasitic capacitance of $0.7\,fF/\mu^2$. The length of the interconnection is $100\,\mu$

4.2   A comparator must be able to detect an input signal with 5 mV accuracy. The parasitic capacitance of the regenerative loop of the latch is 0.1 pF and the gain of the pre-amplifier is 100. Determine the transconductance of the regenerative loop that ensures a metastability error probability equal to 0.001 at 400 MHz.

4.3   The resistive divider used to generate the reference voltages of an 8-bit flash is a serpentine made by 32 unity elements per linear segment. The value of resistances is affected by a gradient by 0.1% per resistance position in both y and x directions. Determine the maximum INL and estimates the harmonic distortion for a full scale input sine wave.

4.4   Determine the transistor sizing of an MOS preamplifier which differential pair uses $300\,\mu A$ bias current and uses $300\,mV$ overdrive. The variance of the offset must be $1\,mV$ with dominant contribution from the tem controlled by the process parameter $A_{VT} = 1.6\,mV/\mu$.

4.5   Estimate the attenuation of the sinusoidal component of the 1/f noise at 2 kHz in an auto-zeroed comparator running at 2 MHz. Solve the problem assuming considering at the input of the offset a normal sine wave with amplitude 1.

4.6   Repeat the Example 4.5 but assume a finite resistance of the switch-on which value is 5 Ω. Estimate the improvement with two equal shunting resistances connecting the two reference voltages to the middle point.

4.7   The overdrive of the input differential pair of a pre-amplifer is 360 mV and its output load is 0.2 pF. Estimate the required bias current that obtains a gain equal to 6 when the preamplifier run at 1 GHz.

4.8   Determine the optimum splitting of the bits in a 10-bit two step converter assuming as quality factor the power consumption. The converter runs at 200 MHz and used $V_{ref} = 2\,V$. The power consumption of a comparator is given by $P_{comp} = 0.3 + 10/\Delta\,[mW]$ where Δ is the resolution required at the input of the comparator. The power of the residue generator that obtains an amplification by $2^{N_{MSB}}$ is $2 + 1.2 \cdot 2^{N_{MSB}}\,mW$.

4.9   Repeat the Example 4.6 for a 5+5 bit converter. The required maximum resolution loss is 0.5-bit and the SFDR must be better than 70 dB.

4.10   Use the simulation file of the Example 4.6 for a 5+5-bit converter. The maximum loss in resolution is 0.5 bits and the SFDR must be better than 70 dB.

4.11   Describe with a behavioral language the non-linear response of a 3-bit folder and determine the output spectrum with a full scale input sine wave. Compare the results with the input-output characteristics of a 3-bit residue generator.

4.12   Assume that the real folding response of Fig. 4.19 (with 4-bit) is approximated by parabolas joining together the 10% terminations of the linear segments with continuous derivative. Unfold the response and use the result to estimate the output spectrum with a half-scale sine wave.

4.13   The input-output transfer function of a preamplifier is modeled by $V_{out} = 2/\pi\,atan(V_{in}/0.1)$. Determine the minimum number of preamplifier that obtains, with interpolation, a 6-bit flash.

4.14   The folding response of a 4-bit folder is linear in the 20% to 80% fraction of each folding segment. The joining parts are part of sine waves. Determine the 1/8, 1/4, and 1/2 interpolation curves and estimate possible errors in the zero crossings.

4.15   Estimate the INL improvement obtained by the interpolation network used in Example 4.7 by averaging the results of 100 simulations. Use interpolating resistances equal to 2 and 1/2 R_{out}.

4.16   Repeat the Example 4.9 studying separately effect of time jitter, random noise in the comparator, and random and systematic errors of the DAC.

4.17   Determine the input-output characteristics of a residue generator that uses a 6-comparator flash with thresholds uniformly distributed on the dynamic range. Assume that the residue is converted by a 5-bit flash and the system uses digital correction. What is the obtained number of bits?

4.18   Repeat the Example 4.10 but use a gain error in the residue generator. Assume that the gain of the three stages are 2.1, 1.96, 2.07 respectively. Use a linear ramp at the input and determine the analog equivalent of the output. Plot the error as a function of the input amplitude.

4.19   Extend the number of stages of the pipeline of Example 4.10 to 6 and convert the residue of the last stage with a 4-bit flash. Use a gain equal to $(2+2/27)$ in the last pipeline stage and plot the input-output characteristics. Try different gain errors and explain the results.

4.20   Use the stage model of Example 4.11 for a 10-bit pipeline made by k stages followed by a (10-k)-bit flash. The pipeline runs at 100 MHz. An empiric equation for estimating the power consumption of the residue generator is $P = (SR + f_T/\beta)/10$ mW while the power consumption of a comparator is 2 mW. Scale the performance requirements along the pipeline so that the SFDR is better than 70 dB. Determine the value of k that minimizes the power consumption.

REFERENCES

Books and Monographs

P. Allen and D. R. Holberg: *CMOS Analog Circuit Design*. Oxford University Press, New York, Oxford, 2002.

M. Gustavsson, J. J. Wikner, and N. N. Tan: *CMOS Data Converters for Communications*. Kluwer Academic Press, Boston, Dordrecht, London, 2000.

Journals and Conference Proceedings

General Issues

H. R. Kaupp: *Waveform degradation in VLSI interconnections*, IEEE Journal of Solid-State Circuits, vol. 24, 1150–1153, 1989.

M. J. McNutt, S. LeMarquis, and J. L. Dunkley: *Systematic capacitance matching errors and corrective layout procedures*, IEEE Journal of Solid-State Circuits, vol. 29, pp. 611–616, 1994.

C. L. Portmann and T. H. Y. Meng: *Power-efficient metastability error reduction in CMOS flash A/D converters*, IEEE Journal of Solid-State Circuits, vol. 31, pp. 1132–1140, 1996.

Flash Converters

J. G. Peterson: *A monolithic video A/D converter*, IEEE Journal of Solid-State Circuits, vol. 14, pp. 932–937, 1979.

A. G. F. Dingwall and V. Zazzu: *An 8-MHz CMOS subranging 8-bit A/D converter*, IEEE Journal of Solid-State Circuits, vol. 20, pp. 1138–1143, 1985.

Geelen, G.: *A 6 b 1.1 GSample/s CMOS A/D converter*, 2001 IEEE International Solid-State Circuits Conference, pp. 128–129, 2001, Digest of Technical Papers. ISSCC. 2001.

M. Choi and A. A. Abidi: *A 6-b 1.3-Gsample/s A/D converter in 0.35-μ m CMOS*, IEEE Journal of Solid-State Circuits, vol. 36, pp. 1847–1858, 2001.

R. J. Van De Plassche and R. E. J. Van Der Grift: *A high-speed 7 bit A/D converter*, IEEE Journal of Solid-State Circuits, vol. 14, pp. 938–943, 1979.

C. Moreland, F. Murden, M. Elliott, J. Young, M. Hensley, and R. Stop: *A 14-bit 100-Msample/s subranging ADC*, IEEE Journal of Solid-State Circuits, vol. 35, pp. 1791–1798, 2000.

J. Mulder, C. M. Ward, C. Lin, D. Kruse, J. R. Westra, M. Lugthart, E. Arslan, R. J. van de Plassche, K. Bult, and F. M. L. van der Goes: *A 21-mW 8-b 125-MSample/s ADC in 0.09-mm^2 0.13-μ CMOS*, IEEE Journal of Solid-State Circuits, vol. 39, pp. 2116–2125, 2004.

Folding Converters

H. Kimura, A. Matsuzawa, T. Nakamura, and S. Sawada: *A 10-b 300-MHz interpolated-parallel A/D converter*, IEEE Journal of Solid-State Circuits, vol. 28, pp. 438–446, 1993.

B. Nauta and A. G. W. Venes: *A 70-MS/s 110-mW 8-b CMOS folding and interpolating A/D converter*, IEEE Journal of Solid-State Circuits, vol. 30, pp. 1302–1308, 1995.

P. Vorenkamp and R. Roovers: *A 12-b, 60-MSample/s cascaded folding and interpolating ADC*, IEEE Journal of Solid-State Circuits, vol. 32, pp. 1876–1886, 1997.

M. P. Flynn and B. Sheahan: *A 400-Msample/s, 6-b CMOS folding and interpolating ADC*, IEEE Journal of Solid-State Circuits, vol. 33, pp. 1932–1938, 1998.

R. Taft, C. Menkus, M. R. Tursi, O. Hidri, and V. Pons: *A 1.8V/1.6 GS/s 8 b self-calibrating folding ADC with 7.26 ENOB at Nyquist frequency*, ISSCC Dig. Tech. Papers, pp. 252–253, 2004.

Time Interleaved Converters

W. C. Black Jr. and D. A. Hodges: *Time interleaved converter arrays*, IEEE Journal of Solid-State Circuits, vol. 15, pp. 1022–1029, 1980.

A. Petraglia and S.K. Mitra: *Analysis of mismatch effects among A/D converters in a time-interleaved waveform digitizer*, IEEE Transactions on Instrumentation and Measurement, vol. 40, pp. 831–835, 1991.

R. Khoini-Poorfard, R. B. Lim, and D. A. Johns: *Time-interleaved oversampling A/D converters: theory and practice*, IEEE Transaction on Circ. and Systems, II, pp. 634–635, 1997.

S. Limotyrakis, S. D. Kulchycki, D. K. Su, and B. A. Wooley: *A 150-MS/s 8-b 71-mW CMOS time-interleaved ADC*, IEEE Journal of Solid-State Circuits, vol. 40, pp. 1057–1067, 2005.

Successive Approximation Converters

K. Bacrania: *A 12-bit successive-approximation-type ADC with digital error correction*, IEEE Journal of Solid-State Circuits, vol. 21, pp. 1016–1025, 1986.

Gardino and F. Maloberti: *High Resolution Rail-To-Rail ADC in CMOS Digital Technology*, IEEE Proc. ISCAS 99, pp. II-339/II-342, 1999.

G. Promitzer: *12-bit Low-power fully differential switched capacitor noncalibrating successive approximation ADC with 1 MS/s*, IEEE Journal of Solid-State Circuits, vol. 36, pp. 1138–1143, 2001.

N. Verma and A. P. Chandrakasan: *A 25 μW 100kS/s ADC for Wireless Micro-Sensor Applications*, IEEE Intern. Solid State Circ. Conf., Vol. 49, pp. 222–223, 2006.

Pipeline Converters

S. H. Lewis and P. R. Gray: *A pipelined 5-Msample/s 9-bit analog-to-digital converter*, IEEE Journal of Solid-State Circuits, vol. 22, pp. 954–961, 1987.

K. Nagaraj, H. S. Fetterman, J. Anidjar, S. H. Lewis, and R. G. Renninger: *A 250-mW, 8-b, 52-Msamples/s parallel-pipelined A/D converter with reduced number of amplifiers*, IEEE Journal of Solid-State Circuits, vol. 32, pp. 312–320, 1997.

I. Mehr and L. Singer: *A 55-mW, 10-bit, 40-Msample/s Nyquist-rate CMOS ADC*, IEEE Journal of Solid-State Circuits, vol. 35, pp. 318–325, March 2000.

B. Min, P. Kim, F. W. Bowman III, D. M. Boisvert, and A. J. Aude: *A 69-mW 10-bit 80-MSample/s pipelined CMOS ADC*, IEEE Journal of Solid-State Circuits, vol. 38, pp. 2031–2039, 2003

Other Converters

P. Li, M. J. Chin, P. R. Gray, and R. Castello: *A ratio-independent algorithmic analog-to-digital conversion technique*, IEEE Journal of Solid-State Circuits, vol. 19, pp. 828–836, 1984

Chapter 5

CIRCUITS FOR DATA CONVERTERS

This book assumes that the reader is familiar with the features and design techniques of basic blocks like OTAs, op-amps and comparators; accordingly, this chapter concentrates on the study of circuits that are specific to data converters. First, we shall study S&H circuits, realized in either bipolar or CMOS technology. After that, we shall study the clock boosting technique used to enhance (or make possible), the switching on of MOS transistors at very low supply voltages. Following this, we shall study the circuit techniques used in folding systems for current and voltage inputs. Since various architectures use V/I converters, we shall review some of the general V/I schemes. Finally, we shall examine the generation of overlapping and non-overlapping phases, as required by data converter controls.

5.1 SAMPLE-AND-HOLD

The *S&H* function, defined in the first Chapter, is realized under the control of two phases, one for sampling the input signal and the other for retaining the signal and making it available at the output during the hold phase. In some cases, the output is also available during the sampling period. These are named track and hold (*T&H*) circuits.

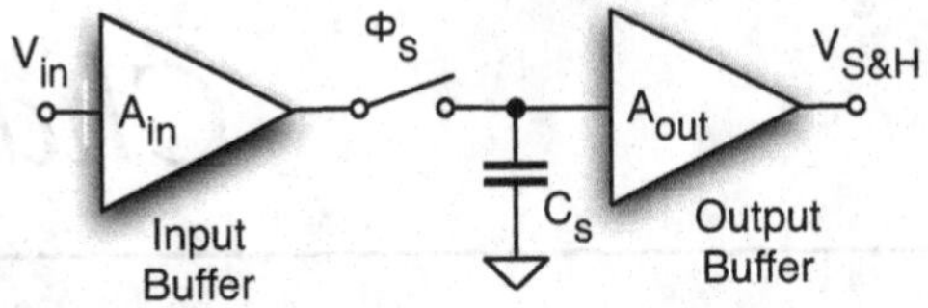

Figure 5.1.   Conceptual block diagram of the Sample and Hold.

The simplest *S&H* scheme is the one shown in Fig. 5.1. When the switch is *on*, the capacitor C_s is charged to the input voltage. Then, when the switch goes *off*, the top plate of the hold capacitor is left floating, and it retains the sampled voltage. In addition, an input buffer is used to reduce the input load and an output buffer to avoid any leakage that would discharge the hold element. The output buffer also provides the result in a volt-metric fashion. Notice that the actual sampling occurs when the switch goes off and freezes the input information on the sampling capacitance. Also, the scheme of Fig. 5.1 is single ended but it can be implemented in a pseudo-differential fashion for applications that require a differential *A/D* conversion.

The basic scheme of Fig. 5.1 can be implemented in both *CMOS* and bipolar technology. Among the solutions that use the bipolar technology we shall also study the diode-bridge and the switched emitter follower; for the *CMOS* based realizations we shall consider configurations based on the two-stages amplifier, and the flip around *S&H*.

5.2   DIODE BRIDGE S&H

The first applications of the *S&H* date back to many decades ago when undersampling was used in oscilloscopes to visualize the waveforms of very fast repetitive signals. The scheme used was the diode bridge *S&H* shown in Fig. 5.2 which is based on a pair of bipolar transistors Q_H and Q_S that form a current switch. When the sampling control is high the current I_1, nominally equal to I_2, flows through the diode bridge. Since all the diodes are *on*, then voltages of nodes A and B are one diode down and one diode up with respect to the input. Due to the symmetry of the circuit, similar voltage drops across D_2 and D_4 ensure that the voltage across the sampling capacitor C_s is equal to the input voltage.

Next Q_S is switched off, and the control of Q_H goes high forcing the current I_1 to flow through Q_H. As a result, the current in the diode bridge goes to zero and the reversely biased diodes D_2 and D_4 isolate the capacitor C_s that stores the sampled signal. The signal is then reproduced at the output by a unity gain

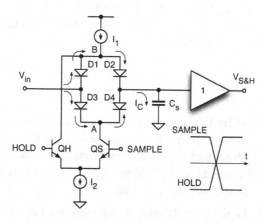

Figure 5.2. Diode bridge sample and hold.

buffer with low output impedance. It is worth noting that during the sampling phase the voltage across C_s tracks the input; therefore the circuit is actually a *T&H*.

Any mismatch between the current sources I_1 and I_2 affects the circuit operation. If $I_1 > I_2$ then the difference, $I_1 - I_2$, flows towards the input node during the sampling phase and causes a current to enter the input generator. During the hold phase the excess current brings the generator I_1 toward triode, forcing the voltage at node B to increase. When node B is one V_D more than the input then the diode D_1 turns on and also D_2 can turn on corrupting the stored information. To avoid this risk it is important to set $I_1 < I_2$ which ensures that when transistor Q_S is off than the node B is pulled down and reversely biases both D_1 and D_2. Thus the mismatch has no influence during hold mode, although there will still be current leakage from the input during sampling mode.

> ### Diodes in the bridge
>
> The speed of the diodes establishes the speed of the diode-bridge S&H
> The use of Sckottly diodes makes the scheme very fast and suitable for medium-resoluton samplers running at multi-GHz frequencies.

5.2.1 Diode Bridge Imperfections

The accuracy of the diode bridge *S&H* is limited by a number of imperfections. The most important are:

- *Aperture distortion*: is the error on the switching-off time caused by the derivative of the input signal. The change of input voltage gives rise to a current through the capacitance C_s of

$$I_C = \frac{1}{C_s} \cdot \frac{dV_{in}}{dt} \tag{5.1}$$

which is provided by the input terminal, and is approximately divided in equal parts as it flows through the *on*-diodes. A positive derivative makes the current through D_1 and D_4 lower than the current in the other pair leading to asymmetrical currents in the bridge. The effect of this asymmetry is a non linear error of the switching instant.

- *Hold Pedestal*: is given by the charge injected from the diodes when they switch from *on* to *off*. Since the charge of the depletion region is non-linear, its injection into C_s causes a non-linear error on the stored voltage.

- *Track-mode Distortion*: is caused by the non-linear impedance of the input buffer that charges C_s through the bridge. Since the non-linear voltage drop is proportional to the input voltage derivative, the distortion increases with the input frequency.

- *Hold feed-through*: is caused by the parasitic coupling between input and output during hold mode. The off diodes are equivalent to parasitic capacitances that connect nodes A and B to the hold capacitance. The equivalent circuit of Fig. 5.3 (a) estimates the hold feed-through whose effect is limited by small decoupling resistance R_A and R_A or small parasitic capacitances.

5.2.2   Improved Diode Bridge

The latter limit can be corrected with the more complex circuit shown in Fig. 5.3 (b) that provides good control of nodes A and B during hold mode which also improves the hold pedestal.

The two extra diodes (D_A and D_B) do not operate during the track mode as they are reverse biased by the drop voltage across the on diodes D_2 and D_4. The two extra diodes are turned on during the hold phase by the extra current generators, I_x, and clamp the voltages of nodes A and B at one V_D up and one V_D down respectively.

Notice that the fixed reverse biasing of D_2 and D_4 gives rise to a constant hold pedestal. Moreover, $V_A = V_H + V_D$ and $V_B = V_H - V_D$ keep D_1 and D_3 turned off provided that the change of V_{in} during the hold period is less than $\pm V_D$. Also, the biasing of nodes A and B through the low impedance of the buffer, and the $1/g_m$ of the diodes reduces the resistances that are represented by R_A and R_B in the circuit of Fig. 5.3 (a) improving the decoupling between input and storing capacitance in the hold mode.

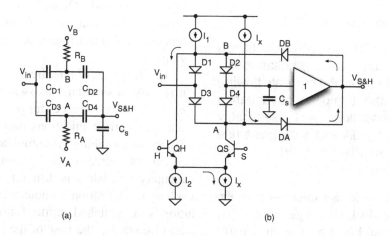

Figure 5.3. (a) Circuit for estimating the hold-mode feed-through. (b) Diode bridge with clamping diodes.

5.3   SWITCHED EMITTER FOLLOWER

The diode bridge must accommodate the diodes shift up and shift down (about $0.7\ V$) plus the voltage room for the top and the bottom current sources (again about $0.7\ V$ for bipolar circuits). Therefore, a bipolar diode bridge requires a supply voltage equal to the input range plus at least $2.8\ V$.

A simplified version shown in Fig. 5.4 (a) reduces this by $0.7\ V$ as the top diodes are removed and the current through D_1 is provided by the input terminal. This circuit can be useful, however a *dc* term in the input current is not suitable for many applications. A more convenient scheme is the one of Fig. 5.4 (b) that

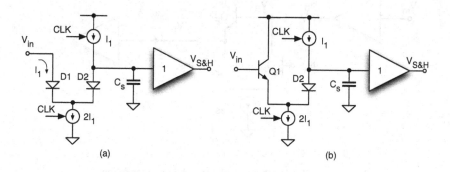

Figure 5.4. (a) Simplified diode bridge. (b) Reduction of the input current using a BJT.

replaces the diode D_1 with a bipolar transistor, thus reducing the input current to I_1/β (β is the transistor current gain).

The drawback of this circuit is that it requires a simultaneous switching *off* of the current sources for preventing spur current injections into C_s. Since the switching synchronization of current sources made by complementary transistors is difficult, it is necessary to use solutions that obtain a self-synchronization without relying on the clock phases generation. The solution is the switched emitter follower shown in Fig. 5.5. The input buffer biases, through R_b, the base of the transistor Q_E, which operates as an emitter follower during the sampling period. The phases Φ_S and Φ_H are high and low respectively in the track mode and change to low and high for switching in the hold mode. It is assumed that the phases change is simultaneous. The current I_{bias} causes a drop across R_b and pulls V_B down by $R_b I_{bias}$. Since the emitter voltage of Q_E is sustained by the hold capacitance, the V_{BE} of Q_E is reversed switching *off* the transistor simultaneously with Q_S.

Observe that the voltage across C_S and the output voltage are a shifted replica of the input and not, strictly speaking, a *S&H* signal. Compensating for this difference with a shift up is normally not necessary for high-speed applications as a *dc* component is not a problem.

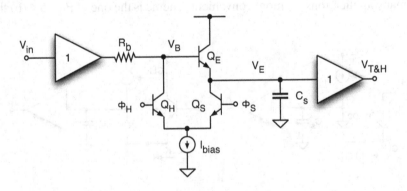

Figure 5.5. Basic scheme of the switched-emitter follower.

5.3.1 Circuit Implementation

The input buffer must be linear, fast and capable of quickly switching from the sample to the hold condition and vice-versa. The circuit in Fig. 5.6 (a) uses a fully differential amplifier whose gain is -1 thanks to the use of equal resistances and equal emitter areas in the bipolar transistors for compensating for the V_{BE} non-linearity. The unity gain makes the circuit very fast but because of the inverting operation, its possible for the *dc* level of one the outputs to become close to its input causing a small V_{CB}. Since in the off-state the two outputs are both pulled down, one of the V_{CB} can turn into a direct bias. The switching "on" of the base-to-collector diode drains current from the input and slows down the next *off–on* transition.

The circuit in Fig. 5.6 (b) avoids this problem by using two simple buffers in a pseudo-differential arrangement. Each buffer is an emitter follower with a diode shift-up that compensates for the shift down of the input transistor. When the output nodes are pulled down by output currents larger than $I_{bias}/2$ the diodes Q_3 and Q_4 turn off, and can quickly switch back when the circuit returns to the track mode. However, the scheme of Fig. 5.6 (b) uses twice the current of the circuit of Fig. 5.6 (a) for equal f_T.

The capacitance of the base-emitter junction of Q_E of Fig. 5.5 causes a hold pedestal whose amplitude depends on the reverse base-to-emitter voltage of Q_E during the hold phase. The lack of control of the voltage V_B makes the reverse biasing of Q_E, and consequently the hold pedestal, unpredictable. The use of the clamping element Q_C of Fig. 5.7, whose base voltage replicates the held

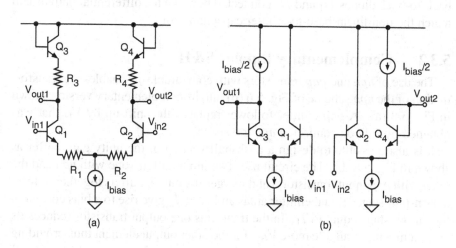

Figure 5.6. (a) Gain stage with -1 gain. (b) Pseudo-differential unity gain buffer.

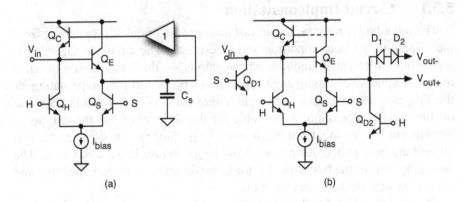

Figure 5.7. (a) Use of auxiliary buffer for clamping the input swing. (b) Circuit solutions for reducing the hold pedestal.

output voltage, obtains a constant reverse biasing of Q_E during the hold phase which is equal to a diode voltage. In the track phase Q_C goes off and does not influence the circuit operation.

Another source of pedestal error is the charge injection from the parasitic base and collector capacitances of Q_S. This error is compensated for by a matched injection of charge provided from dummy transistors Q_{D1} and Q_{D2}, which are driven by complementary phases as shown in Fig. 5.6 (b). Finally, back-to-back diodes D_1 and D_2 connected between the differential outputs can match the non-linear base-to-emitter capacitance of Q_E.

5.3.2   Complementary Bipolar S&H

The use of *npn* and *pnp* transistors with comparable f_T enables the transformation of the *npn* scheme of Fig. 5.6 (b) into its complementary version shown in Fig. 5.8 (a). A *pnp* emitter follower replaces the shift-up by V_D that was obtained by a diode connected *npn* transistor.

It is also possible to design a push-pull version of the unity gain buffer as shown in Fig. 5.8 (b). The circuit uses two input buffers (one with *npn* and the other with *pnp* input transistors) and merges the output buffers together to form a push-pull stage. Equal emitter areas and $I_1 = I_2$ give rise to a bias current in the output stage equal to I_1. In the transients one output transistor reduces its current making available more V_{BE} for the other output element thus providing the current required by the load (*AB* class operation). If one transistor goes off the entire current of the other transistor flows through the output node (*B* class operation).

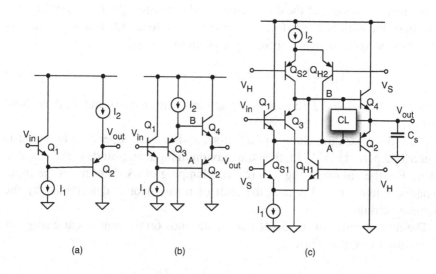

Figure 5.8. (a) Unity gain buffer with complementary BJT. (b) Push-pull unity gain buffer with BJT. (c) Complementary bipolar S&H.

The push-pull buffer of Fig. 5.8 (b) becomes a complementary switched-buffer if the voltages at nodes A and B are suddenly pushed up and pulled down respectively. The circuit of Fig. 5.8 (c) obtains this result by adding two pairs of differential switches (Q_{S1}-Q_{H1} and Q_{S2}-Q_{H2}) which are used to divert I_{B1} and I_{B2} from the emitters of Q_1 and Q_2 to the nodes B and A respectively.

> **Observe that**
>
> Complementary bipolar S&H requires a comparable f_T for the npn and the pnp devices. The feature is not granted by many technologies.

In hold mode the block CL clamps the voltages of nodes A and B at suitable levels above or below the output and also provides the currents I_{B1} and I_{B2}. However during the sampling period the same block becomes ineffective similar to the transistor Q_C of Fig. 5.7 (a).

5.4 FEATURES OF S&Hs WITH BJT

The use of bipolar transistors in *S&Hs* obtains high-speed and good linearity but the dynamic range is limited by the need to keep junctions reversely biased in the off state.

Consider, for example, a differential version of the circuit of Fig. 5.7 (a) driven by the buffer of Fig. 5.6 (b). In the hold state the off-condition is ensured by the reverse biasing of the diode-connected transistor Q_3 (or Q_4) of Fig. 5.6 (b). Since the emitter of Q_3 follows the input (due to the Q_1 follower), there is a limit to the voltage swing of the input given by

$$V_{in}(nT) - V_D < V_{in}(t) < V_{in}(nT) + V_D \qquad (5.2)$$

for $nT < t < nT + T/2$, where nT are the hold instants and $T/2$ the hold period.

Since the maximum change of a full-scale sine wave (operating at Nyquist) over a $T/2$ period is $V_{ref}/2$, then the maximum input amplitude cannot exceed $2V_D$. For the circuit of Fig. 5.8 (c) the clamping block determines the input dynamic range. It is V_D plus the shift of nodes A or B determined by the clamping circuit.

The non-linearity of the S&H mainly depends on the non-linear change of V_{BE} with the emitter current

$$V_{BE} \simeq V_{BE0} + V_T ln \frac{I_E}{I_{bias}} \qquad (5.3)$$

where V_{BE0} is the base-to-emitter voltage at $I_E = I_{bias}$.

Since the emitter current is equal to the bias current plus the current into C_S then

$$I_E = I_{bias} + C_S \frac{dV_{in}}{dt}. \qquad (5.4)$$

With a pseudo differential scheme, the differential output voltages are

$$V_{out+} = V_{in+} - V_{BE0} - V_T ln \frac{I_{bias} + C_S \frac{dV_{in}}{dt}}{I_{bias}} \qquad (5.5)$$

$$V_{out-} = V_{in-} - V_{BE0} - V_T ln \frac{I_{bias} - C_S \frac{dV_{in}}{dt} dt}{I_{bias}}, \qquad (5.6)$$

resulting in a the differential error

$$\delta V_{out,d} = V_T ln \left[\frac{I_{bias} + C_S \frac{dV_{in}}{dt}}{I_{bias} - C_S \frac{dV_{in}}{dt}} \right]; \qquad (5.7)$$

which, for an input sine wave $V_{in} = A sin(\omega_{in} t)$, yields

$$\delta V_{out,d} = V_T ln \left[\frac{I_{bias} + A\omega_{in} C_S cos(\omega_{in} t)}{I_{bias} - A\omega_{in} C_S cos(\omega_{in} t)} \right]. \qquad (5.8)$$

Since the error is an odd function, there are only odd distortion terms whose amplitude is proportional to the input frequency, the sampling capacitance, and the input amplitude. To minimize the error the bias currents should be significantly larger than the current in the sampling capacitance. If, for example, a $4pF$ capacitance samples a $200\ MHz$ sine wave with amplitude $1\ V$ the current in the sampling element can be as large as $5\ mA$. Since an *SFDR* $> 100\ dB$ requires a minimum $I_{bias} = 8\ I_C$, (which will be verified in the following example), then the bias current must be $40\ mA$.

Example 5.1

An emitter follower with bias current 5 mA drives a 2 pF sampling capacitor. Estimate the output spectrum of a 100 MHz input sine wave with peak amplitude 1 V. Determine the SFDR for bias currents ranging from 2 to 10 times the peak current in the capacitor. Verify the SFDR reduction when decreasing the sine wave input amplitude.

Solution

The file Ex5_1.m provides the code for measuring the distortion as described by equation (5.7). The length of the sequence is not impor-tant because there is no quantization and it is unlikely that numerical noise would mask the distortion tones. The sequence of 2^{10} samples of 19 input sine waves is more than enough for the study.

The peak current in the 2 pF capacitance at the input frequency is

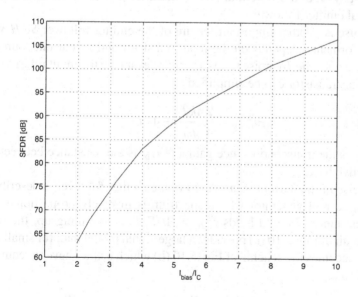

Figure 5.9. SFDR versus the bias normalized current in the emitter follower.

*about 1.25 mA, which corresponds to a quarter of the bias current.
The spectrum of the output voltage shows, as expected, only odd distortion terms with the third tone -82.8 dB below the input amplitude.
The fifth harmonic at -123 dB is much lower. Therefore, the SFDR is
determined by the third harmonic.*

*Fig. 5.9 plots the SFDR as a function of the bias current. When
$I_{bias} = 8I_C$ the SFDR becomes greater than 100 dB. The SFDR also
improves as the input amplitude decreases: with $I_{bias} = 5\ mA$ the
SFDR is 95 dB for A = 0.5 V and 107 dB for A = 0.25V.*

The linearity of the push-pull buffer of Fig. 5.8 (b) and (c) is given by the
linearity of the transistors of the output stage, Q_2 and Q_4. This is because the
dynamic current in the input emitter followers is negligible with respect to the
one in Q_2 and Q_4, which is equal to the dynamic current of the output stage
divided by β.

Since the currents of Q_2 and Q_4 are

$$I_{Q_2} = I_{bias} - C_S \frac{dV_{in}}{dt}; \quad I_{Q_4} = I_{bias} + C_S \frac{dV_{in}}{dt}, \qquad (5.9)$$

then the single ended error is given by equations similar to (5.7) while the
differential error becomes twice the single ended limit. Accordingly, the harmonic distortion of a differential push-pull buffer is 6 *dB* worst than a class *A*
differential emitter follower.

The noise is another important feature of the emitter follower *S&H* whose
noise performance depends on the input transistor and its current source. At
high frequencies the *1/f* noise terms are not relevant, so the input referred noise
of a bipolar transistor can be simplified to

$$v_{n,in}^2 = \frac{2}{3} \frac{4kT}{g_m} + 4kTr_{bb'} \qquad (5.10)$$

where g_m is the transconductance gain and $r_{bb'}$ is the resistance between base
and effective base.

The noise of the current source used in the emitter follower is described by
a random current generator whose spectrum is inversely proportional to the
output resistance, R_0 and holds $i_{n,0}^2 = 4kT/R_0$. Summing up, the single-
ended circuit of Fig. 5.10 (a) provides a large signal model that, for small signal
analysis, becomes the scheme of Fig. 5.10 (b) which is the Thevenin equivalent
of Fig. 5.10 (c)

$$v_{n,eq}^2 = \gamma \frac{4kT}{g_m} = \frac{2}{3} \frac{4kT}{g_m} + 4kTr_{bb'} + \frac{4kT}{R_0 g_m^2} \qquad (5.11)$$

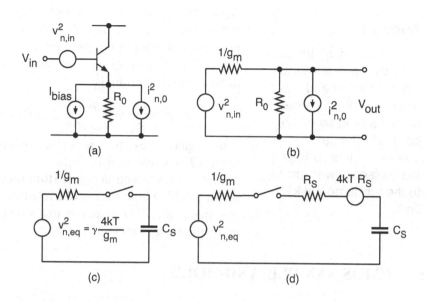

Figure 5.10. Equivalent circuits used to study the emitter-follower noise.

where γ is an excess noise factor estimated using the noise spectrum of a $1/g_m$ resistance as a reference. It becomes its theoretical minimum with $r_{bb'} = 0$ and an ideal current source.

The sampling of $v_{n,eq}^2$ with a capacitance C_S determines a noise voltage across C_S whose total power within the Nyquist interval, similar to that already studied, is

$$V_{n,C_S}^2 = \frac{\gamma kT}{C_S} \tag{5.12}$$

showing that the excess noise factor γ augments the noise with respect to the expected kT/C value. The result shows again that the value of $1/g_m$ is not important as it augment either noise and time constant of the RC filtering network. The two effects, as shown in Chapter 1, compensate one each other.

In many cases the speed of the sampling network is very high or can be purposely made very high. Assume, for example, that I_{bias} is $5\,mA$ and that $C_S = 2pF$. Since $1/g_m = 5\Omega$, the time constant of the sampling network is $10ps$. In these situations it is possible to trade speed for noise reduction by adding a resistance in series with the sampling capacitance as shown in Fig. 5.10 (d). The noise spectra of the emitter follower and series resistance are quadratically superposed while the time constant of the circuit increases due to the addition of R_S and $1/g_m$. The result is that the total noise power across the sampling

Trade-off

A resistance in series with a S&H slows down its speed but reduces the excess noise caused by the $r_{bb'}$ base resistance and the current source, bringing the noise power close to the fundamental kT/C limit.

capacitance is again given by a white spectrum equal to $V_{n,C_S}^2 + 4kTR_S$ that is low-pass filtered with time constant $(R_s + 1/g_m)C_S$. The integration of the consequent colored spectrum is

$$V_{n,C_S}^2 = \frac{kT}{C_S} \cdot \frac{\gamma + g_m R_S}{1 + g_m R_S} \qquad (5.13)$$

which approaches its fundamental minimum, kT/C_S, for $g_m R_S \gg 1$.

Therefore, a reduction in speed to its minimum specified limit the use of a resistance in series to the *S&H* gives rise to the optimum noise-speed trade-off.

5.5 CMOS SAMPLE-AND-HOLD

A *MOS* transistor naturally achieves the function of an analog switch, especially when the current through it is zero or negligible. With V_{GS} below the threshold there is no conductive channel connects source and drain making the transistor an opened connection. In contrast, a V_{GS} exceeding the threshold, V_{Th}, establishes a conductive channel between source and drain terminals. If V_{DS} is low or zero it operates in the triode region, and has an on-resistance equal to

$$R_{on} = \frac{L}{\mu C_{ox} W (V_{GS} - V_{Th})} \qquad (5.14)$$

where μC_{ox}, W and L are well known parameters.

The value of the on-resistance must be such that the time constant of the $R_{on} C_S$ network is much lower than the time allowed to charge the sampling capacitor. Depending on the required accuracy it may be necessary to ensure a number of time constants that range from 7 to 10, or more. Therefore, for high frequency applications the aspect ratio *W/L* of the transistor used in the switch can be fairly large.

Equation 5.14 shows that the on-resistance increases as the overdrive decreases and goes to infinite when $V_{GS} - V_{Th} = 0$. Therefore, if the switched voltage is such that the overdrive is constant (or contained within suitable limits) then switch can be implemented using a single *MOS* transistor. However when the switched voltage varies over large ranges a pair of *n-channel* and *p-channel* transistors realize the switch as shown in Fig. 5.11 (a) and (b).

Complementary devices obtain an on-resistance given by the parallel connection of the two components. Since one conductance increases and the other

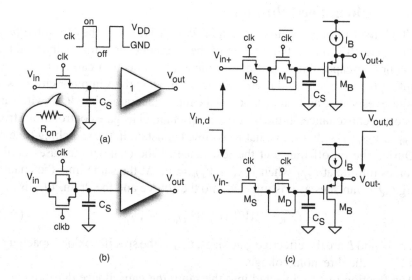

Figure 5.11. (a), (b) CMOS switch with a single and complementary transistors. (c) Simple passive pseude-differential S&H with follower output buffers.

decreases proportionally to the channel voltage, then the total conductance can be kept constant over a wide range of operation if the W/L of the p-type is μ_n/μ_p larger than the W/L of the n-type.

The use of a *MOS* switch to charge a sampling capacitance is the simplest way of obtaining the *S&H* (or better the *T&H*). The charged stored is then preserved by an output buffer which should have a very high input resistance.

For high speed applications the output buffer is normally realized by a simple follower whose input transistor should have its source connected to the substrate to cancel the body effect. A pseudo-differential scheme, as shown in Fig. 5.11 (b), enables the sampling of differential signals.

The dummy transistor M_D, used for clock-feedtrough compensation, has its source and drain shorted and is controlled by a delayed and complemented version of the phase switching the main transistor. Switching M_D *on* creates a conductive channel whose charge, which is almost equal to the charge injected by M_S, compensates for the M_S charge injection.

The scheme of Fig. 5.11 (c), which uses a simple source follower for maximizing the speed, is not able to obtain high resolutions as the linearity of a *MOS* follower is much less than that of its bipolar counterpart. We have seen that with relatively large currents the harmonic tones of an emitter follower can be as low as -100 *dB*. Similar currents used in *CMOS* sub-micron followers obtain speeds in the *GHz* range but the expected resolution is 7-9 *bit* due to the harmonic tones in the -70 *dB* range.

5.5.1   Clock Feed-through

Clock feed-through occurs when a *MOS* transistor switches from the *on* to the *off* state. It is due to the fraction of the channel charge that is injected into the sampling capacitance, and due to the charge injection caused by overlap coupling between gate and drain. Fig. 5.12 models the sampling circuit with the finite resistance and capacitance of the input generator, the *MOS* switch and the storing capacitance. If the transistor is n-channel its gate control drops from $V_{G,on}$ to $V_{G,off}$ with a slope that is assumed constant in the switching time δt.

During the on-off transient the resistance of the channel increases until it becomes infinite at t_{off} when $V_{GS} - V_{Th} = 0$. At the same time the channel charge Q_{ch} vanishes as it is injected into the source and the drain. Q_{ch} is

$$Q_{ch} = (WL)\, C_{ox}(V_{G,on} - V_{Th}) \tag{5.15}$$

where W and L are the effective gate sizes, C_{ox} is the specific oxide capacitance, and V_{Th} is the threshold voltage.

The fraction of Q_{ch} injected into the sampling capacitance depends on the *MOS* parameters, the slope $\alpha = \Delta V_G / \delta t$ of the clock phase and the boundary conditions on both sides of the switch.

The charge injection can be studied with a simplified model which describes the channel and gate as a distributed RC network. The study obtains results as a function of a switching parameter B given by

$$B = V_{ov}\sqrt{\frac{\mu C_{ox} W/L}{|\alpha| C_S}}. \tag{5.16}$$

The numerical solution, for a significant difference between the impedances seen at the two sides of the switch, gives rise to the plots of Fig. 5.13 that show the fraction of injected charge on the sampling capacitance as a function of B and various ratios of C_S/C_{in}. The diagrams for small values of B account for an injected charge equal to half the channel charge, while for large values of B

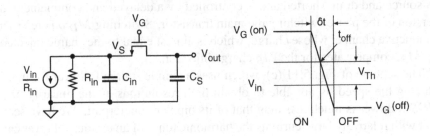

Figure 5.12.   Circuit and clock scheme for studying the clock feed-through.

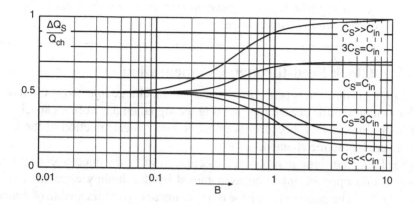

Figure 5.13. Fraction of the charge injected into the source versus B.

(associated with very long fall times), the plots correspond to an equilibrium approached asymptotically, which yields a charge repartition proportional to C_S/C_{in}.

The results are simple for fast and slow fall times, while in the intermediate region the estimation is difficult. Therefore, for a profitable use of the compensation methods described below, it is necessary to operate in predictable regions, typically where the fall-time is fast and the injection is almost half the channel charge.

Example 5.2

The switch of a T&H is an n-channel transistor with W/L = 9μ/0.18 μ and a sampling capacitance of 0.6 pF. The overdrive of the switch is 0.8 V with gate oxide capacitance equal to 1.2 fF/μ². Estimate the offset caused by the clock feed-through with B=0.03.

Solution

The gate capacitance calculated by $C_G = WLC_{ox}$ is 1.94 fF. Since the overdrive is 0.8 V, the charge of the channel is 1.56 fCoul. Since the value of B gives rise to a 50% injection of the charge on the channel into C_S, the caused offset is

$$V_{os} = \frac{1}{2}\frac{C_G}{C_s}V_{ov} = 0.64\,mV \tag{5.17}$$

which is less than 0.1% of the overdrive control.
The calculation does not account for the contribution of the overlap

gate drain which, for a minimum transistor length, is about 10-20%
of the channel contribution.

5.5.2   Clock Feed-through Compensation

Clock feed-through is problematic because the charge injected is not a negligible fraction of the charge representing the signal on the storing capacitance and, more importantly, because its effect is a non linear function of the input voltage applied to the channel.

The clock feed-through is reduced by compensating the injected charge with an equal and opposite injection as obtained by the dummy element M_D of Fig. 5.11 (b). The dummy transistor is a non-linear capacitor capable of matching the non-linear voltage dependence of the charge in the channel of M_S. Since the injected charge is a fraction of the M_S channel, M_D must be smaller (approximately half) than M_S. This asymmetry reduces the effectiveness of the compensation to about $70 - 80\%$.

Another approach used is based on the differential balancing of the charge injections. The method is used in *A/D* converters with differential inputs. Since the signal is differential then common-mode terms are not very important. Therefore, an equal charge injection on both paths gives rise to a common-mode term that is rejected by the fully differential operation.

Consider, for example, the scheme of Fig. 5.14 (a) where two switches reset the capacitors C_S. When the switches turn off they inject charges on the two virtual grounds that will be equal provided that the switches have the same area and the same gate controls. Since the scheme is symmetrical and the accuracy relies only on the matching, the effectiveness of this method can be $80 - 90\%$.

Instead of compensating for the clock feed-through it is possible to ensure a constant charge injection which, in turn, gives rise to an offset. The feature is fully acceptable for any applications which don't allocate information at *dc*.

A constant injection requires equal switching conditions, which means having the same channel charge, and the same channel voltage. These features are obtained by the scheme of Fig. 5.14 (b) which switches the node A between ground and virtual ground. Since the charge injected into the virtual ground depends on the charge injected into A at the opening of S_3, it is essential to ensure the same boundary conditions every clock period. The

Keep in Mind

The effectiveness of the clock feed-through compensation depends on symmetric working conditions and the matching of elements. She expected maximum cancellation is about 90%, the best to do is to obtain a dc (signal independent) term.

contemporary switching at the two sides of C_S is avoided by the phase scheme of Fig. 5.14 (c). The phase $\Phi_{1,d}$, which is slightly delayed with respect to Φ_1, opens the switch S_1 after S_3 opens. Since the plate connected to ground determines the actual sampling, the method is often referred to as bottom plate sampling.

5.5.3 Two-stages OTA as T&H

This *CMOS T&H* is based on the observation that the output of a two-stages *OTA* in the unity gain configuration tracks the input, and that the output of the first stage is the input divided by the second stage gain. Moreover, since the stability requires the use of a capacitor across the input and output of the second stage, the compensation element also stores the input signal and is capable of obtaining the hold if the feedback loop opens.

Fig. 5.15 (a) shows a two-stages OTA in the unity gain configuration and Fig. 5.15 (b) just adds a switch between first and second stage to open the feedback. Also, the name of the compensation capacitance is changed to C_s to outline its double role: as a compensation element during the track phase, and as a sampling element during the hold phase.

Observing that the first differential stage has gain A_1 and the other single ended stage A_2, the voltages at the end of the track period are

$$V_{out}(nT) = \frac{[V_{in}(nT) + V_{os}]A_1 A_2}{1 + A_1 A_2}; \quad V_1(nT) = \frac{V_{out}(nT)}{A_2} \qquad (5.18)$$

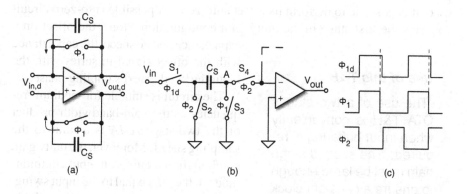

(a)   (b)   (c)

Figure 5.14. (a) Clock feed-through compensation with a fully differential scheme. (b) Bottom plate sampling. (c) Phases scheme.

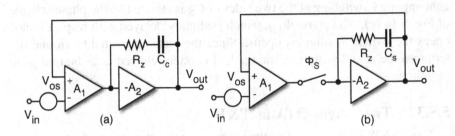

Figure 5.15.   (a) Two stage OTA with Miller compensation. (b) Use of the two stage OTA as a T&H.

showing that with a relatively large gain the output voltage tracks the input plus an input referred offset. Moreover, the charge on the sampling capacitance accounts for the finite gain of the second stage.

When the switch goes off the charge on C_S holds the signal as it just changes because of the clock feed-through

$$V_{out}(nT + T/2) = \frac{[V_{in}(nT) + V_{os}]A_1 A_2}{1 + A_1 A_2} - Q_{ck}C_s \qquad (5.19)$$

Since the voltage of the switch is almost constant $(-V_{in}/A_2)$, the channel charge is independent of the input making the clock feed-through error just an offset. Moreover, since during the hold period the first stage is not used, it is possible to perform its offset auto-zero. A possible auto-zero circuit connects the first stage in the unity gain configuration, stores the offset on a capacitance and connects the capacitance with the offset signal in series with the input during the next sampling phase.

Use of this T&H

The use of a two stages OTA T&H is conveniently when input buffering is required. The second stage gain must be large enough to ensure a constant clock feed-through.

The circuit is suitable for medium frequencies as the gain-bandwidth product of the two-stages *OTA* is a limit to the sampling speed. Moreover, the unity gain configuration requires an input common-mode of the *OTA* equal to the input swing. Since the input voltage is made available at the output terminal during the sampling, the circuit operates as a *T&H*.

5.5.4 Use of the Virtual Ground in CMOS S&H

The use of the virtual ground concept frees the input common mode from the requirement of having a range as wide as the input signal.

Another solution that secure the same benefit is the scheme of Fig. 5.16 (a) operating as a fully-differential amplifier with a gain of -1 that becomes a *S&H* thanks to the switched scheme used. This circuit is known as a charge transferring *S&H* since it pre-charges C_S to the input voltage during the sampling phase Φ_S and then transfers this charge into the hold element, C_H. Also, the hold capacitances C_H are pre-charged to the difference between the common-mode inputs and outputs. In this way, the circuit realizes an offset cancellation, and obtains a possible shift of the common mode output.

The difference between the common modes of input and output of the *OTA* makes the design flexible and allows the use of schemes like a telescopic cascode. Since the left terminals of C_S are tied together during Φ_H the scheme rejects any common mode input components.

The condition $C_S = C_H$ gives rise to a unity gain. However, with different capacitance ratios it is possible to provide gain or attenuation, if needed.

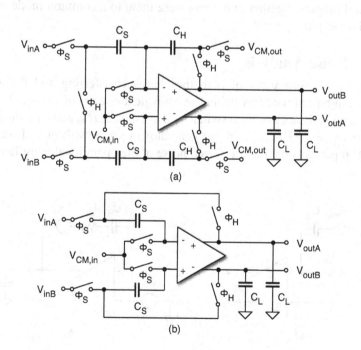

Figure 5.16. (a) Charge transferring S&H. (b) Flip-around S&H.

A more effective solution is shown in Fig. 5.16 (b) as it uses only a pair of differential capacitors for the input sampling during Φ_S, and for the hold during Φ_H. The circuit operation can be viewed as the flipping of the connection of the right terminal of the capacitance; for this reason the scheme is referred to as a flip-around architecture.

The gain of the circuit of Fig. 5.16 (b) can only be 1; moreover, the scheme does not allow the option of having different input and output common-mode voltages. However, the circuit is more power effective than the charge transfer architecture because the feedback factor is 1 (neglecting parasitic capacitances) while the β of the charge transfer scheme is $1/(1 + G)$ (i.e. it is 0.5 for $G = 1$). The more favorable β of the flip-around scheme requires a smaller open-loop gain-bandwidth product with equal sampling frequency thus providing a power saving or enabling a higher sampling rate.

Neither schemes of Fig. 5.16 use the op-amp during the sampling phase as they perform the sampling in a passive way. Therefore, during the sampling period the op-amp is open loop and the output voltage can be stacked near the positive or negative supply, thus requiring a long time to recover and return to the normal region of operation. This possible limit is avoided by shorting the differential outputs together and connecting them to a common mode voltage during the sampling period.

5.5.5   Noise Analysis

The noise performances of both the charge transferring and flip-around schemes can be estimated by using the equivalent circuits of Fig. 5.17. These model the on-resistance of each switch by a resistance, R_{on} and a thermal noise generator $v_n^2 = 4kTR_{on}$. A noise generator in series with one of the input terminals represents the input referred noise of the operational amplifier.

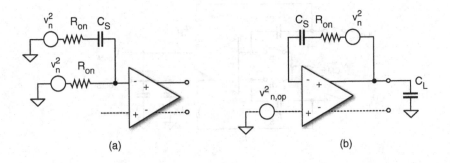

(a)   (b)

Figure 5.17. (a) Noise equivalent circuit of the flip around S&H in the sampling phase and (b) in the hold phase.

The study of the sampled-data noise, measured during the hold phase, uses the same methodology as discussed for the kT/C noise as follows: every noise generator filtered by a low-pass network causes colored noise across each capacitance. When the switches open the noise sampled across the capacitors has noise spectrum whose power is equal to the integral of the colored spectra from zero to infinity. The noise powers generated during the sampling period are transferred to the output (filtered by a possible transfer function) and superposed quadratically if they are uncorrelated, or summed linearly if they are correlated.

The scheme of Fig. 5.17 (a), which models the sampling phase of the flip-around circuit, uses two noise generators to charge C_s through two on-resistances. Since both the noise spectrum and the on-resistance double, then the decrease in the low-pass corner frequency compensates for the higher thermal noise leading to kT/C_S. During the hold phase, the circuit is in the unity gain configuration, as shown by the model of Fig. 5.17 (b), and causes noise at the output. This noise, combined with the sampled term (caused during the sampling phase) are sampled by C_L at the end of the hold phase.

The virtual ground causes the output to be equal to the noise generator $v_n^2 = 4\,kTR_{on}$ until it rolls off due to the finite unity gain frequency, f_T, of the op-amp. Therefore, the white $4\,kTR_{on}$ spectrum gives rise to a colored output spectral density of

$$v_{out}^2 = \frac{4kTR_{on}}{1 + (\omega/\omega_T)^2} \tag{5.20}$$

which, integrated over the infinite frequency interval leads to $kTR_{on}\omega_T$.

Because of the unity gain configuration even the op-amp's noise is f_T band-limited. Therefore, assuming that $v_{n,op}^2 = 4kT\gamma/g_m$ and $\omega_T = g_m/C_L$, the sampled noise across C_L caused by the switches in both the sampling and hold phases, added to the op-amp noise gives

$$V_{n,flip}^2 = \frac{kT}{C_S} + g_m R_{on}\frac{kT}{C_L} + \frac{\gamma kT}{C_L}. \tag{5.21}$$

Since the sampling time constant, $R_{on}C_S$, is typically lower than $1/\omega_T$, $g_m R_{on}/C_L \ll 1/C_S$ making the second term of the above equation negligible.

Studying the noise of the charge transferring scheme is a bit more complicated as it uses two capacitors (assumed equal, $C_S = C_H$), charged by a series of two or three switches. Let us first consider the sampling phase, and observe that the noise charge stored on C_S during the sampling period is then injected into C_H during the hold phase. Therefore, if the charges on the two capacitances are correlated their effect must be superposed linearly. This observation together with the superposition principle shows that since noise generator (3) in Fig. 5.18 (a) imposes equal noise charges on C_S and C_H (as the right and left networks are equal), then it does not contribute to the output noise.

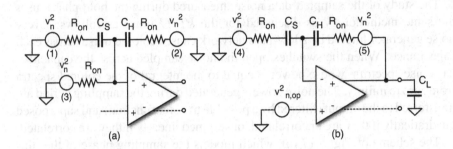

Figure 5.18. (a) Noise equivalent circuit of the charge transferring S&H in the sampling and (b) in the hold phase.

However generators (1) and (2) both establish different charges on the capacitors and their combined effect is dependent on the difference between the two transfer functions from the input to the capacitors terminals

$$H_{eq}(s) = H_{C_S}(s) - H_{C_H}(s) = \frac{1}{1 + 3sR_{on}C_S}; \quad if \ C_S = C_H \quad (5.22)$$

showing a combined filtering action that is better than its passive counterpart which, in turn gives rise to $2kT/(3C_S)$ for the sampling of both C_S and C_H.

For the hold phase it is necessary to account for the contribution of noise generators (4) and (5) and for the op-amp noise as shown in Fig. 5.18 (b). Notice that noise source (4) is amplified by -1 and band-limited at βf_T. Also, since C_H establishes a feedback and the virtual ground voltage is almost zero, then noise source (5) shows up at the output until the band-limit βf_T. Observe that the two switches contribute two white spectra but a lower feedback factor ensures a more effective filtering.

Finally, for the last noise contribution it is worth observing that the feedback network amplifes $v_{n,op}$ by 2, thus amplifying the noise spectrum, $v_{n,op}^2$, by 4, which is, again, band-limitated at βf_T.

The quadratic superposition of the contributions during the sampling and hold phases gives rise to a sampled-data noise whose expressions assume again that the op-amp noise spectrum is $v_{n.op}^2 = 4kT\gamma/g_m$ and $\omega_T = g_m/C_L$. The result, with $\beta = 1/2$, is

$$V_{n,ch_tr}^2 = \frac{2kT}{3C_S} + +g_m R_{on}\frac{kT}{C_L} + \frac{2\gamma kT}{C_L} \quad (5.23)$$

where the two first terms are almost equal to the corresponding terms in equation (5.21) but the last term is two times larger than the flip-around counterpart thus carrying higher noise if the op-amp is the dominant term.

The result shows a slightly worst noise performance than the flip around scheme especially if the γ op-amp noise factor is large. The difference between the two cases with equal design conditions is about 3 dB.

Intuitive Conclusions

The noise power of a sample and hold is inversely proportional to the used capacitances (sampling and load) through the kT/C relationship. The use of noisy op-amp can significantly worsen the noise performance.

Example 5.3

A flip-around S&H is made offset insensitive by pre-charging the sampling capacitance to the input minus the offset. Sketch a possible circuit diagram and determine the noise performance. Use a sampling capacitance equal to 1 pF and $R_{on} = 10\ \Omega$. The model of the OTA has $g_m = 12\ mA/V$, $\gamma = 2$ and $C_L = 2\ pF$.

Solution

A possible offset insensitive S&H is the single-ended solution shown in Fig. 5.19 (a) that places the OTA in the unity gain configuration during the sampling phase, thus making the offset available at the non-inverting terminal. The sampling capacitance is pre-charged to the offset and when it is connected in feedback it obtains the offset cancellation.

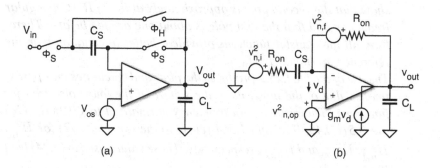

Figure 5.19. (a) Offset insensitive flip-around S&H (b) Noise model during the sampling phase.

The noise behavior of the circuit differs from the flip-around scheme during Φ_S as the action of the right switch not only makes the offset available, but the noise of the op-amp too. The small signal equivalent circuit shown in Fig. 5.19 (b) is described by

$$
\begin{aligned}
v_{out} &= v_{n,i} + v_{n,f} + v_{n,C_S}(1 + 2sR_{on}C_S) \\
v_d &= v_{n,op} - v_{out} + v_{n,f} + v_{n,C_S}sR_{on}C_S \\
g_m v_d &= v_{out} \cdot sC_L + v_{n,C_S}sC_S
\end{aligned}
$$

where v_{n,C_S} is the noise voltage across the sampling capacitance, and $v_{n,i}$ and $v_{n,f}$ are the thermal noise sources of the input and feedback switches respectively.

The solution of the above equations yield

$$
v_{n,C_S} = \frac{v_{n,i}(g_m + sC_L) + v_{n,f}sC_L - v_{n,op}g_m}{g_m + s\{C_L + C_S(1 + g_m R_{on})\} + 2s^2 C_L C_S R_{on}}
$$

showing that the voltage across the sampling capacitor is a combination of the three noise sources, and is properly shaped by different noise transfer functions which have two poles and, for the thermal noise generators, a zero.

Since the noise sources are uncorrelated the voltage spectrum across C_S is given by the quadratic superposition of the various contributions

$$
v_{n,C_S}^2 = v_{n,i}^2 H_{n,i}^2(f) + v_{n,f}^2 H_{n,f}^2(f) + v_{n,op}^2 H_{n,op}^2(f). \tag{5.24}
$$

Fig. 5.20 shows the amplitude of the noise transfer functions obtained using the numerical values of the design, and compared with that of the $R_{on}C_S$ low-pass filter: $H_{n,ref} = 1/(1 + sR_{on}C_S)$. The plots show that the second pole is approximately at the $1/R_{on}C_S$ angular frequency and that the first pole is about one decade before. Therefore, all the transfer functions provide a better attenuation than the reference, particularly $H_{n,op}$.

The use of the file Ex5_2 produces the plots and calculates the approximate value of the integrals of the square of the four noise transfer functions. The results, with frequency normalized to $1/(2\pi R_{on}C_S)$ are 0.51, 0.45, 0.05 and 1.564 (close to the expected $\pi/2$) for $H_{n,i}^2$, $H_{n,f}^2$, $H_{n,op}^2$ and $H_{n,ref}^2$ respectively. These values used in (5.24) lead to

$$
v_{n,C_S}^2 = \frac{4kTR_{on}}{2\pi R_{on}C_S}(0.51 + 0.45) + \frac{4kT\gamma}{g_m 2\pi R_{on}C_S}0.05 = 1.14\frac{kT}{C_S}
$$

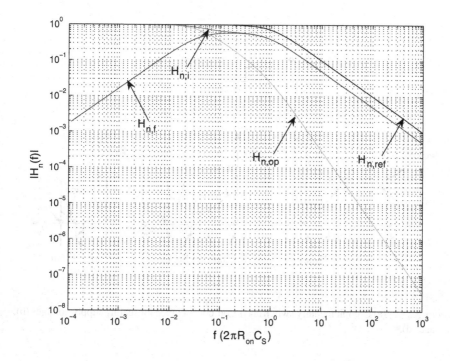

Figure 5.20. Normalized module of the noise transfer functions.

which is a little bit higher than the kT/C_S noise of a flip-around S&H during the sampling phase. The result is due, on one hand, to an increase of the total noise because of the op-amp noise contribution and, on the other hand, to a more effective filtering due to the small feedback factor.

5.6 CMOS SWITCH WITH LOW VOLTAGE SUPPLY

CMOS circuits have benefited significantly from device scaling, as shrinking the device dimensions increases the speed and reduces the parasitic capacitances, resulting in reduced bias currents. However, the gate oxide also shrinks leading to lower and lower supply voltages that ultimately make it difficult to implement key functions required by analog circuits, specifically the switching of *MOS* transistors.

It was observed that a pair of complementary transistors compensates for the increased on-resistance of one device with the reduction of the complementary one. The switch of Fig. 5.21 is *on* as both transistors are driven to

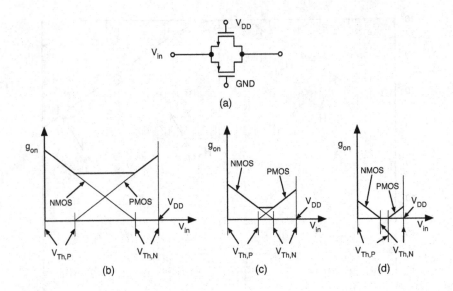

Figure 5.21. (a) Complementary switch and its on-conductance for scaling supply voltages (b), (c), and (d).

their maximum; the n-channel to V_{DD}, and the p-channel to GND. The total conductance is given by

$$G_{on,n} = \mu_n C_{ox} \left[\frac{W}{L}\right]_n V_{ov,n} + \mu_p C_{ox} \left[\frac{W}{L}\right]_p V_{ov,p}, \qquad (5.25)$$

provided that there is some overdrive for the n-channel and the p-channel transistors then

$$V_{ov,n} = V_{DD} - V_{in} - V_{Th,n}; \quad V_{ov,p} = V_{in} - V_{Th,p}. \qquad (5.26)$$

Accordingly, for $V_{in} < V_{Th,p}$ the conductance of the p-channel transistor is zero while the n-channel element goes off for $V_{in} > V_{DD} - V_{Th,n}$.

When the input voltage is such as to turn-on both transistors the condition

$$\mu_n \left[\frac{W}{L}\right]_n = \mu_p \left[\frac{W}{L}\right]_p \qquad (5.27)$$

makes the conductance independent of the input voltage and equal to

$$G_{on,sw} = \mu_n C_{ox} \left[\frac{W}{L}\right]_n (V_{DD} - V_{Th,n} - V_{Th,p}). \qquad (5.28)$$

Therefore, for a given technology, and with aspect ratios expressed by equation (5.27), the minimum on-conductance decreases with the supply voltage.

Furthermore, the input voltage range where the conductance is constant decreases as shown in Fig. 5.21 (b) and (c). When the supply voltage is $V_{Th,n} + V_{Th,p}$ the conductance of the complementary switch can become zero giving rise, from that supply value and below, to a channel voltage range where the switch is not able to open.

The use of low thresholds alleviates this problem, but since the leakage current increases at low thresholds it is necessary to ensure a proper trade-off between analog and digital requirements that inevitably leads to moderate threshold reductions. Technologies with multiple thresholds are possible but they are rare and expensive solutions.

Another possibility is the use of a double supply voltage: one for the analog and the other for the digital part, with the optional addition of using, thin oxide for digital transistors and thick oxide for high voltage devices. However even this method results in an increase in cost as it requires additional processing steps and, more expensive, extra masks.

Circuit solutions can partially avoid the difficulty of closing a switch at low supply voltage. For example, the scheme of Fig. 5.22 is a flip-around S&H whose *OTA* has its input terminals at a low voltage level, near ground. This low *dc* level provides suitable headroom for driving the switch S_b. For the switch connected to the output of the *OTA* the scheme uses a switched op-amp technique that, as shown in the figure, switches off the second stage of the *OTA* at points that are close to V_{DD} and ground. The output node becomes high-impedance, thereby enabling the connection of the left plate of C_S to the input without disconnecting the *OTA* with a switch. The only critical element for the scheme of Fig. 5.22 is the input switch, S_{in}, whose channel voltage range must equal the variable input signal.

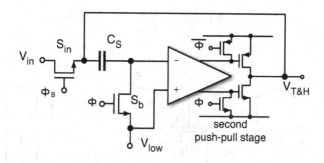

Figure 5.22.   Flip-aroud T&H with input common-mode voltage and switched OTA.

5.6.1 Switch Bootstrapping

Charge pumps provide higher voltages than their supply voltages. However, the use of charge pumps for generating a control that is a multiplied version of V_{DD} is not suitable for clocking *MOS* switches. The reason is that the driving can be adequate when the input signal is close to the supply voltage, but the gate-to-source difference can become too high when the input signal is close to zero. As a matter of fact, high gate-to-source voltages cause failures due to various mechanisms. When short channel lengths are used in deep sub-micron technologies the so-called "hot electrons" or "hot holes" effect degrades the transistor threshold over time. Moreover, high electric fields degrade the oxide breakdown voltage, increasing the local tunneling current that is responsible for oxide lifetime reduction.

The above arguments motivate the switch bootstrapping method which ensures that the stress on the gate is always below its technologic limit. The method is conceptually based on the scheme of Fig. 5.23 that uses a charged capacitor to sustain the gate-to-source voltage of M_S during the on-phase. The switch S_{OFF} grounds the gate of the switch M_S during Φ_{OFF} and, at the same time, switches S_1 and S_2 charge the boosting capacitance C_B to the supply voltage. The switches S_3 and S_4 connect C_B between the source and drain of M_S during Φ_{ON} to obtain gate boosting.

Actually, the voltage at the gate of M_S does not increase by $V_{in} + V_{DD}$ but, due to the initially discharged parasitic capacitance C_p, goes to

$$V_{GS} = (V_{in} + V_{DD})\frac{C_B}{C_B + C_p} \tag{5.29}$$

leading to a gate-to-source voltage equal to

$$V_{GS,M_S} = V_{DD}\frac{C_B}{C_B + C_p} - V_{in}\frac{C_p}{C_B + C_p} \tag{5.30}$$

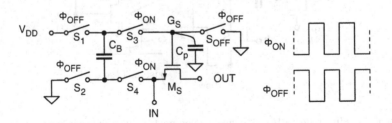

Figure 5.23. Conceptual scheme of switch bootstapping.

which will be lower than V_{DD} and also dependent on the input voltage. This limit is reduced by using relatively large boosting capacitances and small gate areas for all transistors connected to G_S. The on-conductance of M_S becomes

$$G_{on} = \mu_n C_{ox} \left(\frac{W}{L}\right)_{M_S} \left[V_{DD}\frac{C_B}{C_B + C_p} - V_{in}\frac{C_p}{C_B + C_p} - V_{Th,n}\right] \quad (5.31)$$

describing the mentioned reduced boosting effectiveness and a slight on-resistance dependence on the input voltage.

The circuit implementation of the boosting concept illustrated by Fig. 5.23 should ensure the proper driving of the switch transistors, thus avoiding any direct biasing of the channel-substrate diodes, and providing suitable protection for the drains that undergo large voltage swings. Accordingly, it is necessary to satisfy the following conditions:

- S_1 must be able to switch on and off V_{DD};
- the switch S_3 must sustain the boosted voltage during the on phase;
- the switch S_4 must operate under the same conditions as the main switch;
- S_{OFF} must be able to swing from the boosted voltage to zero.

A possible switch boosting scheme is the one shown in Fig. 5.24 which highlights the transistors used to realize the five switches. Since switch S_1 is an n-channel transistor, control of its gate requires a voltage higher than V_{DD} and, for this, the scheme uses a voltage doubler made by the cross pair M_{d1} and M_{d2}, and the charge pumping capacitances C_1 and C_2 driven by Φ_{OFF} and its inverse. Since the charge pump biases the gate of M_1 to V_{DD} during Φ_{ON}

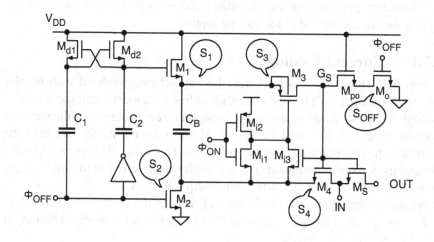

Figure 5.24. Circuit implementation of switch boosting.

and up to $2V_{DD}$ during Φ_{OFF}, the boosting capacitance C_B charges to V_{DD} (because S_2 is also on).

Switch S_3 is realized by the p-channel transistor M_3. The inverter M_{i1} and M_{i2} whose output during Φ_{ON} goes to the voltage of the bottom plate of C_B drives the gate of M_3. Sometimes the transistor M_{i3} takes the control if switch S_4 pushes the source of M_{i1} above the control of the inverter.

The boosted voltage of node G_S has two effects: it switches both S_4 and M_S on, and it gives rise to a high voltage at the drain of M_o, (the transistor realizing S_{OFF}). The high-volage drain protection of M_o is ensured by M_{po} that, during Φ_{ON}, distributes the boosted voltage between M_{po} and M_o.

Observe that the source-to-well connection of M_3 avoids any possible latch-up. Moreover, the use of an n-channel transistor for S_1 requires a voltage doubler but is less problematic than using a p-channel switch, as the proper biasing of a p-channel device would have required switching between V_{DD} during one phase and the boosted voltage during the other phase.

The input voltage of the scheme of Fig. 5.23 can not exceed V_{DD} as the switching on of M_{i3} applies the input voltage to the drain of M_{i2}. Since the well of M_{i2} is at V_{DD} a higher drain voltage would forward bias the well-drain diode. Therefore, input voltages higher than V_{DD} require more complex boosting schemes.

5.7   FOLDING AMPLIFIERS

The input of a folding amplifier can be either a current or a voltage. Ideally, the output is a set of linear segments with alternating slopes equal in module. The circuit implementations can use either bipolar or MOS technology. Here, we study two possible circuits that make use of both technologies, one for current inputs, and the other for voltage inputs.

5.7.1   Current-Folding

Fig. 5.25 shows equivalent current folding schemes realized with bipolar and *MOS* transistors. The first circuit identifies 4 segments, and the second 8 through the use of equal sections each using a current generator of the required fraction ($1/4$ or $1/8$) of the full-scale. If the input current is zero then the current in the follower transistor of each cell is I_E (or I_S). The given additions of currents flowing into the loads R_L, nominally equal, give rise to equal voltage drops and, in turn, produce a differential output of zero.

An input current lower than I_E (or I_S) reduces the current of Q_1 (or M_1) and, consequently, reduces the current in the left load resistor. The output differential voltage becomes

$$V_{out,d} = R_L I_{in}. \tag{5.32}$$

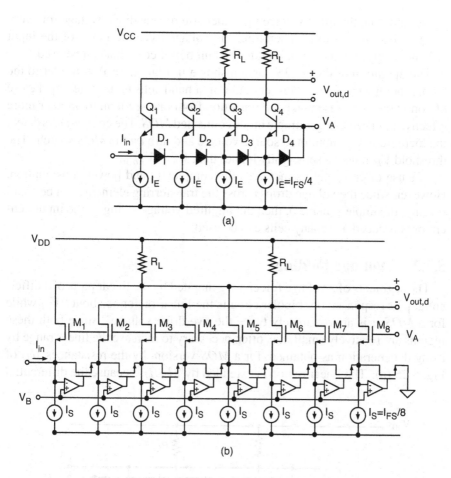

Figure 5.25. (a) Current-folding amplifier with BJT. (b) MOS current-folding amplifier.

When the input current becomes higher than I_E the current in Q_1 goes to zero and the emitter of Q_1 (that is normally one diode voltage below V_A) goes up until it switches diode D_1 *on* and passes $I_{in} - I_E$ to the next cell. The current passed reduces the current of Q_2 which, in turn, decreases the current in the right R_L augmenting the negative output. Therefore, the differential output becomes

$$V_{out,d} = R_L I_E - R_L (I_{in} - I_E) = -R_L I_{in} \qquad (5.33)$$

which is linear but has a negative slope. The differential output decreases until the input current is greater than $2I_E$. At this point $I_{Q,2}$ goes to zero, the diode D_2 switches on, and $(I_{in} - 2I_E)$ is passed to the third cell causing the current in the left load resistor to diminish again starting the third segment and so forth.

Observe that the input voltage must increase by one diode voltage for every cell that becomes active. Therefore, the voltage dynamic range of the input current generator becomes a limit to the number of cells that can be used.

The operation of the MOS version shown in Fig. 5.25 (b) is similar to the bipolar but the transfer of current to the right hand cells is sustained by the use of comparators. These make the transition from one segment to another more effective than just using a simple diode connected MOS. The comparators detect the increase of the monitored source voltage and turn *on* the MOS switch. The threshold V_B must be set slightly higher than $V_A - V_{Th,n}$.

The use of comparators increases the complexity and power consumption. However, since the voltage drops across the transferring elements can be small (even with simple schemes), then the required voltage swing of the input generator is reduced and many cells can be used.

5.7.2   Voltage Folding

The segments of a voltage folder are generated by the linear parts of a differential pair response. For bipolar circuits the linear region is about $2V_T$, while for a MOS it is approximately twice the overdrive voltage. Since both these regions are relatively small, it is often necessary to increase the linear range by using degeneration as obtained, for a MOS version, by the resistances R_D of Fig. 5.26 (a). The linear region is extended by $2I_S R_D$ because the differential

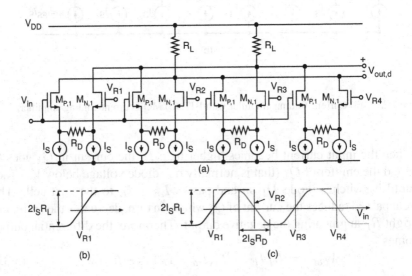

Figure 5.26. (a) MOS voltage-folding. (b) response of a single cell. (c) response of the entire circuit.

pair becomes fully unbalanced with about $\pm(I_S R_D + V_{ov})$ differential input. Fig. 5.26 (b) shows the swing of the output voltage caused by a single cell. The input crosses the reference and the differential output changes by $2I_S R_L$.

The entire folder uses a number of cells working in parallel whose output terminals have alternate connections to the two loads R_L. The first response around V_{R1} combined with the one around V_{R2} obtains the first folding. Combining the second and the third cells together creates the second folding and so forth. The overall result shown in Fig. 5.26 (c) displays rounded parts that are dependent on the sharpness of the saturation of the single response, and on the position of the references.

Since the output swing is $\pm I_S R_L$, then to obtain a 1 V range with $R_L = 1\,k\Omega$ it is necessary to use $I_S = 1\,mA$. The pole established by the load resistance and the effective output capacitance determines the input signal band. If, for example, $R_L = 1\,k\Omega$ and $C_p = 1\,pF$ the frequency of the pole is 159 MHz. For high-speed applications it may be necessary to use a lower resistances with a consequent increase of the power consumption.

The output common-mode depends on the number of cells K used; since the average current in each load resistance is $K I_S$, the common mode output is $V_{DD} - K R_L I_S$. The value of the common mode voltage can be set to a desired value by injecting suitable common mode currents into the two R_L.

5.8 VOLTAGE-TO-CURRENT CONVERTER

The electrical variable normally used in data converters is voltage, but in some cases it is convenient to perform the processing in the current domain. For this purpose it is necessary to use a voltage-to-current converter (*V-I*) to perform the transformation between the two domains. As a matter of fact, a *MOS* transistor in the triode region is a simple voltage-to-current converter that, with some expedients, is also fairly linear. Therefore, for very low resolutions the use of a single *MOS* in saturation can be sufficient. However, since the linearity is not much better than 60 *dB* with few hundreds of *mV* range, a *V-I* with medium-high resolution is a relatively complex scheme.

Fig. 5.27 (a) shows a single-ended *V-I* converter that replicates the input voltage across a resistance and thus obtains a current proportional to the input voltage: $I_{out} = V_{in}/R$. The accuracy of the circuit depends on the offset and input referred noise of the op-amp. Moreover, since the output impedance of the current mirror affects the mirroring factor, it is often necessary to use a cascode configuration.

A benefit of the op-amp used in Fig. 5.27 (a) is that it cancels the shift between source and gate, V_{GS1}. However, since this shift is insignificant for fully differential schemes and the nonlinearities of the V_{GS} partially cancel out, the solution of Fig. 5.27 (a) can be simplified into the one of Fig. 5.27 (b) that

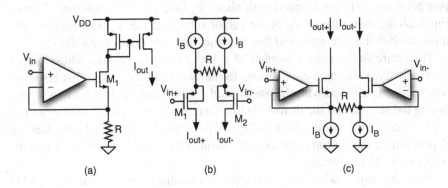

Figure 5.27.   Schemes of Voltage-to-Current converters

employs two *p*-input source followers with wells tied to the source to cancel the body effect. The resistance R across the sources gives rise to a current

$$I_{out\pm} \simeq I_B \pm \frac{V_{in+} - V_{in+}}{R} \tag{5.34}$$

which, of course, is independent of the input common mode voltage.

A limit to this accuracy comes from the finite (small and large signal) transconductance that, in series with the resistance influences the transconductance gain of the *V-I* converter. Therefore, in order to limit the error, it is necessary to use a transconductance that is smaller than the resistance. Since the value of the resistance determines the transconductance gain, the *SFDR* constraint may require the use of a large bias current to minimize the non-linear effect of the transconductance (albeit partially compensated by the fully differential operation) .

In some cases, instead of increasing the *dc* current in the emitter follower, it can be more effective to use the same power (or less) for two dedicated *OTA*, allowing for an active reduction of the transconductance, as shown in Fig. 5.27 (c). This circuit allows the use of *n*-channel transistors as the body effect is reduced by the gain of the *OTA*.

The absolute accuracy of the *V/I* transconductance gain depends on the absolute accuracy of the resistance used and its temperature and voltage coefficients. Since the absolute value of integrated resistances can change by ±15% with process variations, precise applications require using on-chip trimming or calibration or external elements. Another systematic error depends on the *MOS* transistor threshold mismatch that, as already studied before, is inversely proportional to the square root of the gate area and can be as large as few *mV*.

Example 5.4

The V-I converter of Fig. 5.27 (b) has an input voltage range of $\pm 0.5\,V$ and uses bias currents $I_B = 1\,mA$. Determine the value of the resistances required to ensure a SFDR better than 85 dB and 95 dB. Use $V_{ov} = 400\,mV$.

Solution

The output current depends on total resistance and is determined by the series of R with the large signal transconductance of M_1 and M_2, whose values depend on the dc current

$$R_T = R + \frac{V_{ov}}{2(I_B + I_{out})} + \frac{V_{ov}}{2(I_B - I_{out})}$$

which, rearranged gives

$$R_T = R + \frac{V_{ov}I_B}{I_B^2 - (I_{out})^2}.$$

The differential input determines the output current I_{out}, expressed by

$$\Delta V_{in} = RI_{out} + \frac{V_{ov}I_B I_{out}}{I_B^2 - (I_{out})^2}$$

which is a non-linear voltage-current relationship that should be solved to obtain $I_{out} = f(V_{in})$.

Instead of solving the non linear equation $\Delta V_{in} = g(I_{out})$ it can be convenient to just determine the coefficients of its polynomial fitting by using the Matlab file Ex5_3. Since the response is odd there are only odd polynomial coefficients with decreasing amplitude. For example, $R = 10\,k\Omega$ gives rise to:

$$k_2 = -5.95 \cdot 10^{-18}; \quad k_3 = -3.78 \cdot 10^{-04};$$

$$k_4 = -4.02 \cdot 10^{-18}; \quad k_5 = -1.55 \cdot 10^{-06};$$

$$k_6 = -7.83 \cdot 10^{-20}; \quad k_7 = -4.92 \cdot 10^{-08}.$$

If the input voltage is a sine wave at f_{in} with amplitude A_{in} the third harmonic tone (which determines the SFDR) has amplitude $k_3 A_{in}/4$. Therefore, 85 dB or 95 dB SFDR require a distortion coefficient k_3 of less than $2.25 \cdot 10^{-4}$ and $7.11 \cdot 10^{-5}$ respectively. Simulations at different values of R satisfy the above requirements by using $R = 12.5\,k\Omega$ and $R = 18.5\,k\Omega$.

The scheme of Fig. 5.29 obtains the V-I conversion by using a combination of the passive scheme of Fig. 5.27 (b) and the one of Fig. 5.27 (c). The circuit avoids the body effect by connecting source and substrate of M_1 and M_2, and keeps the V_{GS} of the input transistors constant by maintaining a constant current in the input transistors by using local feedback.

The input transistors operate as both the input element of a source follower and the input of a gain stage. The difference between I_{M_1} (or I_{M_2}) and I_B is multiplied by the impedance of nodes A (and B) giving rise to feedback

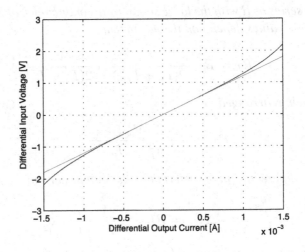

Figure 5.28.   Current-voltage relationship with R = 2 $k\Omega$.

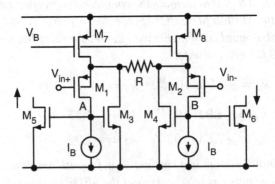

Figure 5.29.   Improved V-I converter using constant currents and source and substrate tighten in M_1 and M_2.

voltages at the gate of M_3 (and M_4). Assuming the current of M_7 and M_8 is constant, then the loop control is such that $-\Delta I_{M_3}$ and ΔI_{M_4} compensate for the current in the resistor. The currents in M_5 and M_6 are output mirrored replicas.

The feedback loop is the cascade of a common source stage and a common gate stage with gain

$$A_{loop} = -g_{M_3} \frac{R/2}{1 + g_{M_1} R/2} \cdot g_{M_1} R_{out,A} \qquad (5.35)$$

where $R_{out,A}$ is the resistance at the drain of M_1.

If R is approximately two times $1/g_{M_1}$, and $g_{M_1} R_{out,A}$ is in the 30-40 dB range, then the linearity improvement by A_{loop} is sufficient for many applications as a good trade-off between speed and complexity.

In some designs the circuit also requires a moderate compensation to ensure stability of the local loops. Moreover, it is necessary to burn some power for biasing M_3 and M_4. Typically the current in these transistors is I_B, doubling the power consumption with respect to the circuit of Fig. 5.27 (b). However this cost is much less than using two complete op-amps although the benefits in terms of speed and linearity is significant.

The inaccuracy of the output current, in addition to the limits caused by the resistance and the thresholds mismatch of the input transistors, is also determined by the inaccuracy of the current mirrors whose mirror factor depends on the accuracy of the μC_{ox} factor, the precision of the aspect ratio and the thresholds' mismatch.

5.9 CLOCK GENERATION

The operation of data converters requires a number of clock phases that are all obtained from a master clock. The logic signals control switches for implementing the conversion algorithm, reconfiguring analog architectures or controlling the transfer of data. The clock phases may require delay, and may need to be overlapping or non-overlapping to ensure specific features. These include the maintaining of feedback during transitions between different states, or avoiding the leakage of charge stored in capacitances. Moreover, transient spurs caused by the switching activity of digital logic must not occur until after the analog signals have been sampled.

The basic schemes used for clock generation are either non-overlapping, or overlapping, where the use of an extra inverter switches from one feature to the other. Fig. 5.30 (a) shows a cross-coupled flip-flop that generates two complementary phases from a single input phase. Fig. 5.30 (a) uses *NOR* gates, and highlights the initial logic state (1 or 0) corresponding to the dashed line of the phase diagram. When the phase goes high, the top *NOR* switches, but

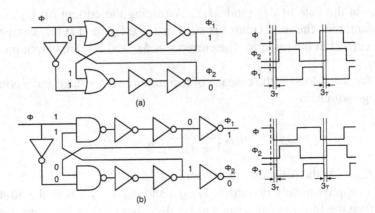

Figure 5.30.   Logic circuits used to generate two non-overlapping phases.

there is a delay of three inverters before the output Φ_1 goes low. Following this transition both inputs of the bottom *NOR* become zero and after a three further delays, Φ_2 becomes logic 1. Therefore, the disoverlap is 3 inverter delays. If needed the circuit can use a chain of 5 inverters (or more) to increase the disoverlap time.

The scheme of Fig. 5.30 (b) works in a similar way, but uses *NAND* gates to obtain overlapping phases. These are then transformed into non-overlapping phases at the output nodes using two additional inverters.

PROBLEMS

5.1 The diodes used in the diode bridge sample and hold are such that the current required to charge C_s flows evenly in the upper and lower parts of the bridge. Determine the variation of the voltages at nodes A and B of Fig. 5.2 for a 0.5 V sine wave input.

5.2 The fall time of switching-off of a diode bridge is 2 ps. Estimate the error due to the aperture distortion with a 600 MHz 1 V peak sine wave at the input.

5.3 Design the scheme of a possible clipping block to be used in the scheme of Fig. 5.8. The swing of the input signal is $\pm 1\,V$.

5.4 The β of the transistors used in Fig. 5.8 (a) is 40 and the output drives a capacitance of 2 pF. Estimate the SFDR for an input sine wave with $\pm 1\,V$ amplitude and $I_2 = 2\,mA$

5.5 Repeat Example 5.1 but use sampling capacitances of 1-2-4-8 pF. The input is a sine wave at 160 MHz with 1 V peak amplitude.

5.6 A switched emitter-follower uses a bias current equal to 5 mA. The excess gain factor is $\gamma = 8$. The sampling capacitance is 2 pF for a 100 MS/s operation. The foreseen accuracy requires a S&H bandwidth that is at least 6 times the sampling rate. Optimize the noise performance.

5.7 Simulate, at the transistor level, the pseudo-differential MOS S&H of Fig. 5.11. Use a bias current of 0.5 mA and the SPICE model of the fastest technology available to the reader. Determine the maximum sampling rate for 8-bit accuracy. Determine suitable sizing of the MOS switches and dummy.

5.8 Determine with a SPICE simulation the transfer functions from all the possible noise inputs in the circuit of Fig. 5.16 and the output. Use for the op-amp a behavioral model with finite gain and bandwidth.

5.9 Repeat Example 5.2 but use a sampling capacitance equal to 4 pF and $R_{on} = 25\,\Omega$. The model of the the OTA has $g_m = 12\,mA/V$, $\gamma = 8$ and $C_L = 4\,pF$.

5.10 Estimate with SPICE, using any MOS model available, the on-resistnace of a complementary switch made by minimum area elements. Use a V_{DD} equal to 3, 2 and 1 times the threshold voltage of the n-channel element.

5.11 Simulate, at the transistor level, the clock boosting circuit of Fig. 5.24. Use the SPICE model of any available CMOS technology. Don't design the voltage doubler but use a suitable signal generator.

5.12 Simulate, at the transistor level, a four-cell MOS voltage-folder. Use the SPICE model of any available CMOS technology. The full scale range is 1 V, $I_s = 200\ \mu A$ and the degeneration resistance is 1 kΩ

5.13 Estimate, using the SPICE model of any available CMOS technology, the linearity of the schemes of V-to-I converter given in Fig. 5.27 (b). Use W/L = 50, $I_B = 1/, mA$ and $R = 10\,k\Omega$. The peak amplitude of the input sine wave is 1 V.

5.14 Design, using SPICE, the voltage-to-current converter of Fig. 5.29. The range of the input signal is $\pm 1\ V$ and its bandwidth is 50 MHz. The desired SFDR is 80 dB.

REFERENCES

Books and Monographs

R. Gregorian and G. C. Temes: *Analog MOS Integrated Circuits for Signal Processing*, John Wiley & Sons, Inc., 1986.

D. A. Johns and K. Martin: *Analog Integrated Circuit Design.* John Wiley & Sons, New York, 1997.

F. Maloberti: *Analog Design for CMOS-VLSI Systems.* Kluwer Academic Press, Boston, Dordrecht, London, 2001.

P. R. Gray, P. J. Hurst, S. H. Lewis, and R. G. Mayer: *Analysis and Design of Analog Integrated Circuits.* John Wiley & Sons, New York, 2001.

P. E. Allen and D. R. Holberg: *CMOS Analog Circuit Design.* Oxford University Press, New York, Oxford, 2002.

Journals and Conference Proceedings

Sample & Hold

K. Poulton, J. J. Corcoran and T. Hornak: *A 1-GHz 6-bit ADC System*, IEEE Journal of Solid-state Circuits, vol. 22, pp. 962–970, 1987.

P. Vorenkamp and J. Verdaasdonk: *Fully Bipolar, 120-Msample/s 10-b Track-and-Hold Circuit*, IEEE Journal of Solid-state Circuits, vol. 27, pp. 988–992, 1992.

F. Murden and R. Gosser: *12b 50MSample/s two-stage A/D converter*, 1995 IEEE International Solid-State Circuits Conference, pp. 287–279, ISSCC. 1995.

C. Fiocchi, U. Gatti, and F. Maloberti: *A 10 b 250 MHz BiCMOS track and hold*, IEEE International Solid-State Circuits Conference, pp. 144–145, ISSCC. 1997.

C. Moreland, F. Murden, M. Elliot, J. Young, M. Hensley, and R. Stop: *A 14-bit 100-Msample/s Subranging ADC*, IEEE Journal of Solid-state Circuits, vol. 35, pp. 1791–1798, 2000.

P. J. Lim and B. A. Wooley: *A High-Speed Sample-and-Hold Technique Using a Miller Hold Capacitance*, IEEE Journal of Solid-state Circuits, vol. 26, pp. 643–651, 1991.

U. Gatti, F. Maloberti, and G. Palmisano: *An accurate CMOS sample-and-hold circuit*, IEEE Journal of Solid-State Circuits, vol. 28, 120–122, 1992.

CMOS Switches and Capacitors

B. J. Siew and C. Hu: *Switched Induced Error Voltage on a Swithced Capacitor*, IEEE Journal of Solid-state Circuits, vol. 19, pp. 519–525, 1984.

W. B. Wilson, et al.: *Measurement and Modeling of Charge Feedthrough in N-Channel MOS AnalogSwitches*, IEEE J. Solid-StateCircuits, vol. SC-20, pp. 1206–1213, 1985.

G. Wegmann, E. A. Vittoz, and F. Rahali: *Charge Injection in Analog MOS Switches*, IEEE Journal of Solid-State Circuits, vol. 22, 1091–1097, 1987.

C. Eichenberger and W. Guggernbuhl: *Charge Injection in Analog CMOS Switches*, IEE Proceedings-G , vol. 138, pp. 155–159, 1991.

F. Maloberti, F. Francesconi, P. Malcovati, and O. J. A. P. Nys: *Design Considerations on Low-Voltage Low-Power Data Converters*, IEEE Transactions on Circuits and Systems-I: Fundamental Theory and Applications, vol. 42, no. 11, pp. 853–863, 1995.

A. Baschirotto and R. Castello: *A 1-V 1.8-MHz CMOS switched-opamp SC filter with rail-to-rail output swing*, IEEE J. Solid-State Circuits, vol. 32, no. 12, pp. 1979–1986, 1997.

J. R. Naylor and M. A. Shill: *Bootstrapped FET Sampling Switch*, United States Patent 5,172,019, December, 15, 1992.

A. M. Abo and P. R. Gray: *A 1.5-V, 10-bit, 14.3-MS/s CMOS pipeline analog-to-digital converter*, IEEE J. Solid State Circuits, vol. 34, pp. 599–606, 1999.

D. Aksin, M. Al-Shyoukh, and F. Maloberti: *Switch Bootstrapping for Precise Sampling Beyond Supply Voltage*, IEEE J. Solid State Circuits, vol. 41, pp. 1938–1943, 2006.

Folding and V-I Converter

B. D. Smith: *An unusual electronic analog-digital conversion method*, IRE Transaction on Instrumentations, vol. 5, pp. 155–160, 1956.

A. Abel and K. Kurtz: *Fast ADC*, IEEE Transaction Nuclear Sciences, vol. NS-22, pp. 446–451, 1975.

M. P. Flynn and B. Sheahan: *A 400-Msample/s, 6-b CMOS folding and interpolating ADC*, IEEE Journal of Solid-State Circuits, vol. 33, pp. 1932–1938, 1998.

B. Fotouhi: *All-MOS voltage-to-current converter*, IEEE Journal of Solid-State Circuits, vol. 36, pp. 147–151, 2001.

R. R. Torrance, T. R. Viswanathan, and J. V. Hanson: *CMOS voltage to current transducers*, IEEE Transaction on Circuits and Systems, vol. CAS-32, pp. 1097–1104, 1985.

J. J. F. Rijns: *CMOS low distortion high-frequency variable-gain amplifier*, IEEE J. Solid-State Circuits, vol. 31, pp. 1029–1035, 1996.

Chapter 6

OVERSAMPLING AND LOW ORDER ΣΔ MODULATORS

Oversampling converters, initially used for audio-band and high-resolution applications, are now widely used in systems requiring video-band and medium-resolution. The technique, as we shall study, benefits from both noise-shaping and oversampling to give an optimum trade-off between speed and resolution. This chapter recalls the basic principles of the oversampling method and discusses first and second order architectures providing the basis for the study, made in the next chapter, of high order sigma-delta architectures, continuous-time solutions and sigma-delta DAC.

6.1 INTRODUCTION

The key advantage of oversampling is that the signal band occupies a small fraction of the Nyquist interval making it possible to use digital cancellation on the relatively large fraction of the quantization noise that is outside the band of interest. The use an ideal digital filter after the A/D conversion removes the noise from f_B to $f_s/2$ and significantly reduces the quantization noise power by a factor of $f_s/(2f_B)$ leading to

$$V_{n,B}^2 = \frac{\Delta^2}{12} \cdot \frac{2f_B}{f_s} = \frac{V_{ref}^2}{12 \cdot 2^{2n}} \cdot \frac{1}{OSR} \tag{6.1}$$

where V_{ref} is the reference voltage and n is the number of bits of the quantizer.

The definition of the equivalent number of bits shows that an oversampling by OSR potentially improves the number of bits from n to

$$ENOB = n + 0.5 \cdot log_2 \cdot (OSR) \tag{6.2}$$

showing that every increase of the OSR by a factor four potentially improves the converter resolution by 1-*bit*. The advantage is not so important as, for example, for gaining 5-*bit* it is necessary to use $OSR = 1024$. Nevertheless, when oversampling is used to relax the anti-aliasing specifications, the additional benefit of obtaining extra bits is obviously positive.

The oversampling, whose benefits can be enhanced by noise shaping, is only appropriate in the analog world; in the digital domain, oversampling is not a suitable feature as storing and transmitting digital data is effective when using the lowest sampling rate that is compatible with the signal band. Moreover, since the power consumed by digital circuits is proportional to the clock frequency, using large oversampling also means wasting power. Because of the above drawbacks the sampling-rate of digital over-sampled signals is normally reduced by the use of a suitable decimation filter.

Decimation by k means using one sample out of k and corresponds to a down-sampling that, as already studied, gives rise to the aliasing of spectral components from the high frequency region of the original Nyquist interval down to the reduced by k band-base. However, since obtaining the $ENOB$ increase requires the removal of the quantization noise from high frequency regions, then the same filter that secures this OSR benefit can also effectively operate as the anti-aliasing filter required for the decimation.

Fig. 6.1 shows the processing chain and the spectra obtained after each step: spectrum #1 represents the sampled-data analog signal whose upper limit is much lower than f_N; spectrum #2 shows that the quantization noise (which is spread over the entire Nyquist interval) is mostly outside f_B; spectrum #3

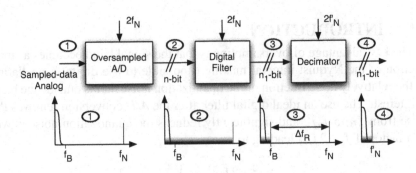

Figure 6.1. Out-of-band noise rejection and decimation of an over-sampled signal.

shows the effect of a digital filter that improves the *SNR* by rejecting the Δf_R fraction. Note that this filter occurs before the decimator and must run at the same frequency as the *ADC* ($f_s = 2f_N$). Spectrum #4, after decimation, has the same noise spectrum level, but occupies a smaller Nyquist interval.

6.1.1 Delta and Sigma-Delta Modulation

Historically, oversampling was not used for stretching the quantization noise over a wide frequency range, but for increasing the effectiveness of a pulse code modulation (*PCM*) transmission. The key to this scheme was to use a high sampling rate to transmit the change (*delta*) between successive samples rather than the actual sample. The block diagram of a delta modulator is shown in Fig. 6.2 (a) where the difference between the signal and its estimation is quantized by a 1-*bit ADC* or a multi-bit *ADC*, while the estimation of the input is the integration of the digital output transformed into analog by a *DAC*.

This method is called delta-modulation if 1-*bit* quantization is used, while it is referred to as differential *PCM* for multi-bit conversions. Fig. 6.2 (b) shows the input and output for 1-*bit* quantization. Observe that it is necessary to ensure that the sampling frequency and *DAC* quantization step are large enough to permit the input tracking.

Since a *dc* input signal does not produce any significant information at the delta modulators output, the circuit has a high-pass response. Moving the integrator of Fig. 6.2 (a) into the position shown in Fig. 6.3 (a) obtains an equivalent system provided an additional derivative block is positioned at the input terminal. The natural evolution of this system (of Fig. 6.3 (a)) is to remove the input derivative to obtain the scheme of Fig. 6.3 (b). The difference between this and the delta modulator is that the integrator operates on the error difference and not on the estimation of the signal, thus changing the modulator response

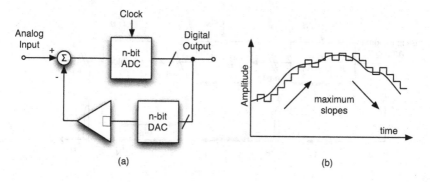

(a) (b)

Figure 6.2.   (a) Delta modulator or differential PCM. (b) Input and output with delta modulation.

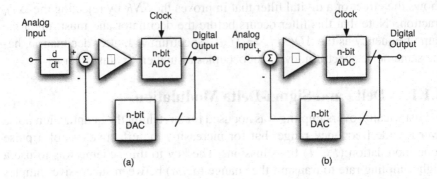

Figure 6.3. (a) Block diagram equivalent to the delta modulator. (b) Sigma-Delta modulator.

from high-pass to low-pass. Since the scheme of Fig. 6.3 (b) is the integration (*sigma*) of the difference (*delta*) the modulator is named sigma-delta ($\Sigma\Delta$) modulator. More specifically, the scheme shown is a first-order $\Sigma\Delta$ modulator since it uses only one integrator around the loop.

An important property of the $\Sigma\Delta$ method, studied in detail in the next sections, is the spectral shaping of the noise, which greatly enhances the benefits of oversampling on *A/D* converters. Therefore, the initial studies in effective *PCM* transmission gave rise to a new and now widely used category of data converters.

6.2   NOISE SHAPING

The oversampling method becomes more effective if the noise spectrum is lowered in the signal band, possibly at the expenses of an increase of the out-

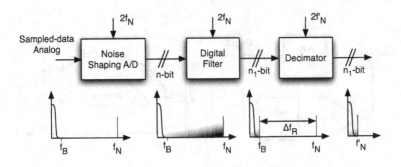

Figure 6.4.   Out-of-band noise rejection and decimation of a noise shaped signal.

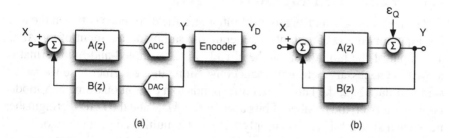

Figure 6.5. Incorporating the quantizer in a feedback loop obtains noise shaping.

of-band portion, thereby changing the white spectrum of the quantization noise into a shaped spectrum. Having more noise in high frequency regions is not problematic as the digital filter used after the *ADC* (Fig. 6.4) removes it.

Incorporating the quantizer in a feedback loop as shown in Fig. 6.5 gives rise to the desired in-band noise reduction, also called noise shaping.

The scheme has a sampled data input that, after the processing block *A(z)*, is converter into digital. For closing the loop it is necessary to generate the analog representation of the converted signal as done by the DAC. A second processing block *A(z)* is used before the subtracting element. The linear model of Fig. 6.5 (b) represents the quantization error with the additive noise ϵ_Q that is a second input of the circuit.

By inspection of the scheme it results

$$[X - Y \cdot B(z)] A(z) + \epsilon_Q = Y, \tag{6.3}$$

whose solution yields

$$Y = \frac{X \cdot A(z)}{1 + A(z)B(z)} + \frac{\epsilon_Q}{1 + A(z)B(z)}. \tag{6.4}$$

The above equation shows that signal and quantization noise pass through two different transfer functions

$$Y = X \cdot S(z) + \epsilon_Q \cdot N(z) \tag{6.5}$$

named signal transfer ($S(z)$) function and noise transfer function ($N(z)$) respectively. For a low pass data converter and for securing a beneficial noise shaping, $S(z)$ should be low pass and $N(z)$ high pass.

Since often the processing block $B(z)$ is not used ($B(z) = 1$), $A(z)$ must be integrator-type for obtaining the desired responses.

6.3   FIRST ORDER MODULATOR

The block diagram of Fig. 6.3 (b) integrates the difference between the analog input and the *DAC* output to generate the sampled-data input of the *ADC*. The input of the modulator can be already in the sampled-and-hold format or a *S&H* is necessary before the data conversion. In the former case we have a sampled-data ΣΔ, the latter case corresponds to a continuous-time ΣΔ modulator. Moreover, the number of bits used by the *ADC* and the *DAC* distinguishes between a single-bit ΣΔ (or simply ΣΔ) and a multi-bit ΣΔ modulator.

Fig. 6.6 (a) shows the block diagram of sampled-data ΣΔ that uses the transfer function

$$H(z) = \frac{z^{-1}}{1 - z^{-1}} \tag{6.6}$$

to realize the analog sampled-data integration.

Since the quantization associated with the n-bit *ADC* is equivalent to the addition of a quantization error, ϵ_Q, and the *DAC* just converts the digital output into a quantized analog signal, then the diagram of Fig. 6.6 (a) can be modeled with the analog scheme of Fig. 6.6 (b) which is a linear sampled-data circuit with two inputs, X and ϵ_Q, and one output, Y. The encoder provides a digital representation of the quantized variable Y.

The equation describing the circuit is

$$Y(z) = \{X(z) - Y(z)\}\frac{z^{-1}}{1 - z^{-1}} + \epsilon_Q(z) \tag{6.7}$$

leading to

$$Y(z) = X(z) \cdot z^{-1} + \epsilon_Q(z)(1 - z^{-1}) \tag{6.8}$$

Notice that the signal is just delayed by one clock period while the noise is passed through $(1 - z^{-1})$. This shows that the signal and the quantization noise are processed differently by the modulator. To generalize, we can state that signal passes through the signal transfer function *STF(z)* and the quantization noise through the noise transfer function *NTF(z)*

$$Y(z) = X \cdot STF(z) + \epsilon_Q(z) \cdot NTF(z). \tag{6.9}$$

The noise transfer function of the first order ΣΔ is high-pass as is evident by its estimation on the unity circle: $z \rightarrow e^{j\omega T}$

$$NTF(\omega) = 1 - e^{-j\omega T} = 2je^{-j\omega T/2}\frac{e^{j\omega T/2} - e^{-j\omega T/2}}{2j}$$

$$NTF(\omega) = 2je^{-j\omega T/2}sin(\omega T/2) \tag{6.10}$$

The result shows that the white spectrum of the quantization noise is amplified by 4 but is shaped by $sin^2(\omega T/2)$ giving rise to a significant attenuation of the low-frequency components of the spectrum.

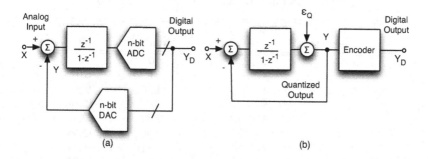

Figure 6.6. (a) Sampled-data first-order sigma delta modulator and (b) its linear model.

If a digital filter removes any noise power outside the signal band, then the resulting square noise voltage is the integral of the shaped spectrum extended from 0 to f_B

$$V_n^2 = v_{n,Q}^2 \int_0^{f_B} 4 \cdot sin^2(\pi fT)df \simeq v_{n,Q}^2 \frac{4\pi^2}{3} f_B^3 T^2 \qquad (6.11)$$

obtained using the approximation $sin(x) \simeq x$, valid for $\omega_B T/2 << \pi/2$.

Since $V_{n,Q}^2 = v_{n,Q}^2 f_s/2$ and $T = 1/f_s$, then equation (6.11) can be rewritten as

$$V_n^2 = V_{n,Q}^2 \frac{\pi^2}{3} \left[\frac{f_B}{f_s/2}\right]^3 = V_{n,Q}^2 \frac{\pi^2}{3} \cdot OSR^{-3}. \qquad (6.12)$$

If the *ADC* uses k thresholds, the *DAC* generates $k+1$ levels in the reference range $0 - V_{ref}$

$$V_{DAC}(i) = i\frac{V_{ref}}{k}; \quad i = 0\cdots k. \qquad (6.13)$$

The cascade of *ADC* and *DAC* gives rise to a quantizer whose quantization interval is $\Delta = V_{ref}/k$, and has $n_Q = log_2(k+1)$ bits. Moreover, since the power of the quantization noise and that of a full-scale sine wave are respectively

$$V_{n,Q}^2 = \frac{V_{ref}^2}{k^2 \cdot 12}; \quad V_{sine}^2 = \frac{V_{ref}^2}{8}, \qquad (6.14)$$

then the maximum *SNR* of the first order $\Sigma\Delta$ modulator is given by

$$SNR_{\Sigma\Delta,1} = \frac{12}{8} \cdot k^2 \cdot \frac{3}{\pi^2} \cdot OSR^3. \qquad (6.15)$$

Assuming $n' = log_2 k$, equation (6.15), in *dB*, becomes

$$SNR_{\Sigma\Delta,1}|_{dB} = 6.02 \cdot n' + 1.78 - 5.17 + 9.03 \cdot log_2(OSR) \qquad (6.16)$$

Table 6.1 - SNR improvement with Multi-level Quantizers

ADC Thresholds	DAC Levels	n_Q	n' extra bits	ΔSNR [dB]
1	2	1	0	0
2	3		1	6.02
3	4	2	1.58	9.54
4	5		2	12.04
5	6		2.32	13.97
6	7		2.58	15.56
7	8	3	2.81	16.84
8	9		3	18.03
15	16	4	3.91	23.52

where the first term accounts for the *SNR* improvement due to a multi-level quantizer, and 5.17 *dB* (equal to $\pi^2/3$ in *dB*), is a fixed cost required to secure an *SNR* improvement of 9.03 *dB* for every doubling of the oversampling ratio. Accordingly, every doubling of the sampling frequency improves the *ENOB* by 1.5-*bit*.

Table 6.1 summarizes the improvement of the *SNR* when using a multi-level quantizer. Notice that with more than 6-8 thresholds the number of gained bits n' almost equals n_Q.

Since the *ADC* output is binary, the number of bits sent to the digital filter is the rounding of $log_2(k+1)+1$; therefore, using a 4-threshold quantizer (which gains 2-*bit*) requires 3-*bit* processing at the input stage of the digital filter.

Example 6.1

Simulate the features of a first order sigma-delta modulator that uses 3-bit quantization and $V_{FS} = 1V$. Use sine wave and dc input signals. Analyze the different spectra of the output bit stream with different inputs and measure the noise level at the Nyquist frequency.

Solution

The files Ex6_1.m *and* Ex6_1_Launch *enable the study of the first order sigma-delta modulator illustrated in Fig. 6.7. The input is made of a sine wave and a* dc *signal that are weighted (or set to zero) by the parameters* Ksine *and* Kdc. *The output of the modulator is sent to the workspace where a suitable number of samples are used to calculate the FFT. The function* plot_spectrum *uses windowing for the FFT calculation but normalizes the result to the power of a sine wave whose peak amplitude is 1 V.*

The simulations with an input sine wave show the expected noise shaping with a zero at dc. *This is shown in Fig. 6.8 which gives*

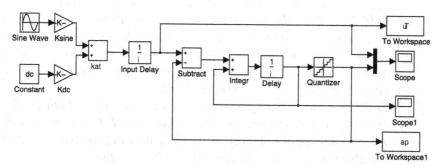

Figure 6.7. Behavioral scheme of the first order sigma-delta modulator.

the output spectrum for a 0.85 V input sine wave corresponding to a signal power of 0.36 V^2 (−1.4 dB).
The quantization noise of the 3-bit quantizer with $\Delta=1/8$ V gives rise to a $\Delta^2/12=0.0013$ V^2 power that, for an fft of a 16384 point

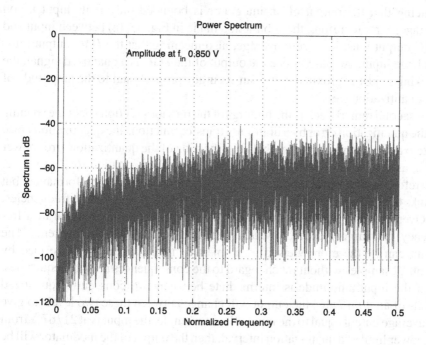

Figure 6.8. FFT of a 16384 sequence of the output bit-stream made by 131 sine waves periods.

sequence, results in a noise power equal to $1.6 \cdot 10^{-7} V^2$ in each of the 8196 bins. Since the NTF at Nyquist corresponds to an amplification by 2 (4 in power) then the expected power in the bins around Nyquist is $6.4 \cdot 10^{-7} V^2$, 65 dB below the power of a 1 V peak sine wave (1/2 V^2). Even though the spectrum of Fig. 6.8 is very noisy the level near Nyquist ranges between –60 and –70 dB.

Simulations at various frequencies, number of bits and amplitudes obtain spectra that, occasionally, are not as well shaped as Fig. 6.8. If the number of bits diminishes the noise floor augments as expected but also tones become apparent above the noise floor.

The use of the dc signal shows that the spectrum is only well shaped for some input amplitudes. For a number of critical values the spectrum is made by big tones with some shaping in between.

6.3.1   Intuitive Views

With simple remarks it is possible to intuitively understand some of the important characteristics of the $\Sigma\Delta$ modulator. A first feature derives from noticing that the output of an integrator is bounded only if its input is, on average, zero; therefore, the subtraction made in Fig. 6.6 (a) between input and *DAC* output must have zero average, also meaning that the *DAC* output must track the input. Actually, since the output of the *DAC* is a quantized signal, the tracking is just approximate with an accuracy proportional to the amplitude of the quantization step.

A second remark recalls the findings of the previous section about the shaping of the quantization error by a high-pass transfer function: the quantization noise is zero at *dc*, but results in a global amplification of the quantization error power by 2, as can be verified by extending the integral of equation (6.22) to $f_s/2$. Therefore, a first order modulator degrades the global noise performance, but thanks to the shaping, pushes a significant part of the power to high frequencies.

Oversampling improves the *ENOB* by sampling the input signal at a frequency that largely exceeds what is required by the Nyquist theorem. The result can be therefore viewed as a smart dynamic averaging performed by plentiful samples without much regard to the consequences at high frequencies.

If the input amplitude is intermediate between two consecutive quantized levels of an *n-bit DAC*, the output will change between these two levels to give an average output equal to the input. If, for example, the input is at $21/67\Delta$ from the lower limit of a quantization interval, then the output of the modulator will be in average 21 times the higher code out of 67 clock periods. The pattern does not repeat itself exactly as the signal changes during the conversion; nevertheless, the result can be viewed as an interpolation between the lower and the upper level of the relevant quantization interval. Therefore, the modulator virtually

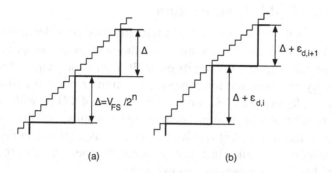

Figure 6.9. (a) Staircase representing the increase of the ENOB. (b) Same case but with a DNL affecting the DAC.

adds extra steps in the staircase representing the static input-output transfer characteristics, as shown in Fig. 6.9 (a).

Assume now that non linearity errors affect the *DAC* causing variations in the amplitudes of the contiguous large steps. Following the previous intuitive view, the interpolating staircase of Fig. 6.9 (a) obtained by oversampling will join consecutive big steps of the *DAC* as shown by Fig. 6.9 (b). The result is that the resolution increases but the linearity does not improve. Moreover, since increasing the oversampling ratio results in smaller quantization steps, then the linearity of the *n-bit DAC* should be increased if an *INL* of less than 1 *LSB* is to be maintained.

The last remark concerns the linearity and noise specifications of the *ADC* and the *DAC*. The digital signal generated by the *ADC* is such that any limit affecting it, namely the error and the noise on the threshold voltages, are relaxed by the feedback loop: indeed, the *ADC* error must be referred to the input of the integrator and then referred to the input of the modulator. The result of these two operations leads to a division of the error by the transfer function of the integrator. Since at low frequency (which is the region of interest) the integrator has a very large gain (infinite at zero frequency), the error is greatly attenuated in the signal band. The same benefit does not apply to the *DAC* as its error is injected directly at the input of the modulator together with the input.

> **Warning!**
>
> The feedback of a $\Sigma\Delta$ modulator does not relax the *DAC* linearity. Remind that the method greatly reduces the number of DAC levels but not their accuracy requirement!

6.3.2   Use of 1-bit Quantization

The previous sub-section showed just with intuitive considerations that the most critical block of the $\Sigma\Delta$ modulator is the *DAC* as the shaping and the feedback does not help in reducing its error. It is therefore required to ensure a *DAC* linearity good enough for complying with the requested *ENOB*. Since $\Sigma\Delta$ converters often target 14-*bit* or more the design of a *DAC* with such high linearity is problematic and often becomes the bottleneck of the modulator design. Special techniques that resolve the linearity issue thus enabling multi-bit $\Sigma\Delta$ modulators are studied in a successive chapter; nevertheless, the request of high *DAC* linearity is a serious design problem.

A simple way to overcome the linearity requirement ensues from the following basic observation: a line that connects many points is typically a broken line but if only two points are used then the interconnecting line is certainly a straight line. Therefore, if the input-output characteristics of the *DAC* are made by only two voltages the linearity problem does not exist. Having a 2-*level DAC* means using a single bit *ADC* (such as a comparator) and two voltage generators, 0 and V_{ref} or $-V_{ref}$ and $+V_{ref}$, for the *DAC*.

Although this solution answers well to the frequent requests of designing data converters with inaccurate technologies, it is problematic for two reasons. The first is that the quantization step is as large as the entire dynamic range. Therefore, since the converter relies only on noise shaping for obtaining its resolution, the oversampling ratio must be very high. The second is that the condition that foresees a large number of quantization steps for justifying the modeling of the quantization error with a white noise is not verified. It may occur that the power of the quantization error is concentrated at some frequency, giving rise to tones which can fall in the signal band.

Fig. 6.10 (a) shows the circuit diagram of a 1-*bit* sigma-delta modulator. The integrator is a switched capacitor structure which samples the input with the capacitance C_1 during Φ_1, and then during Φ_2, C_1 injects a charge proportional to the difference between input voltage and *DAC* output. The *ADC* is just a single comparator whose output is conventionally assumed to be ± 1. The *DAC* is made by two switches that connect the output to $+V_{ref}$ or $-V_{ref}$ depending on the comparator control. The feedback capacitance C_2 is nominally equal to C_1 to obtain the required unity gain of the *SC* integrator.

Fig. 6.10 (b) shows the ± 1 $\Sigma\Delta$ output bit stream and its generating input sine wave with amplitude equal to 0.634 *V*. Observe that the output is mainly $+1$ when the input is close to its maximum and mainly -1 when the input is close to its minimum; also, when the input is close to zero the number of $+1$ and -1 are almost equal. Looking at the output bit stream plot it can be noted that, albeit completely different from the input, the average of the bit stream follows the input. Moreover, the rapid switching from one level to the complementary one results in a high frequency spur as pointed out in the previous study.

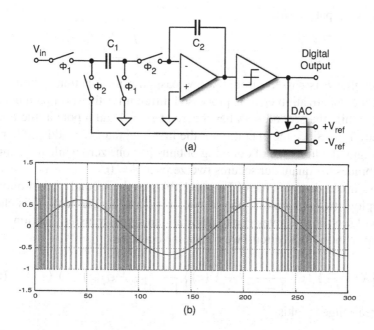

Figure 6.10. (a) First order 1-bit $\Sigma\Delta$ modulators. (b) Input and ± 1 bit-stream output.

6.4 SECOND ORDER MODULATOR

With a first order $\Sigma\Delta$ modulator the number of bits increases by 1.5 for every doubling of the sampling frequency. This result is interesting, however, in order to secure high resolutions using a 1-*bit ADC* it is necessary to use a fairly large *OSR*. Moreover, as was observed by the computer simulations of Example 6.1, the output spectrum is, in some cases, poorly shaped and has large tones that could fall in the signal band.

Better performances and features are secured by using two integrators around the loop thus forming a second-order modulator as shown in the conceptual scheme of Fig. 6.11 (a). Since the use of two integrators in a feedback loop can cause instability, it is necessary to dump one of the two integrators by using one of the two options represented by the dotted lines of Fig. 6.11 (a). One of them makes a conventional approximated integrator, while the other uses a longer path and includes the quantizer in the dumping loop. A study of the second integrator specifies the difference between these two solutions. The short loop path gives

$$R = \frac{P - R}{s\tau} \quad \rightarrow \quad Y = R + \epsilon_Q = \frac{P}{1 + s\tau} + \epsilon_Q \qquad (6.17)$$

The long loop path yields

$$\frac{P - Y}{s\tau} = Y - \epsilon_Q \quad \rightarrow \quad Y = \frac{P}{1 + s\tau} + \frac{s\tau\epsilon_Q}{1 + s\tau} \qquad (6.18)$$

Notice that P is passed through a low pass transfer function in both cases, but that the quantization error is processed differently: the short path does not alter ϵ_Q, while the second path has a zero at $s = 0$ and a pole at the angular frequency $1/\tau$. Since the operation of the first integrator will add another zero in the origin, the first type of dumping obtains just one zero while the dumping that embraces the quantizer secures two zeros at $s = 0$.

The sampled-data scheme of Fig. 6.11 (b) uses the second type of dumping and employs two integrators, one of which has a delay element. This choice, as we will see shortly, gives rise to an optimum signal transfer function.

By inspection of the circuit we obtain the equation

$$\left\{ [X(z) - Y(z)] \frac{1}{1 - z^{-1}} - Y(z) \right\} \frac{z^{-1}}{1 - z^{-1}} + \epsilon_Q(z) = Y(z) \qquad (6.19)$$

which rearranged yields

$$Y(z) = X(z) \cdot z^{-1} + \epsilon_Q(z)(1 - z^{-1})^2 \qquad (6.20)$$

showing that the signal transfer function is just a delay and the noise transfer function is $(1 - z^{-1})^2$, the square of the result obtained by a first order $\Sigma\Delta$ modulator.

Observe that the simple delay of the signal transfer function is a result of the cancellation of the terms $Y(z)(-2z^{-1} + z^{-2})$ with the multiplication of $Y(z)$ by $(1 - z^{-1})^2$. This result exceeds its requirement as a flat response within the signal band (with a gain of 1) would have been sufficient. If the integrators have a

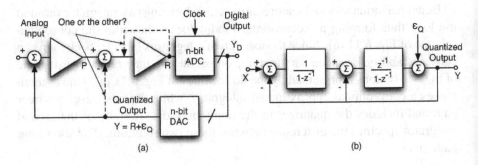

Figure 6.11.   (a) Conceptual continuous-time second-order $\Sigma\Delta$ modulator. (b) Possible sampled-data block diagram.

gain error the generated terms do not exactly cancel $Y(z)(-2z^{-1}+z^{-2})$ giving rise to a parasitic denominator in both the signal and noise transfer functions, whose effect, with small gain errors, is negligible.

The *NTF* calculated on the unity circle is given by

$$NTF(\omega) = (1 - e^{-j\omega T})^2 = -4e^{-j\omega T}\{sin(\omega T/2)\}^2 \tag{6.21}$$

Assuming again that a digital filter completely removes the noise outside the signal band, the noise power remaining in the frequency interval $0\text{-}f_B$ is

$$V_n^2 = v_{n,Q}^2 \int_0^{f_B} 16 \cdot sin^4(\pi fT)df \simeq v_{n,Q}^2 \frac{16\pi^4}{5} f_B^5 T^4 \tag{6.22}$$

which, again, uses the approximation $sin(x) \simeq x$ valid for $\omega_B T/2 << \pi/2$. Moreover, since $V_{n,Q}^2 = v_{n,Q}^2 f_s/2$ and $T = 1/f_s$, the noise power becomes

$$V_n^2 = V_{n,Q}^2 \frac{\pi^4}{5} \left[\frac{f_B}{f_s/2}\right]^5 = V_{n,Q}^2 \frac{\pi^4}{5} \cdot OSR^{-5} \tag{6.23}$$

giving rise to the *SNR* of a second order $\Sigma\Delta$ modulator

$$SNR_{\Sigma\Delta_2} = \frac{12}{8} \cdot k^2 \cdot \frac{5}{\pi^4} \cdot OSR^5 \tag{6.24}$$

which, in *dB*, is

$$SNR_{\Sigma\Delta_2}|_{dB} = 6.02n' + 1.78 - 12.9 + 15.05 \cdot log_2(OSR). \tag{6.25}$$

This shows a $12.9\ dB$ loss as a result of the $5/\pi^4$ term; nevertheless, the second order modulator secures an increase by $15\ dB$ ($2.5\text{-}bit$) for every doubling of the oversampling ratio.

6.5 CIRCUIT DESIGN ISSUES

The design of any sigma-delta modulator (including the high-order and the cascade architectures that will be studied shortly) requires knowledge of a number of design issues both at the architecture and the circuit level. This section studies the limits and the solutions related to the use of real basic blocks. The next section will discuss issues related to the architectural design.

The most important limits of the basic blocks used are:

- Offset of the op-amp (or *OTA*).
- Finite op-amp gain.
- Finite op-amp bandwidth.
- Finite op-amp slew-rate.
- Non-ideal operation of the *ADC*.
- Non-ideal operation of the *DAC*.

6.5.1   Offset

The offset of an integrator's op-amp can be described as an input referred voltage generator. The offset of the first integrator is added to the input signal and gives rise to an equal offset at the digital output.

The offset of the second integrator is referred to the input of the modulator by dividing it by the transfer function of the first block, which for a sampled data scheme is $(1 - z^{-1})/k$ (k is the gain of the integrator and a possible delay is disregarded). Therefore, since the offset is a *dc* signal the high-pass transfer function cancels out the effect.

The offset of the *DAC* is added to the input and causes, similar to the offset of the first integrator, an offset in the digital output. In contrast, the offset of the *ADC* is referred to the input by dividing it by the transfer function of one or more integrators and does not limit the *dc* operation of the modulator. This feature is interesting as it provides the flexibility of positioning the *ADC* thresholds around a more convenient voltage level.

6.5.2   Finite Op-Amp Gain

The finite gain of an op-amp reduces the *dc* response of the integrator from infinite to the value of the op-amp gain. The scheme of Fig. 6.12 describes a typical implementation of the integrators used in modulators. Its analysis leads to

$$C_2 V_{out}(nT + T)\left(1 + \frac{1}{A_0}\right) = C_2 V_{out}(nT)\left(1 + \frac{1}{A_0}\right) +$$

$$+C_1\left[V_1(nT) - V_2(nT + T) - \frac{V_{out}(nT)}{A_0}\right] \qquad (6.26)$$

which, in the z-domain, becomes

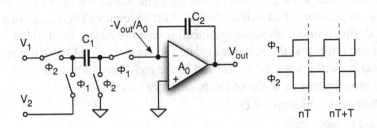

Figure 6.12.   Switched capacitor integrator that uses an op-amp with finite gain.

$$V_{out}(z-1)\left(1+\frac{1}{A_0}\right) = \frac{C_1}{C_2}\left[V_1 - zV_2 - \frac{zV_{out}}{A_0}\right] \quad (6.27)$$

yielding, after rearrangement

$$\frac{V_{out}}{V_1 - z^{-1}V_2} = \frac{C_1}{C_2}\left[\frac{A_0}{A_0 + 1 + C_1/C_2}\right]\frac{z^{-1}}{1 - \frac{(1+A_0)C_2}{C_1+C_2+A_0C_2}z^{-1}} \quad (6.28)$$

which shows a gain error equal to $A_0/(1 + A_0)$ and a shift of the pole from $z = 1$ to a point inside the unity circle $z_p = (1 + A_0)/(1 + A_0 + C_1/C_2)$.

The effect of the gain error only slightly affects the signal transfer function, however the shift of the pole causes an equal shift of the *NTF* zeros and, for a second order modulator, leads to

$$NTF \simeq (1 - z_{p1}\cdot z^{-1})(1 - z_{p2}\cdot z^{-1}) \quad (6.29)$$

where z_{p1} and z_{p2} represent the shift caused by the first and the second integrator.

At *dc* $(z = 1)$ the *NTF* is not zero but equal to $(1 - z_{p1})(1 - z_{p2})$. If the two gains A_0 are equal and the capacitances used are also equal the *NTF* becomes

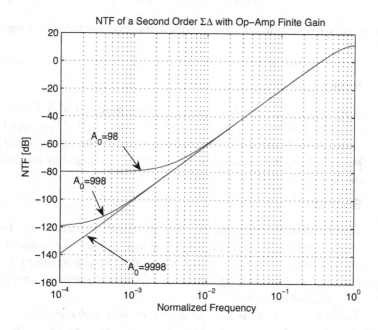

Figure 6.13. NTF of a second order modulator with three different finite gains of the op-amps.

$$NTF = \left(1 - z^{-1}\frac{A_0 + 1}{A_0 + 2}\right)^2 \tag{6.30}$$

The result, illustrated in Fig. 6.13 for three different values of the *dc* gain, shows that the finite gain only affects the shaping below a corner frequency f_c. If the signal band is larger than f_c then the noise power within the signal band is only minimally dependent on the flat region as the frequencies near the bandwidth limit dominate the calculation of the spectrum integral.

The corner frequency is such that

$$e^{s_p T} = \frac{A_0 + 1}{A_0 + 2} \tag{6.31}$$

which is on the real axis of the *s*-plane at

$$f_c = \frac{f_s}{2\pi}ln\left\{1 - \frac{1}{A_0 + 2}\right\} \simeq \frac{f_s}{2\pi(A_0 + 2)} \tag{6.32}$$

The finite gain does not affect the SNR if $f_B >> f_c$; therefore both gain and oversampling must be set to satisfy the condition

$$\pi(A_0 + 2) >> OSR \tag{6.33}$$

which is a very relaxed op-amp gain request for modulators with a medium *OSR*.

Example 6.2

A second-order 1-bit $\Sigma\Delta$ modulator uses op-amps with 40 dB gain. Plot the SNR as a function of the OSR with a $-10\,dB_{FS}$ input sine wave. Assume that the output bit stream passes through an ideal low-pass digital filter capable to reject all the noise out of the band of interest.

Solution

The Simulink description Ex6_2.mdl and the file Ex6_2Launch.m model the $\Sigma\Delta$ modulator. It uses a first integrator without delay and a second one with delay. The finite gain that determines the integrator response (6.28) gives rises to a gain error and a delay error accounted for by two suitable functions.

The launcher runs the simulation twice; once with a gain of 100 and the other with a gain of 100 k and plots the two spectra on the same diagram (Fig. 6.14). The Nyquist frequency is 1 MHz and the corner frequency obtained is about 3 kHz in excellent accordance with (6.32).

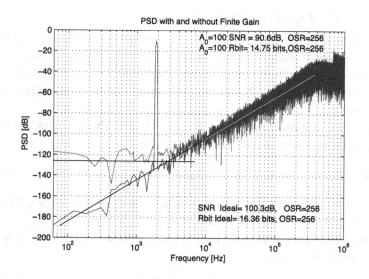

Figure 6.14. Output spectra with ideal and finite gain ($A_0 = 100$) op-amp.

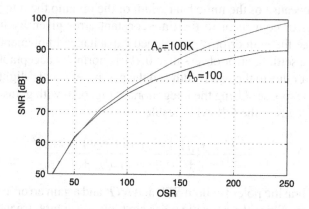

Figure 6.15. SNR at different OSR and op-amp gains.

The condition of (6.33), verified by estimating the SNR at various OSR, provides the result shown in Fig. 6.13. Notice that equation (6.33) requires an OSR << 320 and with OSR = 50 the SNR with 40 dB gain does not differ from the ideal case. In contrast, at OSR = 250 the SNR drops by about 10 dB.

6.5.3   Finite Op-Amp Bandwidth

It is well known that the poles of the transfer function determine the bandwidth and the phase margin of the op-amp. The first dominant pole should produce a 20 *dB/dec* roll-off until the gain is lower than 0 *dB*. The other poles, named non-dominant poles, should occur after the 0 *dB* crossing, and affect the phase margin.

If the effect of the non-dominant poles is negligible, the step response of the integrator is an exponential that, for large finite gain A_0, is

$$V_{out}(t + nT) = V_{out}(nT) + V_{step}U(nT)(1 - e^{-t\beta/\tau_d}) \qquad (6.34)$$

where β is the feedback factor. For a switched capacitor integrator with input element C_1, and C_2 in feedback, then $\beta = C_2/(C_1 + C_2)$.

Since the integration phase only lasts $T_s/2$, the output does not reach its final value which causes an error of

$$\epsilon_b = V_{step}e^{-T_s\beta/(2\tau_d)} \qquad (6.35)$$

that is proportional to the input signal.

Therefore, because of the finite bandwidth of the op-amp the integrator displays a gain error equivalent to the time-constant error given by its passive elements. Since the gain error of a switched capacitance implementation is a fraction of %, a settling error of less than 0.1% is normally acceptable.

The gain error due to the bandwidth and the passive elements slightly affects the modulator response. Using the integrators of Fig. 6.16 with gains $(1 - \epsilon_{b,1})$ and $(1 - \epsilon_{b,2})$ leads to the following result

$$V_{out}(z) = \frac{V_{in}z^{-1}(1 - \epsilon_{b,1})(1 - \epsilon_{b,2}) + \epsilon_Q(1 - z^{-1})^2}{1 - z^{-1}(\epsilon_{b,1} + 2\epsilon_{b,2} - \epsilon_{b,1}\epsilon_{b,2}) + z^{-2}\epsilon_{b,2}} \qquad (6.36)$$

which shows parasitic poles for both *STF* and *NTF* and a gain error in the signal transfer function. Since the parasitic poles are typically at high frequency, they do not alter the operation in the signal band.

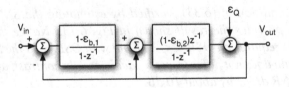

Figure 6.16.   Block diagram of a second order modulator using op-amps with finite bandwidth.

6.5.4 Finite Op-Amp Slew-Rate

Finite slew-rate combined with a finite bandwidth can be a significant limit. The charge injected by an *SC* structure gives rise to a slewing period that, obviously, must be smaller than $T_s/2$; and is followed by an exponential settling for the remaining time that can be incomplete.

Since the effect of finite slew-rate and bandwidth of an op-amp was studied in Chapter 3 for an *SC* amplifier, it is not necessary to repeat the analysis here. Just recall that the set of equations describing the transient is non-linear, and that a longer slewing period reduces the time left for exponential settling. Therefore, since the error becomes a non-linear function of the input, then its estimation is not possible in the z domain but must be calculated using time domain simulations.

The use of the model described in Chapter 3 enables the calculation of the error at the end of the injection period without doing a full transient analysis. Indeed, an input signal equal to $-V_{in}$ would result in an ideal output step equal to $\Delta V_{out} = V_{in}C_1/C_2$. In contrast, a real op-amp has a slewing time of

$$t_{slew} = \frac{\Delta V_{out}}{SR} - \tau. \tag{6.37}$$

From t_{slew} the output voltage, differs to the final value by

$$\Delta V = SR \cdot \tau; \tag{6.38}$$

and evolves exponentially in the remaining fraction of the injection time-slot, $T/2 - t_{slew}$. The output voltage error at $T/2$ equals

$$\epsilon_{SR} = \Delta V e^{-(T/2 - t_{slew})/\tau} \tag{6.39}$$

The use of the above set of equations in a behavioral simulator or described with a behavioral language helps in speeding up the study of combined finite bandwidth and limited slew-rate.

Since the error must be added to the integrator output, it is shaped by the noise transfer function from the injection point to the output. For this reason the error in a second order $\Sigma\Delta$ modulator the second integrator is less critical than the first one. This feature can be used in defining the speed specification of op-amps for optimizing the power consumption.

Example 6.3

Determine by computer simulations the minimum slew-rate required for the op-amps used in a single-bit second-order $\Sigma\Delta$ modulator that uses the cascade of two delayed integrators with gains 1/2 and 2. Study the combined effect of slew-rate and finite bandwidth (use an *equivalent bandwidth of 100 MHz) on the output spectrum. Use ± 1 V reference voltages, $f_s = 50\,MHz$ and $V_{in} = -6\,dB_{FS}$.*

Solution

The integrators used in the Matlab file Ex6_3 are described by a behavioral model based on the equations (6.37), (6.38) and (6.39). The model, in addition to the slew-rate and the equivalent bandwidth ($B_{EQ} = \beta f_T$), also uses the parameter $\alpha = 1 - 1/A_0$ and the hard saturation of the output voltage.

A preliminary simulation with ideal parameters determines the amplitude of the input signals which lead to maximum output changes of first and second integrator equal to 0.749 V and 3.21 V respectively. If the bandwidth is very large, the exponential settling is negligible; therefore, the slew-rates must be at least

$$SR_1 > \frac{\Delta V_{out,1}}{T/2} = 74.9\,V/\mu s; \ SR_2 > \frac{\Delta V_{out,2}}{T/2} = 321\,V/\mu s$$

The ideal case gives an SNR = 72 dB with OSR = 64 and $f_{in} = 140$ kHz. The use of $SR_2 = 325\,V/\mu s$ and $SR_1 = 78\,V/\mu s$ does not change the SNR significantly. The use of $SR_1 = 74\,V/\mu s$ also has little affect on the SNR, but the reduced saturation gives rise to third and fifth harmonics distortion tones as shown in Fig. 6.17.

The combined limits due to bandwidth and slew-rate obviously require some margin with respect to the calculated minimum slew rates. If the second integrator is ideal a first integrator with $SR = 150\,V/\mu s$

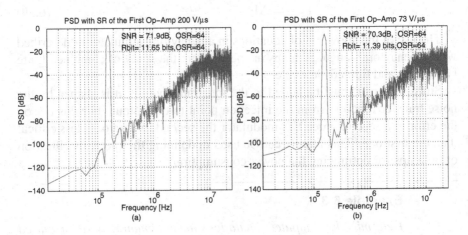

Figure 6.17. (a) PSD with very large bandwidths, $SR_2 = 325\,V/\mu s$ and $SR_1 = 78\,V/\mu s$. (b) Same as (a) but $SR_1 = 74\,V/\mu s$.

and $B_{EQ} = 100\,MHz$ causes a third order tone at $-80\ dB_{FS}$. The tone goes down to $-111\ dB_{FS}$ if both parameters are doubled.

With simulations it is possible to verify that, thanks to better shaping, the performance degradation of the second integrator has less influence on the SNR and distortion. A first ideal integrator and a second integrator with $SR = 250\,V/\mu s$ and $B_{EQ} = 100\ MHz$ diminishes the SNR by few dB, but no tones appear on the spectrum. This verifies that the limit due to the second integrator is less critical than the first one.

The reader can use the file Ex6_6 controlled by Ex6_6Launch to perform a more detailed study of these limits.

6.5.5   ADC Non-ideal Operation

The static and dynamic limitations of a real *ADC* degrade the modulator performances. However, since the signal at the output of the *ADC* is a digital representation of its input plus both quantization error and *ADC* error

$$ADC_{out} = V_{in,ADC} + \epsilon_Q + \epsilon_{ADC}, \tag{6.40}$$

then the shaping of the $\Sigma\Delta$ modulator acts on the addition of ϵ_Q and ϵ_{ADC}. Therefore, if $\epsilon_{ADC} < \epsilon_Q$ the *ADC* limit does not hamper the circuit performance.

This condition, which requires a *DNL* and *INL* of less than 1 *LSB*, is easily verified since the number of thresholds is small and the dynamic range is large: having an *LSB* of many tens of mV makes it easy to design *ADC*'s up to hundreds of *MHz*.

6.5.6   DAC Non-ideal Operation

It was already observed that the *DAC* errors are not shaped by the *NTF*. Instead they are added to the input of the modulator and transferred to the output through the *STF*. The stringent linearity requirements encourage the use of a 1-*bit* quantizer, or make careful design necessary to ensure the *INL* meets the overall resolution requirements.

The *DAC*s used in sampled-data $\Sigma\Delta$ modulators are switched capacitor schemes and possibly include the input signal injection. In some cases the capacitance is divided into parts to realize a multi-level *DAC*. This scheme, and the methods used for digital error correction are discussed shortly, and again in another chapter. Here, we will only consider the limit associated with the *kT/C* noise.

As studied previously, the *kT/C* is the integral over the Nyquist interval of the white spectrum voltage across any capacitance *C* that is sampled through a switch. The data conversion of the noise and the digital filter used to remove the out of band noise rejects part of the white spectrum. The remaining fraction must satisfy the condition

$$v_{n,kT/C}^2 = \frac{kT}{OSR \cdot C_{in}} < \frac{V_{ref}^2}{8 \cdot 2^{2n}} \qquad (6.41)$$

which determines the minimum usable input capacitance.

High order modulators use extra DACs to realize multiple feedback from the output to internal analog points. The additional DACs inject extra-noise and errors. However, the high-pass transfer function from an internal node to the output reduces the low frequency components of errors and, in the case of noise, shapes the spectrum. Therefore, since the linearity and noise requirements are less stringent than the ones of the input DAC the capacitances used can be scaled.

6.6   ARCHITECTURAL DESIGN ISSUES

The design of a $\Sigma\Delta$ architecture, in addition to the obvious requirement of satisfying the desired functions, also requires optimization of the voltage swing at the output of the op-amps (or OTA), proper analysis and control of the noise, and that a tone free output spectrum is ensured.

6.6.1   Integrator Dynamic Range

The voltage swing at the output of an integrator depends on signal amplitude and quantization noise. Therefore, the dynamic range of both operational amplifiers and quantizer can need to be larger than the reference to accommodate both the signal and the noise.

When the integrator output exceeds the dynamic range of the op-amp (or OTA) the signal is clipped to a saturation level resulting in a loss of feedback control. This occurs because the inverting terminal of the op-amp is no longer tightened to the non-inverting terminal, and is free to go up or down giving rise to an incomplete transfer of the input signal.

The effect is studied by using the switched capacitor model of Fig. 6.18 (a) that locks the output to the saturation voltages $\pm V_{sat}$ when they try to exceed their limits. The capacitance C_1 charged to the input voltage during Φ_2 is connected to the virtual ground during the next phase. Assuming that the voltage of the left terminal falls to zero at a constant pace, then the op-amp and the feedback loop maintain control over the virtual ground until the output voltage of the op-amp reaches one of its saturation limits. At this time the clipping generator takes control and the output voltage remains constant. If the input capacitance is still charged to Q_{res}, the virtual ground starts moving and when the left terminal of C_1 reaches zero the fraction $Q_{res}C_2/(C_1+C_2)$ is transferred into C_2 leaving a fraction $Q_{res}C_1/(C_1 + C_2)$ in the input capacitance. The input referred voltage error becomes

$$\epsilon_s = \frac{Q_{clip}}{C_1 + C_2} \tag{6.42}$$

Since the error depends on both the closeness of the output to one of the saturation limits before every injection, and on the amplitude and the sign of the input signal, noisy inputs and noisy outputs make the saturation error almost unpredictable. Therefore, the error can be modeled by an input referred noise source whose amplitude is proportional to the probability of saturation occuring.

Even a quantizer input that exceeds the limits is problematic: if the input of the flash *ADC* is above or below the first or last *ADC* threshold by more than $\pm\Delta/2$, then the first and the last *DAC* outputs cannot properly quantize the input. The error, similar to the one due to the op-amp saturation, is added to the quantization error. Even the quantization over-range error looks like a random variable and is modeled by a noise, $\epsilon_{s,Q}$.

Summing up, the saturations of the two op-amps and the over-range of the quantizer give rise to three noise terms which limit the performance of a second-order modulator as shown in Fig. 6.11 (b). The error caused by the first op-amp saturation, $\epsilon_{s,1}$, added to the input signal, is transferred to the output multiplied by the signal transfer function z^{-1}; $\epsilon_{s,Q}$, injected at the same input point as the quantization noise ϵ_Q, is shaped by the noise transfer function, while $\epsilon_{s,2}$, being injected at the input of the second integrator, is shaped by a first-order high-pass transfer function (a simple calculation can verify this statement). Therefore, including the quantization error, it results

$$Y = Xz^{-1} + \epsilon_{s,1}z^{-1} + \epsilon_{s,2}(1 - z^{-1}) + (\epsilon_Q + \epsilon_{s,Q})(1 - z^{-1})^2 \tag{6.43}$$

Observe that the four error sources are uncorrelated; therefore, the total noise power in the signal band is their quadratic superposition. The contributed

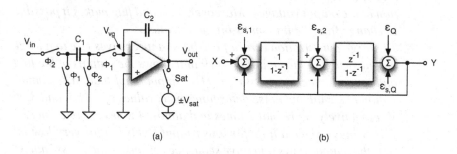

(a)   (b)

Figure 6.18. (a) Switched capacitor integrator with op-amp saturation. (b) Second-order modulator model including the saturation effects.

powers are the integrals extended from zero to f_B of the spectra passed through the square of the corresponding transfer function. The results, assuming that the spectra $\epsilon_{s,1}^2$, $\epsilon_{s,2}^2$ and $\epsilon_{s,Q}^2$ are white, is

$$V_n^2 = \frac{V_{n,1}^2}{OSR} + V_{n,2}^2 \frac{\pi^2}{3 \cdot OSR^3} + \left[V_{n,Q}^2 + \frac{\Delta^2}{12} \right] \frac{\pi^4}{5 \cdot OSR^5} \qquad (6.44)$$

where $V_{n,1}^2 = \epsilon_{s,1}^2 f_B$, $V_{n,2}^2 = \epsilon_{s,2}^2 f_B$ and $V_{n,Q}^2 = \epsilon_{s,Q}^2 f_B$.

Equation (6.44) shows that the noise sources modeling saturation have very different effects on the output. For example, with $OSR = 64$ the noise power $V_{s,2}^2$ is reduced by 64, $V_{s,1}^2$ by $79,682$, and $V_{s,Q}^2$ by as much as $5.51 \cdot 10^7$. Therefore, the most critical limit is the saturation of the first integrator followed by the saturation of the second one. The quantizer over-range only matters for errors that are comparable with Δ.

In order to avoid clipping it is therefore necessary to ensure a suitably large dynamic range of the op-amps used in the integrators. Alternatively, a given op-amp output range determines the maximum integrator swing and, in turn, establishes the maximum usable reference.

Example 6.4

Determine the voltages at the output of the first and second integrators of the architecture shown in Fig. 6.11 (b). Use a 1-bit and 3-bit quantizer with different input amplitudes and $V_{Ref} = \pm 1$ V. Estimate the resolution and determine the loss caused by the hard saturation of the op-amp outputs and the quantizer.

Solution

The files Ex6_4 and Ex6_4launch are the basis for the study of this Example. A flag enables the analysis of unconstrained or hard saturation of the output voltages. Moreover, a second flag makes it possible to change from 1-bit to multi-bit quantization.

The large quantization error of a 1-bit quantizer gives rise to a large output swing as shown in Fig. 6.19 that shows the outputs of the first and second integrators with $-6\,dB_{FS}$ input amplitude. The combination of signal and noise determines peak values of 2.18 V and 3.96 V respectively, more that 2 times and almost 4 times the reference.

The swings reduce a little for lower input levels but for very low inputs the output shows repetitive sequences of spikes and a less "noisy" behavior, denoting an inaccurate modeling of the quantization error with noise.

The SNR does not reach the maximum value predicted by equation (6.25) because large signals hamper the proper operation of the *modulator. However, with an input amplitude equal to – 10 dB$_{FS}$ and OSR = 64 the SNR is 67.6 dB, close to the value predicted by using equation (6.25): 69.2 dB.*

A – 10 dB$_{FS}$ input sine wave gives rise to a first integrator swing equal to 1.91 V, and a second integrator swing as large as 3.1 V. Fig. 6.20 shows how the saturations affect the power spectral densities. The spectra correspond to the ideal case and three saturation cases. The ideal situation gives rise to the expected 40 dB/decade increase of the noise spectrum. The clipping of the first integrator output to 1.85 V causes extra white noise whose floor at– 100 dB limits the SNR to 64.4 dB. A clipping of the second integrator output voltage to 2.5 V also causes extra noise but its spectrum is first order shaped as shown by the 20 dB/decade spectrum slope at low frequency. The limit gives rise to a negligible SNR loss (67.5 dB) and a barely visible third order tone. The spectrum with both integrators in saturation is dominated by the white term, however the noise floor increases and the SNR drops to 60.2 dB.

The file Ex6_4launch also measures the power of the total quantization error. The result using ideal integrators and a –10 dB$_{FS}$ input is 39% larger than the expected $\Delta^2/12$. This extra noise is due to the

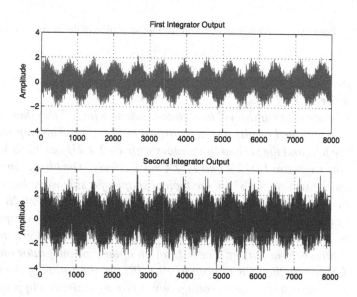

Figure 6.19. Outputs of the first and the second integrators of a $\Sigma\Delta$ modulator with a – 6 dB$_{FS}$ input amplitude.

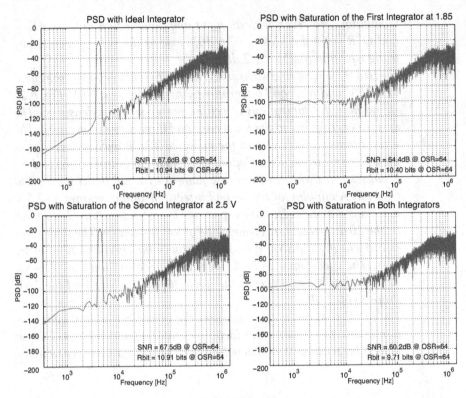

Figure 6.20. Power spectral densities with and without saturation of the intergrator's output.

over-range of the quantizer: apparently its effect is to increase the noise by 1.32 dB.

A study of a multi-level modulator is done with a 7 thresholds ADC. The obtained results are summarized in Fig. 6.21. The amplitudes of the first and the second integrators with a $-2.4\,\mathrm{dB}_{FS}$ (0.758 V) input sine wave are 1.037 V and 1.17 V respectively. The histograms of the output waveforms show the typical shape of a sine wave but extends beyond 0.758 V because of the quantization noise. The 94 dB SNR is in good accordance with the predicted value: 93.6 dB corresponding to 76.8 dB for 1-level at $-2.4\,\mathrm{dB}_{FS}$ input plus 16.84 dB granted by a 7-threshold quantizer. The saturation of the first integrator output to 1 V causes the SNR to drop to 79.1 dB; the saturation of the second integrator to the same voltage gives rise to a higher clipping error but, thanks to the first order shaping, the SNR is 79.4 dB. The clipping of both outputs results in a reduction of the SNR to 77.3 dB. Moreover,

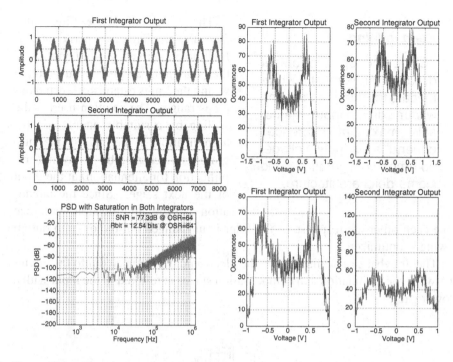

Figure 6.21. Simuation results with a 7-levels quantizer and $-2.4\ dB_{FS}$ input sine wave.

the spectrum shown in Fig. 6.21 reveals harmonic tones with about −100 dB amplitude.

The histograms of Fig. 6.21 compare the output occurrences without (top diagrams) and with saturation (bottom diagram). The effect of saturation is to spread out the occurrence distributions to compensate for the reduced output ranges. The effect also reduces the correlation between the output and the input sine wave.

6.6.2   Dynamic Ranges Optimization

We have seen that using a suitable dynamic range in the op-amps (or *OTA*s) is essential for preserving the *SNR* and avoiding harmonic distortion. The most critical is the first integrator, as its saturation error is not shaped by any transfer function. However, as verified in the previous example, even the saturation of the second integrator and the saturation of the quantizer are limits to the modulator performances. It is therefore essential to carefully estimate the expected voltage swings and to keep them within limits that, on one hand, are not large enough to cause saturation but, on another hand, are not so low as to become comparable with the electronic noise.

Keep in Mind

Since the op-amp output dynamic range must accommodate the integrators swing with maximum input, the usable maximum references depend on supply voltage, op-amp configuration and sigma-delta architecture.

The design of active filters also faces the problem of properly controlling the output swing of integrators as their maximum amplitude must remain within the limits established by the op-amps range. The solution used is the scaling by a suitable attenuation (or amplification) at the input of the integrator compensated by an inverse amplification (or attenuation) at the input of the next stage. Since the attenuation and the amplification neutralize one each other the response of the filter is unchanged but the op-amp output swing is optimized.

The drawback of scaling is that the electronic noise injected into the circuit between the attenuation and compensation stages is referred to the input with a lower gain.

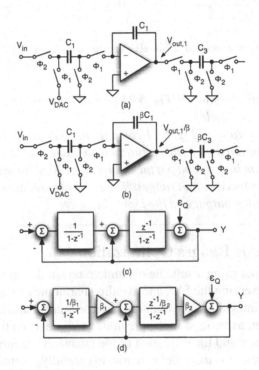

Figure 6.22.   Use of scaling for reducing the swing of the integrators.

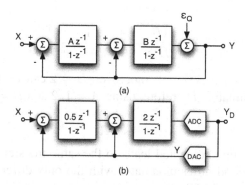

Figure 6.23. (a) Block diagram of a second order $\Sigma\Delta$ modulator with delayed integrators. (b) Optimum gains.

Suppose, for example, that it is necessary to reduce the swing of $V_{out,1}$ in the switched-capacitor integrator of Fig. 6.22 (a). Suppose also that $V_{out,1}$ is the input of a second integrator via the *SC* structure C_3. Increasing the integrating capacitance C_2 by β, as shown in Fig. 6.22 (b), reduces the output voltage by β but an equal increase of the sampling capacitance C_3 gives rise to the same injected charge into the virtual ground of the next stage.

The use of the scaling method in the second order $\Sigma\Delta$ of Fig. 6.22 (c) gives rise to the scheme of Fig. 6.22 (d) which provides an unchanged operation and equal quantization noise performance.

The scaling technique applied to the second integrator would require amplification at the input of the quantizer. However this request can be transferred to the pre-amplifiers used in the flash by scaling down all the *ADC* thresholds by β_2. Obviously a single threshold quantizer does not require any scaling as it only detects the zero threshold.

Observe that the first integrator of the architecture of Fig. 6.11 (b) is without delay, whereas the second integrator block has a delay element. The scheme of Fig. 6.23 shows a possible realizion of a sampled-data second-order $\Sigma\Delta$ modulator with two delayed integrators whose gains are A and B for the first and second position respectively. In addition to the benefit of an extra clock period available for the feedback loop implementation, the use of proper gains realizes appropriate signal and noise transfer functions and, also, obtains a scaling of the first integrator.

The equation describing the circuit is

$$\left[(X - Y)\frac{Az^{-1}}{1 - z^{-1}} - Y\right]\frac{Bz^{-1}}{1 - z^{-1}} + \epsilon_Q = Y, \tag{6.45}$$

giving rise to

$$Y = \frac{X \cdot ABz^{-2} + \epsilon_Q(1 - z^{-1})^2}{1 - (2 - B)z^{-1} + (1 - B + AB)z^{-2}}. \qquad (6.46)$$

The signal gain is 1 if $AB = 1$; moreover, $B = 2$ cancels the z^{-1} and z^{-2} terms of the denominator. Therefore, using $A = 1/2$ and $B = 2$ (Fig. 6.23 (b) results in

$$Y = Xz^{-2} + \epsilon_Q(1 - z^{-1})^2 \qquad (6.47)$$

which is the optimum choice and leads to the same transfer functions as the already studied second order modulator with the only difference of an extra delay in the input transfer function.

Since the gain of a the first integrator scheme of Fig. 6.23 (b) is $\frac{1}{2}$ its output swing is reduced accordingly, while the gain by 2 of the second integrator compensates for the first integrator attenuation. Moreover, the path from the DAC to second integrator input obtains the required dumping.

Example 6.5

Determine by computer simulations the histograms of the output voltages of the two integrators used in the modulator of Fig. 6.23. Utilize a 2-level (1 threshold) quantizer, OSR = 64, $V_{in} = -10\,dB_{FS}$, $A = 1/2$ and two values for B: 2 and 1/2.

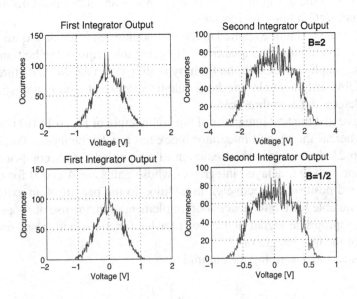

Figure 6.24. Histogram of the integrators output voltage for B = 2 and B = 1/2.

Solution

This Example uses the description and the launcher given in the files Ex6_5 and Ex6_5Launch that are a modification of the files used for the previous Example. The simulations show, as expected, that the scaling reduces the first integrator swing by approximately a factor of 2. The second integrator swing is almost unchanged with respect to the scheme that cascades one integrator without delay and one with delay (that is 3.1 V). The scaling of B from 2 to 1/2, as shown in Fig. 6.24, simply reduces the dynamic range of the second integrator by 4 without affecting other performances.

The Example verifies simple expected results, and also provides files that the reader can use for studying the saturation of integrators, and analyzing the difference between single threshold and multiple threshold quantizers.

Using feed-forward paths is an effective way of reducing the op-amp dynamic range in multi-level architectures. Before describing the uses of the feed-forward method, let us consider again the modulator of Fig. 6.23 (b) whose response is $Y = Xz^{-2} + \epsilon_Q(1 - z^{-1})^2$. The output P of the first integrator is

$$P = \frac{(X - Y)z^{-1}}{2(1 - z^{-1})} = X\frac{z^{-1}(1 + z^{-1})}{2} + \frac{\epsilon_Q z^{-1}(1 - z^{-1})}{2}. \qquad (6.48)$$

With a multi-level quantizer the amplitude of P is dominated by the first term of (6.48). Indeed, the second term, equal to the quantization error attenuated by 2 and passed through a first order high-pass transfer function, is at most equal to Δ.

The feed-forward path represented by the dotted line transforms the scheme of Fig. 6.23 (b) into the one of Fig. 6.25. The additional branch, when referred

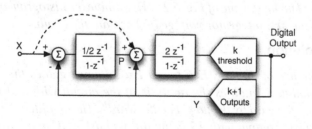

Figure 6.25. Second order modulator with the feed-forward path.

to the input, corresponds to $2 \cdot X(1 - z^{-1})/z^{-1}$. Therefore, the output becomes

$$Y = X\left[z^{-2} + 2 \cdot z^{-1}(1 - z^{-1})\right] + \epsilon_Q(1 - z^{-1})^2 \qquad (6.49)$$

which, when used to estimate the output of the first integrator, yields

$$P = \frac{(X - Y)z^{-1}}{2(1 - z^{-1})} = X\frac{z^{-1}(1 - z^{-1})}{2} + \frac{\epsilon_Q z^{-1}(1 - z^{-1})}{2}. \qquad (6.50)$$

Comparing the above equation with (6.48) shows a significant reduction of the X contribution: X passes through a high-pass $(1 - z^{-1})$ filter that, in the signal band, gives rise to a large attenuation.

Observe that the additional branch changes the signal transfer function from a simple double delay into

$$STF = z^{-2} + 2(1 - z^{-1}); \qquad (6.51)$$

however, the additional high pass term is negligible and, if necessary, can be compensated for in the digital domain.

The use of the feed-forward branch in the modulator of Fig. 6.11 (b) obtains a similar result without the factor 2 of equations (6.48) and (6.51).

The benefit of the feed-forward path in reducing the op-amp swing can be explained by the following intuitive view. Since the second block is an integrator, its output is bounded if its input is, on average, zero. In turn, since the input of the second integrator is made by three terms, $-Y$, P and X, it must be verified that $-Y + P + X \simeq 0$. Recall now that the output follows the input with a difference that is in the order of the quantization error. Accordingly, the condition $P \simeq Y - X$, indicates that P is also in the order the quantization error, and has a fairly small amplitude for multi-level quantization.

Example 6.6

Verify, by computer simulations, the effect of the additional branch used in the scheme of Fig. 6.25. Determine the histogram of the output of the first integrator voltage and explain the result.

Solution

The use of the files Ex6_6 and Ex6_6launch guides this study. The simulation enables the analysis of the circuit with and without the feed-forward by the flag "Feedforward". The amplitude of the first integrator output with a 7 comparators flash ADC and $-3\,dB_{FS}$ input sine wave diminishes from 0.84 V down to 0.14 V. The SNR at OSR = 64 is equal to 93 dB, almost unchanged by the feed-forward path. The file sinusx that calculates the DFT at a defined frequency estimates the amplitude of the output component at the input frequency.

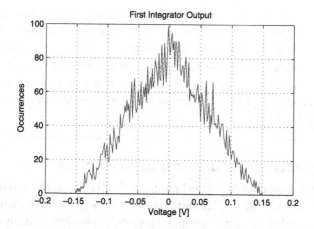

Figure 6.26. Histogram of the first integrator output.

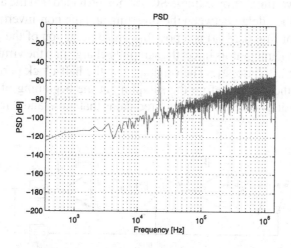

Figure 6.27. Spectrum of the first op-amp output.

As expected the result is higher than 1; however, very close to the bandwidth limit ($f_s/128.3$) the input amplification is fairly small, just 0.02 dB.

The histogram of the first integrator output, shown in Fig. 6.26, looks like a triangle. The result, quite different from the histogram of a sine wave, indicates, on one hand, a very low input sinusoidal term (as is shown in the spectrum of Fig. 6.27), and on the other hand, verifies the result of equation (6.50) foreseeing a first order shaping of the quantization error.

For this matter, remember that the probability distribution function of ϵ_Q is uniform from $-\Delta/2$ to $\Delta/2$; therefore, the probability of having x as result of the subtraction between two successive samples of ϵ_Q is the probability of having the first sample larger than $-\Delta/2+x$; in contrast, the probability of having $-x$ requires a the second sample larger than $-\Delta/2+x$. The result is a triangle that goes to zero at $\pm\Delta/2$.

6.6.3   Sampled-data Circuit Implementation

The switched-capacitor (*SC*) technique is the basis for the circuit implementations of sampled-data ΣΔ modulators. As known with switches and capacitors, controlled by non-overlapped phases, is possible to design integrators with delay or without delay, as required by modulator architectures. The subtraction of signals can be obtained with two SC structures, one inverting and the other non-inverting. or by a single SC scheme provided that the architecture enables half a clock delay between the inverting and the non-inverting terms.

The scheme of Fig. 6.28 (a) is a possible implementation of the diagram of Fig. 6.23 (b). Both integrators inject the sampling charge into the virtual grounds at the beginning of phase Φ_1. The op-amps use the half-clock period defined by Φ_1 to settle the output before the sampling at the beginning of phase Φ_2. The capacitor ratios determine the required gains that are $\frac{1}{2}$ and 2 respectively.

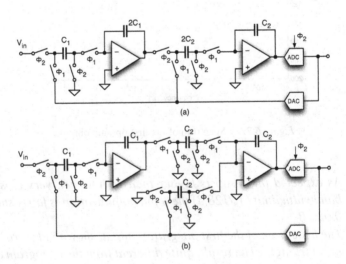

Figure 6.28.   (a) Circuit implementation of the architecture of Fig. 6.23 (b). (b) Implementation of the solution of Fig. 6.11 (b).

The two input input structures obtains the subtraction of the input and the *DAC* feedback by pre-charging the input signal or the output of the first integrator during Φ_2, thus obtaining a non-inverting operation. The signal from the *DAC*, on the contrary, determines an inverting integration.

The architecture foresees one clock period delay around the loop that goes from the output to the second integrator input and two clock period delays for the other loop. By inspection of the scheme it is easy to verify that the obtained delays match the architecture.

The scheme of Fig. 6.28 (b), implementing the architecture of Fig. 6.11 (b), reduces the delay of the loop from the DAC to the input node by one clock period. The injection of the *DAC* output into the first integrator gives rise to a signal that is immediately sampled and injected into the second integrator by the top *SC* structure at the input of the second integrator. The second *SC* structure is used to inject the *DAC* signal.

The latches of the *ADC* are activated by the rising edge of phase Φ_2 for leaving the entire half-period of Φ_2 for the digital conversion and the pre-setting of the *ADC* that must provide its output during the next Φ_1.

A limit of the scheme (caused by the architecture) is that during Φ_1 the two op-amps are one cascaded to the other leading to an overall step response equal to the convolution of the responses of the two op-amps. The resulting slowing-down limits the usable clock frequency or requires using more power to speed-up the circuit operation. By contrast, the feedback factors of the two integrators are both $\frac{1}{2}$ while for the scheme of Fig. 6.28 (a) the feedback factors equal to $\frac{2}{3}$ and $\frac{1}{3}$ requires using op-amps with different gain-bandwidth.

6.6.4 Noise Analysis

The noise of any $\Sigma\Delta$ architecture is caused by the switching off of capacitors and the noise of the op-amps. The calculation of the total noise follows an approach similar to what was previously done for a flip-around *S&H*. Namely, since the scheme has two different configuration during both sampling and injection phase, it is necessary to identify all the separate sub-networks formed, to calculate the noise charges caused by the noise sources of those sub-networks on the capacitors at the end of the considered phase and to estimate their colored spectra. Then, the successive sampling determines a almost white spectra produced by the folding into the band base.

The samples of noise charge on capacitors give rise to output voltage contributions because of the transfer functions from the capacitors charge to the output terminal. For this calculation it is necessary to multiply the noise spectra by the square of the transfer functions and to integrate the result over the signal band. The superposition of all the integrals obtains the total noise power.

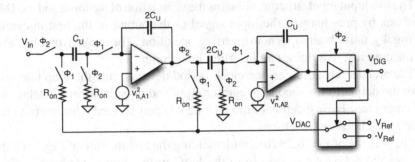

Figure 6.29.   Circuit diagram of a 1-bit second order Sigma-Delta modulator.

The above described methodology can be used to study the noise of the circuit in Fig. 6.29. It is a single-ended second order $\Sigma\Delta$ modulator with two delayed integrators in a feedback loop with gains of 0.5 and 2 respectively. The scheme includes the on-resistance of the pair of switches that close together and the input referred noise generators of the op-amps of which we consider only the white part of the spectrum

$$v_{n,A1}^2 = \gamma_{A1}\frac{4kT}{g_{m,A1}}; \qquad v_{n,A2}^2 = \gamma_{A2}\frac{4kT}{g_{m,A2}} \qquad (6.52)$$

During Φ_2 the input capacitance C_U samples the input voltage through R_{on} and its thermal noise $v_{n,R}^2 = 4kTR$. As known from Chapter 1 the low pass filter $R_{on}C_U$ colors the noise charge spectrum and the folding give rise to the sampled noise power kT/C_U.

The output of the first integrator charges the input capacitance of the second stage $2C_U$ through two switches. Since the first *OTA*, as shown by the noise equivalent circuit of Fig. 6.30 (a), is in the unity gain configuration (thanks to the feedback established by the integrating capacitance), then the equivalent noise model used is the one in the bottom part of the figure. Obviously, $g_{m,A1}$ is the transconductance gain of A_1.

The noise spectrum $v_{n,A1}^2$ is filtered by the square module of a two pole network to give rise to colored noise across $2C_U$ given by

$$H_{A1,in2} = \frac{v_{n,C_{in2}}}{v_{n,A1}} = \frac{1}{1 + s(\tau_0 + \tau_0 2C_U/C_L + \tau_R) + s^2\tau_0\tau_R} \qquad (6.53)$$

where $\tau_0 = C_L/g_m$ and $\tau_R = 2C_U R_{on}$.

If $2C_U/C_L < 1$ then the poles of the noise transfer function are controlled by τ_0 and τ_R with the f_T of the op-amp the one at lower frequency. Since the

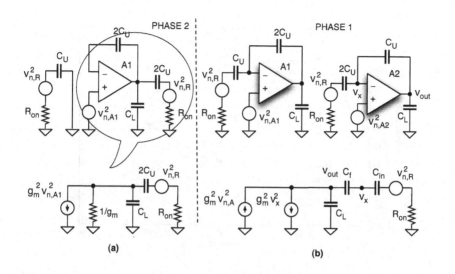

Figure 6.30. (a) Noise model during phase Φ_2. (b) Noise model during phase Φ_1.

effect of the second pole is almost negligible, $v_{n,A1}^2 = \gamma_{A1} 4kT/g_m$ gives rise to the noise power

$$V_{n,A1,in2} = \gamma_{A1} \frac{kT}{C_L} \qquad (6.54)$$

If $2C_U/C_L > 1$ the first pole moves to a slightly lower frequency and improves the noise shaping. However, in practical cases, the benefit is not larger than 1 dB.

The noise spectrum $v_{n,R}^2$ of the bottom scheme of Fig. 6.30 (a) is filtered by a transfer function with two poles and one zero

$$H_{R,in2} = \frac{1 + s\tau_0}{1 + s(\tau_0 + \tau_0 2C_U/C_L + \tau_R) + s^2 \tau_0 \tau_R} \qquad (6.55)$$

If $2C_U/C_L < 1$ the zero almost cancels out the first pole and the time constant $\tau_R = R_{on} C_U$ determines the noise transfer function. The noise power becomes

$$V_{n,R,in2} = \frac{kT}{2C_U} \qquad (6.56)$$

If, in contrast, $2C_U/C_L > 1$ the two poles are separated giving rise to a 20 dB/dec roll-off, a flat region and a second 20 dB/dec roll-off as shown by one of the curves of Fig. 6.31 for $f_T = 200\,MHz$, $C_L = 1\,pF$, $2C_U = 0.5\,pF$ and $R_{on} = 100\,\Omega$. The figure also shows the single-pole responses of the *OTA* and the $2C_s R_{on}$ network. Observe that the effect of the second pole can help a little

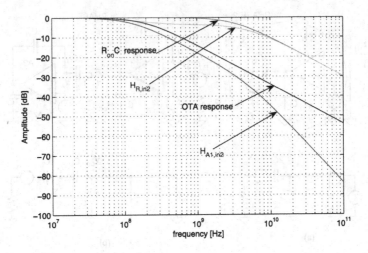

Figure 6.31. Input referred noise transfer functions for second integrator input capacitance.

in obtaining improved noise performances. However, the benefit is, in practical cases in the 1-2 *dB* range.

The study of the noise during phase Φ_1 follows the same procedure and uses the equivalent circuit of Fig. 6.30 (b) described by equations

$$v_x = \frac{v_{out}(R_{on} + \frac{1}{sC_{in}}) + v_{n,R}\frac{1}{sC_f}}{R_{on} + \frac{1}{sC_f} + \frac{1}{sC_{in}}} \tag{6.57}$$

$$g_m(v_{n,A} - v_x) = v_{out}sC_L + (v_{out} - v_x)sC_f \tag{6.58}$$

$$v_{C_{in}} = \frac{C_L(v_x - v_{out})}{C_{in}} \tag{6.59}$$

where proper input and feedback capacitances must be used for analyzing the first or the second stage.

Solving equations (6.57), (6.58) and (6.59) yields

$$v_{C_{in}} = \frac{C_L}{C_f} \frac{-v_{n,A} + (1 + s\tau_0)v_{n,R}}{1 + (\tau_0/\beta + \tau_0 C_{in}/C_L + \tau_R)s + \tau_0\tau_R s^2} \tag{6.60}$$

where $\tau_0 = C_L/g_m$, $\tau_R = C_{in}R_{on}$ and $\beta = C_{in}/(C_{in} + C_f)$.

Equation (6.60) shows that even during Φ_1 a two poles transfer function filters the op-amp noise generator, while a transfer function with two poles and one zero shapes the thermal noise generator of the on-resistance.

Therefore, with a single pole the op-amp noise gives rise to $\gamma_A kT/C_L$ while the on-resistance causes about kT/C_{in}. Again, the extra pole and the possible

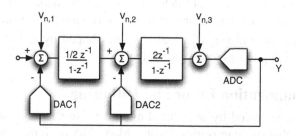

Figure 6.32. Noise model of the second-order modulator.

zero improves the noise filtering and obtains a lower noise power of, in practical cases, about 1–2 *dB*.

The second integrator output, which is sampled by the quantizer at the rising edge of Φ_2, contributes with two noise terms; one because of the noise of the second op-amp, and the other due to the sampling on the *ADC* capacitance, C_{ADC}.

The above accurate study is useful to understand the limit of approximate expressions that can be obtained by inspection of the circuit and, for the modulator of Fig. 6.29, give rise to the noise terms listed in Table 6.2.

Since the power corresponding to each column is uniformly spread over the Nyquist interval, then the noise can be described by three white noise generators

$$v_{n,1}^2 = 2T_s \left[2\frac{kT}{C_U} + \gamma_{A1}kT/C_L \right]$$

$$v_{n,2}^2 = 2T_s \left[\frac{kT}{C_U} + \gamma_{A1}kT/C_L + \gamma_{A2}kT/C_L \right] \qquad (6.61)$$

$$v_{n,3}^2 = 2T_s \left[\frac{2kT}{C_{ADC}} + \gamma_{A2}kT/C_L \right]$$

as used in Fig. 6.32 at the input of the two integrators and the quantizer.

The quadratic superposition gives rise to the output spectrum

$$v_{n,out}^2 = v_{n,1}^2 |z^{-2}|^2 + v_{n,2}^2 |2z^{-1}(1-z^{-1})|^2 + v_{n,3}^2 |(1-z^{-1})^2|^2 \qquad (6.62)$$

Table 6.2 - Noise Power Terms of the Second Order $\Sigma\Delta$ Modulator

Phase	Source	V_{n1}^2 $[V^2]$	V_{n2}^2 $[V^2]$	V_{n1}^3 $[V^2]$
Φ_2	$4kTR_{on}$	kT/C_U	$kT/(2C_U)$	kT/C_{ADC}
Φ_2	$\gamma_{Ai}4kT/g_m$	–	$\gamma_{A1}kT/C_L$	$\gamma_{A2}kT/C_L$
Φ_1	$4kTR_{on}$	kT/C_U	$kT/(2C_U)$	–
Φ_1	$\gamma_{Ai}4kT/g_m$	$\gamma_{A1}kT/C_L$	$\gamma_{A2}kT/C_L$	–

The different injection points determine different shaping of the spectra. Since $v_{n,1}^2$ is not shaped its contribution is only reduced by the used oversampling ratio. In contrast, $v_{n,2}^2$ and $v_{n,3}^2$ are shaped by a first-order and a second-order high pass response respectively.

6.6.5   Quantization Error and Dithering

The benefits provided by oversampling and noise shaping enable the use of a small number of quantization levels. Many $\Sigma\Delta$ designs even use binary quantization to avoid the trouble of obtaining very linear *DACs*. However, the use of a small number of quantization intervals is in direct contrast with the approximations of Chapter 1, which stated that if the quantization error was to be modelled as noise, then a large number of quantization intervals was required. Since this is not properly verified, a study of the possible implications is required.

Suppose, for example, that the input of a first order 1-*bit* $\Sigma\Delta$ modulator is a *dc* signal. If the input amplitude is $\Delta n/m$ with Δ the quantization interval, n and m integer numbers and $n < m$, then the output of the modulator is a pattern made by n ones that repeats itself every m clock periods. The quantization error is also a repetitive pattern, like the one shown in Fig. 6.33 corresponding to $V_{in} = 23/37$ and $V_{ref} = \pm 1$. The repetitive pattern produces a spectrum with tones at f_s/m and its multiples. These kinds of spectral lines are often referred to as *idle channel tones* or *pattern noise*.

Different input amplitudes and higher order modulators make the situation less critical, especially when the input is an irrational fraction of the reference or, as it normally happens, is a sine wave or a band-limited spectrum. However,

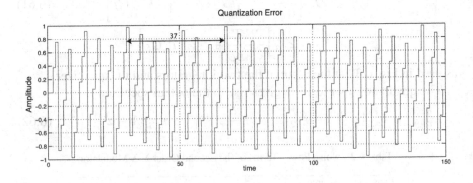

Figure 6.33.   Quantization error of a first order $\Sigma\Delta$ with $V_{in} = 23/37$ and $V_{ref} = \pm 1$.

the risk of having tones in the signal band remains even with busy signals, especially for low order architectures and 1-*bit* quantization.

The occurrence of tones means that the power of the quantization error is concentrated at specific frequencies instead of being spread over the Nyquist interval. The high-pass noise transfer function eventually reduces the amplitude of the tones that fall in the signal band. However, the situation remains problematic because a large quantization noise power causes big tones that, despite the attenuation, can be comparable with small sine wave components of the input.

Although a small sine wave corrupted by noise can be made visible by reducing the bandwidth of the measure filter (as it happens when using a long *fft* sequence), in contrast however, the power of tones does not diminish with the bandwidth of the band-pass filter across it and can often mask the small sine wave located at the same or at near frequencies.

Tones in the spectrum indicate limit cycles in the state-variable space. In contrast to this, noise gives rise to a chaotic path in the state-variable space. Therefore, for favoring noise and removing tones it is necessary to perturb any possible limit cycles and favor the chaotic behavior. Accordingly, since the limit cycles are due to correlated behaviors not destroyed by the features of the input signal, it is necessary to use an auxiliary input capable of breaking the cycles. The method is named *dithering* and the auxiliary signal is the *dither*.

Obviously the dither must be effective against the tones and should not alter the signal. For this there are two possibilities: the first is to inject a sine wave or a square wave whose frequency is out of the signal band. The digital filter used to cancel the out-of-band noise then removes the effect of the dither. The amplitude of the dither must be as low as possible as its amplitude reduces the input dynamic range.

The second method utilizes a noise-like signal whose contribution does not degrade the *SNR*. A noise source that takes the role of the dither can be the thermal noise of the electronics. If this is not sufficient then it is necessary to specifically inject a random signal whose spectrum can be shaped for reducing its negative effect in the signal band.

The dither is typically a bipolar signal $\pm V_{dith}$ with constant amplitude and sign controlled by a pseudo-random (PN) bit-stream generator. The schemes of Fig. 6.34 shows two possible points of injection: the input of the modulator or before the quantizer. In the former case the transfer function of the dither is the same as the input signal, so it is therefore necessary

> **Call up**
>
> Use multi-bit quantizers or dithering to destabilize the tonal behavior of $\Sigma\Delta$ modulators especially when the input may contain a dominant dc component.

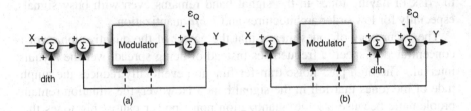

Figure 6.34. (a) Injection of dither at the modulator input. (b) Use of input equivalent shaped dither.

to properly shape the dither spectrum by passing the *PN* bit-stream through a $(1 - z^{-1})^p$ filter before generating its $\pm V_{dith}$ analog representation. In the latter case, the same transfer function that shapes the quantization noise also modifies the spectrum of dither injected before the quantizer. Therefore, since with a *PN* gaussian distribution the power of the dither is $V_{dith}^2/12$, it is just necessary to use a dither amplitude such that $V_{dith} < \Delta$.

6.6.6   Single-bit and Multi-bit

Until now we have not considered in much detail the advantages and the disadvantages of using single-bit or multi-bit quantizers. It was only outlined that 1-*bit* conversion intrinsically ensures a linear response and that the multi-bit approach results in an obvious improvement of the resolution. This section provides some design considerations for suitably choosing the number of quantization levels.

Obtaining high *SNR* with a 1-*bit* $\Sigma\Delta$ entails the use of high order modulators with consequent difficulties in the design of a stable architecture or using high oversampling ratios. Since the bandwidth of the op-amps (or *OTAs*) must be obviously higher than the clock frequency, then the technology speed constrains, or a limited power budget make the high *SNR* 1-*bit* solutions normally suitable for audio-bands or for instrumentation applications.

The usable reference voltages of 1-*bit* modulators, for the reason that the op-amp output swings are fairly large, is a small fraction of the supply voltage. Assume that the output dynamic-range of the op-amp is αV_{DD} and that a $-6\,dB_{FS}$ sine wave gives rise to a $\pm\,\beta_{swing}V_{ref}$ maximum swing at the first integrator. The reference voltage that is at the limit of op-amp saturation is

$$|V_{ref}| < \frac{\alpha V_{DD}}{2\beta_{swing}} \qquad (6.63)$$

For low supply voltages α can be 0.7 and β_{swing} about 2, obtaining $V_{ref} = \pm 0.175/V_{DD}$. If the supply voltage is 1.8 V then the reference voltage can be as low as $\pm 0.3V$ which is problematic because of the constrains on the kT/C noise and the op-amp thermal noise ($\gamma kT/C_L$, especially critical for the first stage). Therefore, using 1-*bit* quantization is normally convenient only for medium-high supply voltages.

The slew-rate of the op-amp must ensure, together with the gain-bandwidth product, an accurate settling of the integrators. The input of the first integrator is the difference between the analog input and the *DAC* output whose value, as is known, follows the input with a closeness that depends on the *DAC* resolution and the bandwidth of the input. Since it is reasonable to assume that the maximum difference is about 2Δ, then for 1-*bit* quantization ($\Delta = V_{Ref}$) the maximum amplitude of the first integrator input becomes $2V_{Ref}$ (value corresponding to the previously performed simulations). The slew rate that brings the output voltage close to the final value by the overdrive of the input stage (value at which the op-amp starts its exponential settling) in a fraction γ of the injection period $T_s/2$ is therefore

$$SR = \frac{2G(V_{Ref} - V_{ov})}{\gamma T_s/2}. \tag{6.64}$$

where G is the gain of the integrator.

The corresponding op-amp output current, including the one required by the capacitive load C_L, becomes

$$I_{out} = \frac{2V_{Ref}(C_{in} + C_L)}{\gamma T_s/2}. \tag{6.65}$$

If the clock frequency is 20 MHz and $V_{ref} = 1 V$, the required slew-rate is $400 V/\mu s$ with $\gamma = 0.1$; I_{out} is $0.8 mA$ with $(C_{in} + C_L) = 2 pF$. Since the use of multi-bit reduces the integrator input by approximately the number of quantization levels the above figures would diminish by the same factor.

The use of multi-level quantization improves the equivalent number of bits but consumes additional power for the *ADC*. However, for a second order modulator increasing the resolution by 2.5-*bit* requires a doubling of the clock frequency; therefore, the optimum use of the power depends on the trade-off between increased speed of the op-amps or more comparators in the quantizer. As a rule of thumb assume that the power of a comparator is approximately 1/20 the power of an *OTA* operated at the same clock frequency.

Also notice that addition more comparators means increased complexity, multi-bit digital processing in the first stage of the decimation filter and extra digital logic for the digital calibration or matching of elements. The use of all the above considerations leads to multi-bit $\Sigma\Delta$ modulators with 3-15 comparators.

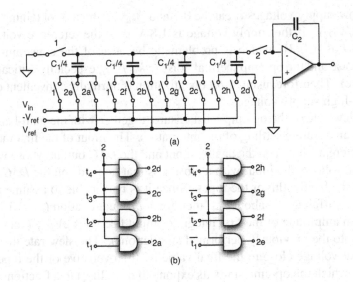

Figure 6.35. (a) Integration of the input minus the DAC signal. (b) Logic network generating the phases for the DAC control.

Even the multi-bit analog section is more complicated than the single bit counterpart. The multi-bit *DAC* can use one of the various architectures studied before but normally the capacitive MDAC is the preferred scheme. Furthermore, normally the subtraction and the *DAC* functions are obtained by the same capacitive array working as an *MDAC* as shown in Fig. 6.35 for 2-*bit*. The input capacitance C_1 is divided into 4 equal elements, pre-charged to the input signal during phase Φ_1 and, during phase Φ_2, under the control of the thermometric codes $t_1, \cdots, t_4$, connected to $+V_{ref}$ or $-V_{ref}$.

The sharing of the same array for the input injection and the *DAC* function (also used for a single-bit architecture) is positive because the feedback factor of the *OTA* is half the value than using separate capacitances. However, the charge that the reference voltage generator is required to deliver is a non-linear function of the input signal. If the control of the *DAC* is $k(n) \simeq V_{in}(n-1)/\Delta$, $k(n)$ capacitances already charged to $V_{in}(n)$ are connected to the reference voltage. Therefore

$$Q_{Ref}(n) = k(n) \left[V_{Ref} - V_{in}(n) \right] \qquad (6.66)$$

that, using the expression of $k(n)$, gives a quadratic term in V_{in}. Therefore, the output resistance of the reference generator must be very small for avoiding significant distortion in the *DAC* response.

PROBLEMS

6.1 Estimate the increase of the effective number of bits for an 8-bit ADC that uses oversampling ratios equal to 2^k with $k = 4 \cdots 12$

6.2 Use in the feedback loop of a first order $\Sigma\Delta$ modulator a block with transfer function $1.2z^{-2}/(1 - 0.95z^{-1})$. Determine the power of the quantization noise in the signal band $f_B << f_s$.

6.3 Repeat Example 6.1 but use triangular and square wave input waveforms. Plot the spectrum of the input signal and the spectrum of the modulator output. Are the results satisfactory?

6.4 Simulate a first order modulator with a 1-bit quantizer and different dc input amplitudes. Accumulate the output bit stream for 2^{12} clock periods and interpret the obtained results.

6.5 Repeat the previous problem with a non-ideal 8-level quantizer. Use a random (fixed) error with 5% σ and plot the accumulated output values as a function of the dc input amplitude.

6.6 Use a behavioral model to simulate the second order modulator of Fig. 6.11 (b). Observe the signals at the output of the two integrators and determine the peak value as a function of the input amplitude and a 1, 2, 3-bit quantizer.

6.7 Use the behavioral model of a second order modulator with a 1-bit quantizer to determine the output bit-stream for different input amplitudes. Use, for example, an input of 3/12 of the full scale or other fractional values. Identify any possible cycles by studying the response in the time domain or in the frequency domain.

6.8 Use the scheme of the previous example to study the effect of random noise added to the dc input signal. Use one of the cases that show tones in the output spectrum.

6.9 Estimate the equivalent number of bits of a low-pass fifth-order modulator that uses a 3-comparator quantizer and OSR = 20. In order or ensure stability the NTF has three poles at z = 0.8 and at the conjugate positions z = 0.6±j0.7.

6.10   Repeat Example 6.2 but use a soft saturation modeled by cascading the integrator with the function $atan(x)$ where x is the ratio between the input signal and a saturation level. Compare the results with an equivalent case that uses hard saturation.

6.11   Use the simulation setup of the Example 6.2 to determine the histograms of the output of the two integrators with and without soft saturation. Determine the histogram of the quantization noise.

6.12   The dynamic range of the integrators used in the second order sigma-delta modulator of Fig. 6.11 (b) is $\pm 0.5 V_{ref}$ and the input range is $\pm 0.5 V_{ref}$. The ADC uses 5 comparators. Scale the design and determine the value of the comparator thresholds.

6.13   Use the feed-forward method to reduce the swing of the second integrator of the scheme of Fig. 6.25. The ADC uses three comparators and by sharing the charge of capacitances realizes a possible subtraction. Draft a circuit implementation using the switched-capacitor method.

6.14   Study the effect of the input offset on a second order modulator with hard saturation in the integrator response. Consider separately the effect of the offset on the first, second integrator, and quantizer.

6.15   Repeat Example 6.5 but assume different gain in the two op-amps. Plot the SNR loss as function of the two gains ranging from 30 to 60dB. The OSR is 64.

6.16   Repeat Example 6.6 using the sigma-delta scheme with no delay in the first integrator. Modify the scheme so that the first integrator output is scaled by a factor of 3. Compare the results with the ones of the example.

REFERENCES

Books and Monographs

J. Candy and G. Temes: *Oversampling Delta-Sigma Data Converters: Theory, Design and Simulation*, New York: IEEE Press, 1992.

S. R. Norsworthy, R. Schreier, and G. C. Temes: *Delta-Sigma Data Converters Theory, Design, and Simulation*, New York, NY: IEEE Press, 1997.

R. Schreier, and G. C. Temes: *Understanding Delta-Sigma Data Converters*, New York, NY: J. Wiley & Sons, NJ, 2005.

Journals and Conference Proceedings

Delta and Sigma-Delta

F. De Jager: *Delta modulation, a method of PCM transmission using the 1-unit code*, Philips Research Rep., no. 7, pp. 442–466, 1952.

J. B. O'Neal and R. W. Stroh: *Differential PCM for speech and data signals*, IEEE Trans. on Communications, vol. COMM-20, pp. 900–912, 1972.

J. C. Candy and O. J. Benjamin: *The Structure of Quantization Noise from Sigma-Delta Modulation*, IEEE Trans. Communications, vol. COM-29, pp. 1316–1323, 1981.

J. C. Candy: *A Use of Double Integration in Sigma Delta Modulation*, IEEE Trans. on Communications, vol. COM-33, pp. 249–258, 1985.

R. Koch, B. Heise, F. Eckbauer, E. Engelhardt, J. A. Fisher, and F. Parzefall: *A 12-bit sigma-delta analog-to-digital converter with a 15-MHz clock rate*, IEEE Journal of Solid-State Circuits, vol. 21, pp. 1003–1010, 1986.

S. H. Ardalan and J. J. Paulos: *An analysis of nonlinear behavior in delta – sigma modulators*, IEEE Trans. Circuit Syst., vol. CAS-34, pp. 593–603, 1987.

S. R. Norsworthy, I. G. Post, and H. S. Fetterman: *A 14-bit 80-kHz sigma-delta A/D converter: Modeling, design and performance evaluation*, IEEE Journal of Solid-State Circuits, vol. 24, pp. 256–266, 1989.

R. M. Gray, W. Chou, and P. W. Wong: *Quantization Noise in Single-Loop Sigma-Delta Modulation with Sinusoidal Inputs*, IEEE Trans. Communications, vol. COM-37, pp. 956–968, 1989.

R. M. Gray: *Spectral Analysis of Quantization Noise in a Single-Loop Sigma-Delta Modulator with dc Input*, IEEE Trans. Communications, vol. COM-37, pp. 588–599, 1989.

B. P. Brandt, D. E. Wingard, and B. A. Wooley: *Second-order sigma-delta modulation for digital-audio signal acquisition*, IEEE Journal of Solid-State Circuits, vol. 26, pp. 618–627, 1991.

Dynamic Range Optimization

B. E. Boser and B. A. Wooley: *The design of sigma-delta modulation analog-to-digital converters*, IEEE Journal of Solid-State Circuits, vol. 23, pp. 1298–1308, 1988.

J. Silva, U. Moon, J. Steensgaard, and G. Temes: *Wideband low-distortion delta-sigma ADC topology*, Electron. Lett., vol. 37, pp. 737–738, 2001.

A. A. Hamoui and K. W. Martin: *High-order multibit modulators and pseudo data-weighted-averaging in low-oversampling $\Sigma\Delta$ ADC's for broadband applications*, IEEE Transaction on Circuits Syst. I, vol. 51, pp. 72–85, 2004.

K. Nam, S. Lee, D. K. Su, and B. A. Wooley: *A low-voltage low-power sigma-delta modulator for broadband analog-to-digital conversion*, IEEE Journal of Solid-State Circuits, vol. 40, pp. 1855–1864, 2005.

Circuit Implementations

L. Schuchman: *Dither signals and their effect on quantization noise*, IEEE Trans. Communications Tech., vol. COMM-12, pp. 162–165, 1964.

K. Nagaraj, T. R. Viswanathan, K. Singhal, and J. Vlach: *Switched-capacitor circuits with reduced sensitivity to amplifier gain*, IEEE Trans. Circuit Syst., vol. CAS-34, pp. 571–574, 1987.

L. Le Toumelin, P. Carbou, Y. Leduc, P. Guignon, J. Oredsson, and A. Lindberg: *A 5-V CMOS line controller with 16-b audio converters*, IEEE Journal of Solid-State Circuits, vol. 27, pp. 332–342, 1992.

P. J. Hurst, R. A. Levinson, and D. J. Block: *A switched-capacitor delta-sigma modulator with reduced sensitivity to op-amp gain*, IEEE Journal of Solid-State Circuits, vol. 28, pp. 691–696, 1993.

J. W. Fattaruso, S. Kiriaki, M. de Wit, and G. Warwar: *Self-calibration techniques for a second-order multibit sigma-delta modulator*, IEEE Journal of Solid-State Circuits, vol. 28, pp. 1216–1223, 1993.

P. Ju, K. Suyama, P. F. Ferguson Jr., and W. Lee: *A 22-kHz multibit switched-capacitor sigma-delta D/A converter with 92 dB dynamic range*, IEEE Journal of Solid-State Circuits, vol. 30, pp. 1316–1325, 1995.

V. Peluso, M. S. J. Steyaert, and W. Sansen: *A 1.5-V-100-μW $\Sigma\Delta$ modulator with 12-b dynamic range using the switched-opamp technique*, IEEE Journal of Solid-State Circuits, vol. 32, pp. 943–952, July 1997.

W. Wey and Y. Huang: *A CMOS delta-sigma true RMS converter*, IEEE Journal of Solid-State Circuits, vol. 35, pp. 248–257, 2000.

Chapter 7

HIGH-ORDER, CT ΣΔ CONVERTERS AND ΣΔ DAC

After studying the basic principles of the oversampling and low order sigma-delta architectures this chapter analyzes high-order modulators employing either single bit or multi-bit quantizers. In addition to single stage architectures we shall study cascaded solutions normally named as MASH. Then we shall consider the continuous-time counterpart of the already studied sampled-data ΣΔ modulators before discussing band-pass implementations and, briefly ΣΔ DAC.

Even if the digital filter is an essential parts of the architecture we shall consider it as a black box without entering into its design details.

7.1   SNR ENHANCEMENT

In the previous Chapter we learnt that using many quantization levels augments the *SNR* of a sigma delta modulator. However, the internal *DAC* can have a limited number of unity elements because the techniques used to improve the linearity becomes too expensive or ineffective with many *DAC* levels. This observation is the basis of *SNR* enhancement techniques described in this section, usable with any kind of ΣΔ modulator.

The *SNR* enhancement methods exploit the fact that the linearity of a $\Sigma\Delta$ modulator does not depend on the precision of the *ADC* but on the linearity of the *DAC*. Therefore, the goal is to obtain a good *SNR* by using many levels in the *ADC*, but to reduce the *DAC* resolution to just 2-levels (or to the small number that the digital matching techniques can properly handle).

Consider the scheme of Fig. 7.1 (a) that conceptually uses an *n-bit ADC* and an *m-bit DAC*. Actually, the reduction of resolution from *n* to *m* is simply obtained by truncating the longer digital word and discarding the *n-m LSB*. A suitable processing of the full word and its truncated version give rise to the digital output. The linear model, given in Fig. 7.1 (b), uses two quantization errors, $\epsilon_{Q,n}$ and $\epsilon_{Q,m}$, for the *n-bit* and *m-bit* quantization. The equations describing the circuit are

$$Y_1 = X \cdot STF + \epsilon_{Q,m} \cdot NTF \tag{7.1}$$

$$Y_2 = Y_1 - \epsilon_{Q,m} + \epsilon_{Q,n} \tag{7.2}$$

that, eliminating $\epsilon_{Q,m}$, yields

$$Y_2 \cdot NTF + Y_1(1 - NTF) = X \cdot STF + \epsilon_{Q,n} \cdot NTF. \tag{7.3}$$

showing that a proper combination of Y_1 and Y_2 obtains the signal plus the noise shaping of the error $\epsilon_{Q,n}$ rather than $\epsilon_{Q,m}$. Therefore, the *SNR* corresponds to an *n-bit* quantization although the *DAC* only uses 2^m levels.

The method relies on a well predicted *NTF* as determined by the analog circuit. Possible mismatches between the actual transfer function, say *NTF'*, and the presumed ideal expression *NTF* used in (7.3) gives rise to a residual fraction *(NTF-NTF')* of $\epsilon_{Q,m}$ affecting the output. Since this error must be kept lower than $NTF \cdot \epsilon_{Q,m}$, it is necessary to ensure that the matching between the ideal and the real *NTF* is better than $2^{-(n-m)}$.

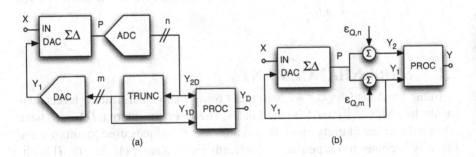

Figure 7.1.   (a) Conceptual block diagram for the SNR enhancement. (b) Linear model.

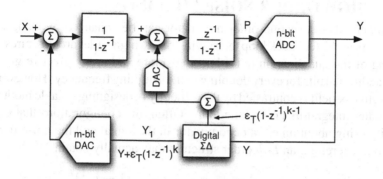

Figure 7.2. Use of a digital sigma-delta and error cancellation to enhance the SNR.

Rather than just using truncation it is possible to reduce the number of bits from n to m using more sophisticated approaches. Fig. 7.2 shows a possible technique applied to a second order modulator. The scheme uses of a digital sigma-delta between ADC and DAC. The output of the digital $\Sigma\Delta$ whose input is the main output Y results

$$Y_1 = Y + \epsilon_T (1 + z^{-1})^p \qquad (7.4)$$

where ϵ_T is the quantization error of the truncation from n to m bits.

Since $Y = P + \epsilon_{Q,n}$, the signal feed back at the input of the sigma delta is

$$Y_1 = P + \epsilon_{Q,n} + \epsilon_{Q,m}(1 + z^{-1})^p \qquad (7.5)$$

which is the quantization noise of an n-bit modulator that is then shaped by the *NTF* plus the extra noise term $\epsilon_{Q,m}(1 + z^{-1})^p$ whose effect on the output is modified by the *STF*. The output of the modulator becomes

$$Y_2 = X \cdot STF + \epsilon_{Q,n} \cdot NTF + \epsilon_{Q,m} \cdot STF \cdot (1 + z^{-1})^p. \qquad (7.6)$$

Notice that the contribution given by $\epsilon_{Q,m}$ is negligible if p is higher than the order of the modulator. Moreover, the solution avoids the processing required by equation (7.3) whose main disadvantage is to increase the word-length of the digital output and, consequently, to increase the complexity of the digital filter used for the noise rejection before the decimation.

A possible improvement, also shown in Fig. 7.2, includes a second digital input to the secondary *DAC*. It is equal to the truncation error multiplied by a suitable function that makes the effect of the secondary input equal to the contribution of the shaped truncation error passed through the first integrator. Therefore, the extra injection cancels out the noise caused by truncation provided that the actual transfer function of the first integrator equals $1/(1 - z^{-1})$.

7.2 HIGH ORDER NOISE SHAPING

The benefit of a second order modulator can be enhanced by using many integrators in the feedback loop thus giving rise to high-order architectures. The shaping of the quantization noise becomes more and more effective granting many additional bits for every doubling of the sampling frequency. However, the perspective benefit is contrasted by the difficulty in designing a stable modulator with many integrators around the loop. Often, the configurations that ensure stability bring about an extra denominator in the signal and the noise transfer function. Therefore, an *L-th* order structure may actually obtain

$$STF(z) = \frac{N(z)}{D(z)}; \quad NTF(z) = \frac{(1-z^{-1})^L}{D(z)}. \tag{7.7}$$

We will see that only few architectures guarantee stability with $D(z) = 1$, nevertheless, for those cases the noise transfer function is just $(1 - z^{-1})^L$. Therefore, after calculations similar to what was done before, it results

$$V_n^2 = V_{n,Q}^2 \frac{\pi^{2L}}{2L+1} \left[\frac{f_B}{f_s/2} \right]^{2L+1} = V_{n,Q}^2 \frac{\pi^{2L}}{2L+1} \cdot OSR^{-(2L+1)} \tag{7.8}$$

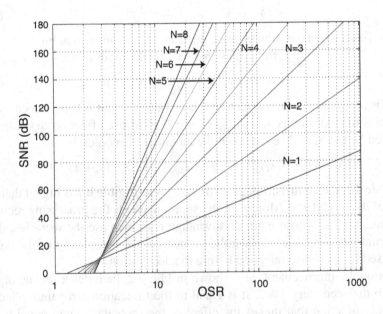

Figure 7.3. SNR versus the oversampling ratio for different order modulators with 1-bit quantization.

yielding

$$SNR_{\Sigma\Delta,L} = \frac{12}{8}k^2 \cdot \frac{2L+1}{\pi^{2L}} \cdot OSR^{2L+1} \qquad (7.9)$$

which, in *dB*, is

$$SNR_{\Sigma\Delta,L} = \left[1.78 + 6.02n'\right] +$$

$$-10log\frac{\pi^{2L}}{2L+1} + 3.01(2L+1) \cdot log_2(OSR). \qquad (7.10)$$

Fig. 7.3 plots the result of equation 7.10 for a 1-*bit* modulator of order ranging from 1 to 8. Observe that the oversampling benefit is excellent for high orders. For example, a sixth order modulator achieves *SNR* = 100 *dB* with just *OSR* = 16. However, the result does not account for $D(z)$ whose zeros must be placed inside the unity circle to ensure stability; moreover, the zeros are typically at frequencies pretty close to the signal band. The estimation of $D(z)$ in the signal band gives rise to a small number that reduces the shaping benefit. The next example provides a calculation for a possible case without and with denumerator.

Example 7.1

The signal and the noise transfer function of a high order sigma-delta modulator have two poles that are located at $f_{p1} = -4f_B$ and $f_{p2} = -8f_B$ in the frequency domain. The numerator $N(z)$ results in an STF that is flat (with gain 1) until $2f_B$ and past that frequency the denominator $D(z)$ controls the response. Determine the effect of $D(z)$ on the NTF at z = 1. The OSR is 64.

Solution

Since the frequency f_B is $f_s/128 = 1/(128 \cdot T_s)$, the two poles are at $f_{p1} = -1/(32 \cdot T_s)$ and $f_{p2} = -1/(16 \cdot T_s)$. The $s \to z$ mapping determines the position of the two poles on the z-plane

$$z_{p1} = e^{-\pi/16} = 0.822, \quad z_{p2} = e^{-\pi/8} = 0.675. \qquad (7.11)$$

The obtained denominator is $D(z)=(1-0.822z^{-1})(1-0.675z^{-1})$. For z=1 it results $D(1)=0.05785$ whose inverse is 17.28 corresponding to a loss of 24.8 dB.

The effect of the denominator causes at low frequency an upward shift of the noise transfer function. Possibly, the loss diminishes at high frequency as can be observed in Fig. 7.4 where the loss is 3 dB less at 0.1 f_s. This does not help because the shift up is at its maximum just in the band of interest.

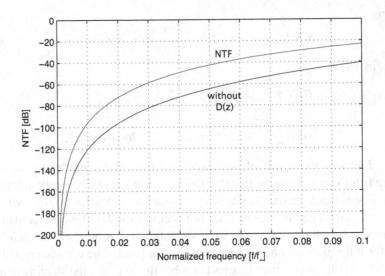

Figure 7.4.   Noise transfer function and its numerator contribution.

7.2.1   Single Stage Architectures

The use of many integrators in a $\Sigma\Delta$ architecture obviously obtains high-order noise transfer functions but, at the same time, poses special challenges for designing a stable architecture. Even the well known stability criteria studied in detail in linear feedback networks, are not definite conditions for $\Sigma\Delta$ modulators. It may happen that a stable linear architecture becomes unstable when the quantizer is inserted into the loop, especially if it uses a small number of bits. The reason for this is that the error caused by the non-linearity associated with quantization can trigger instability making it necessary to use more stringent conditions than the ones required by its linear counterpart (i.e. with $\Delta = 0$).

Before going into more detail of the stability analysis let us first consider the class of architectures that is most commonly used when obtaining high order noise shaping. Fig. 7.5 (a) is the generic scheme of that class featuring single or multi-feedback branches, and no feed-forward. The main feedback ensures that the quantized output tracks the input. Other feedbacks can be used to adjust the signal transfer function and control the stability. The architecture of Fig. 7.5 (a) can be transformed into the linear model of Fig. 7.5 (b) by moving the internal feedbacks to the input thus giving rise to a single feedback embracing the entire quantizer. The signal and the noise transfer functions are

$$STF = \frac{H(z)}{1 + H(z)}; \quad NTF = \frac{1}{1 + H(z)} \qquad (7.12)$$

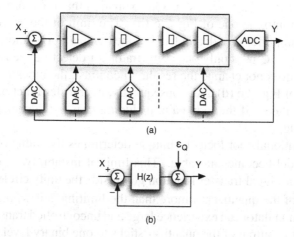

(a)

(b)

Figure 7.5. (a) Conceptual diagram of a high order modulator. (b) Solution with a single feedback.

where the *STF* should be flat with a gain of 0 *dB* in the signal band and the *NTF* generating multiple zeros in the signal band.

Assume that $H(z)$ is given by

$$H(z) = \frac{P(z)}{Q(z)} \tag{7.13}$$

The *STF* and the *NTF* become

$$STF = \frac{P(z)}{P(z) + Q(z)}; \quad NTF = \frac{Q(z)}{P(z) + Q(z)}. \tag{7.14}$$

The circuit implementations of the single feedback scheme start from a filter topology (either continuous-time or sampled data) which can then be modified to allow for suitable placement of the *ADC* and for implementing the feedback paths with one or more *DAC*s. Among the many possible filter schemes the ones frequently used for $\Sigma\Delta$ are the chain of integrators with weighted feedback summation, and the chain of integrators with distributed feedback. These solutions will be studied shortly after the following general study of stability.

7.2.2 Stability Analysis

The study of stability determines the constrains for the modulator parameters and, also, the minimum number of quantization bits that can be used. Since the most critical situation occurs when using binary quantization, then this study is done for that case. The result can then be extended to multilevel quantizations.

The use of the noise model is an approximation that transforms the scheme of Fig. 7.6 (a) into the one of Fig. 7.6 (b). The limit of the approximation is evident if an amplifier is added in front of the comparator as shown in Fig. 7.6 (c). Since the *ADC* uses only one comparator for detecting the zero crossing, then a gain k does not change the result. In contrast, the block k modifies the linear model of Fig. 7.6 (d) and, consequently, its transfer functions. As know from stability theory if the loop gain, including k, is not adequate the system can be unstable.

The conventional root locus technique determines the value of k at which the linear model becomes unstable. This limit of instability, $\tilde{k}$, occurs when the poles of the signal transfer function go outside the unity circle. Therefore, if the "gain" of the quantizer is more than the limiting gain $\tilde{k}$, the unbounded nodes of the modulator can experience large and uncontrolled transients. It may happens that the output of the quantizer sticks to one binary level permanently of for many clock periods or gives rise to low frequency oscillations with an alternation of long strings of 1's and 0's that cause low frequency tones likely to be located in the signal band.

The key point is to find a meaningful definition of the quantizer gain. Since only two levels represent the quantization output, it can be assumed that if the absolute input amplitude is smaller than V_{ref} the quantizer "amplifies" the input, and if the absolute input amplitude is larger than V_{ref} then the quantizer "attenuates" the input. This view can be quantified with an accurate statistical study that obtains the value of the quantizer gain as the level that minimizes the variance of a suitable function of input and output. The theory behind this quantitative definition is not studied here as, for the present scope, an intuitive view is sufficient.

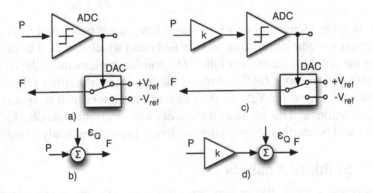

Figure 7.6. (a) Single-bit quantizer and (b) its linear model. (c) Modification of the single-bit quantizer by adding a fictitious gain and (d) its linear model.

Observe that since the output of a bipolar quantizer is its input V_x plus the noise N, then the gain k is a loose measure of the ratio $(V_x + N)/V_x$, very large for V_x close to zero. Assuming that for ensuring stability the mean square average of V_x is within the $\pm V_{ref}$ limits, then the expected average value of k is more than 1. Therefore, the filter used to obtain the modulator by adding a quantizer must include a fictitious loop gain $k > 1$ in the stability analysis. After defining the filter that gives rise to a stable high-order modulator it is worth verifying with time-domain simulations that that parameters ensure a stability region that is large enough to accommodate the alterations caused, in the worst case, by non-ideal behavior of the active and passive element used in the architecture.

It is also necessary to study the stability of multi-level quantizers for two reasons. The first is that the *DAC* clipping can give rise, in addition to an *SNR* reduction, to non-linear terms that may trigger unstable conditions. The second is that a finite number of quantization levels gives rise to a fictitious gain similar to the 2-*level* case. The estimation of this "gain" is again the basis of the study.

The minimum dynamic range of the quantizer, which in turn determines the required number of quantization levels for a given quantization step and maximum input amplitude, is normally determined by behavioral simulations. A first simulation with an ideal quantizer and full-scale input determines the number of levels N_{max} required. The result typically exceeds $2^N = V_{Ref}/\Delta$ and in some cases by a good fraction of 2^N. The extra quantization levels are what at the best ensures stability and full-scale operation. However, the number of levels can be reduced at the expenses of some *SNR* worsening. Indeed, a clipping in the quantizer enable reducing the number of levels until the simulation shows the stability limit.

> **Warning!**
>
> The study of the stability of a high-order modulator must be followed by extensive time-domain simulations with different amplitudes and frequencies of the sine wave input.

7.2.3   Weighted Feedback Summation

The scheme of Fig. 7.7 is a chain of sampled-data integrators with or without delay whose outputs are weighted and summed up to obtain the input of the quantizer generating the feedback signal Y. Observe that since it is necessary to have at least one delay around every loop then $k_1 > 0$. The transfer function of the *p-th* order architecture, $H_p(z)$, is

$$H_p(z) = \frac{z^{-k_1}}{1 - z^{-1}}a_1 + \frac{z^{-(k_1+k_2)}}{(1 - z^{-1})^2}a_2 + \cdots + \frac{z^{-\sum_1^p k_i}}{(1 - z^{-1})^p}a_p \qquad (7.15)$$

which used in (7.12) gives rise to the signal and noise transfer function.

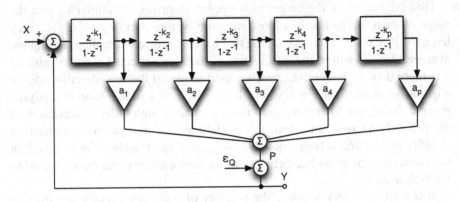

Figure 7.7. Modulator that uses a chain of integrators with weighted feedback summation.

Observe that the architecture provides a number of degrees of freedom which are normally used by the designer to realize the desired transfer functions, ensure stability and to optimize the dynamic range of the integrators. Representing $H(z)$ again as

$$H_p(z) = \frac{P(z)}{(1 - z^{-1})^p} \tag{7.16}$$

and $P(z)$ as

$$P(z) = \alpha_1 z^{-1} + \alpha_2 z^{-2} + \cdots + \alpha_p z^{-p} = \sum_1^p \alpha_i z^{-i} \tag{7.17}$$

it results

$$STF = \frac{\sum_1^p \alpha_i z^{-i}}{\sum_1^p \alpha_i z^{-i} + (1 - z^{-1})^p}$$

$$NTF = \frac{(1 - z^{-1})^p}{\sum_1^p \alpha_i z^{-i} + (1 - z^{-1})^p}. \tag{7.18}$$

Recall that the zeros of the polynomial

$$D(z) = \sum_1^p \alpha_i z^{-i} + (1 - z^{-1})^p = 1 + \beta_1 z^{-1} + \beta_2 z^{-2} + \cdots + \beta_p z^{-r} \tag{7.19}$$

determine the poles of the *STF* and *NTF*. Moreover, the stability requires that the poles be inside the unity circle, therefore

$$D(z) = \prod_1^r (1 - z_i z^{-1}) \quad |z_i| < 1. \tag{7.20}$$

Assuming that $D(z)$ almost equals $D(1)$ in the signal band, it results

$$NTF = \frac{NTF_{id}}{\prod_1^r(1 - z_i)} = \frac{NTF_{id}}{K_p}. \qquad (7.21)$$

The poles required to ensure stability give rise to a pole gain factor $K_p < 1$. The effect is a reduction of the *SNR* with respect to the expected value granted by an ideal $(1 - z^{-1})^p$ *NTF*

$$SNR|_{dB} = SNR_{ideal}|_{dB} - 20log_{10}K_p. \qquad (7.22)$$

Example 7.2

Use Simulink to study a third order modulator with weighted feed-back summation. Determine the weights to be used in the architecture whose integrators are all delayed and estimate the number of comparators and ADC dynamic range necessary to ensure proper operation of the circuit.

Solution

The use of equation (7.15) determines the expression of P(z) that, used in (7.19), yields the denominator of the signal and noise transfer functions

$$1 + (a_1 - 3)z^{-1} + (a_2 - 2a_1 + 3)z^{-2} + (a_1 - a_2 + a_3 - 1)z^{-3}$$

which becomes equal to 1 for $a_1 = 3$, $a_2 = 3$ and $a_3 = 1$.
The use of the files Ex7_2 and Ex7_2Launch provide the basis for the study of the example. A single comparator and 2-level DAC gives rise to instability as can be verified by setting the flag MultiBit to zero. In contrast, the use of a multi-level quantizer without any clipping of the quantized output (that possibly exceeds the dynamic range) obtains the expected noise shaping that leads to an increasing value of the SNR as the quantization step Δ diminishes.
Observe that the number of comparators to be used should be larger than the amount necessary to cover the $\pm V_{ref}$ range as the swing at the output of the last integrator can be higher than the $\pm V_{ref}$ interval. In order to avoid saturation it is therefore necessary to use extra comparators in the ADC and extra levels in the DAC. If, for example, the ADC uses quantization levels separated by $\Delta = 0.25V$, then the swing at the input of the quantizer is around the $\pm 1.5\ V_{ref}$ interval when using a $-3\ dB_{FS}$ input sine wave. Accordingly, the output of the DAC uses 12 levels as shown in Fig. 7.8: 8 levels are

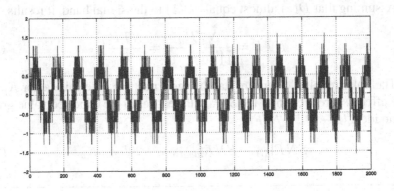

Figure 7.8. Output voltage of the DAC for a $-3dB_{FS}$ input singe wave and $\pm 1V_{ref}$.

> *for the ±1 range while two extra levels are used above and below the dynamic range limits.*
>
> *It can be verified that if $\Delta = 0.5$ V the SNR with OSR = 64 decreases from about 120 dB by 6 dB but the required number of levels diminishes to only 10 as the dynamic range at the input of the quantizer becomes a little more than $\pm 2 V_{ref}$.*
>
> *A clipping of the quantized output can lead to instability: for instance with $\Delta = 0.25$ V it is necessary to use at least 10 quantized level to ensure proper circuit operation as a smaller number of levels causes an unbounded increase of the outputs of the ideal integrators.*

7.2.4   Modulator with Local Feedback

If the zeros of the *NTF* are all at $z = 1$ then the noise attenuation is very good at low frequency giving rise to an optimum noise shaping for large *OSR*s. However, relatively large signal bands do not require a strong localized noise reduction but necessitate a low noise level over the entire band as it is the integral of the spectrum over the signal band that matters, and not the value in a restricted frequency interval. Moreover, since for high order modulators whose zeros are all at $z = 1$ the spectrum increases with f^{2L}, and so the noise power is dominated by the region near the edge of the signal band, f_B.

The use of complex conjugate zeros on the z-circle obtains low noise around the zeros at the expense of a less effective shaping at $z = 1$. Therefore, for a given *SNR* a suitable shift of a pair of zeros from $z = 1$ to complementary positions of the unity circle can increase the usable frequency range. Fig. 7.9 (a) compares the *NTF* of a third order modulator that has all its zeros at $z = 1$

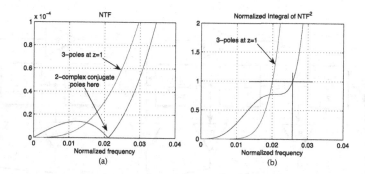

Figure 7.9. (a) NTF of a modulator with three zeros at z = 1 and NTF of a modulator with one zero at z = 1 plus two conjugate zeros. (b) Normalized power voltage versus the signal band.

with the *NTF* of the same modulator with one zero at $z = 1$ and the two other zeros moved to complex conjugate positions corresponding $\pm j\, 0.022\, f_N$. If the oversampling ratio is 50 ($f_B = 0.02 f_N$) the noise power of the first case is higher than when using complex conjugate zeros as that case would enable a bandwidth $f_B = 0.026\, f_N$ with an equal *SNR*. The result is verified by the curves of Fig. 7.9 (b) that give the square integral of the *NTF*s. At low frequency the multiple zeros at $z = 1$ give rise to an excellent noise shaping, but for a band around $0.02\, f_N$ the solution with complex zeros becomes more effective. The zeros cause a flat that extends the usable band for medium-high *SNR*.

The use of local feedback obtains complex conjugate zeros as is achieved by the fifth-order modulator of Fig. 7.10 that uses negative feedback around the pairs of integrators 2–3 and 4–5 with gains equal to g_1 and g_2 respectively. The transfer function of the loop around the first pair of integrators is

$$[X_1(z) - g_1 Y_1(z)] \frac{z^{-1}}{(1 - z^{-1})^2} = Y_1(z) \qquad (7.23)$$

that yields

$$\frac{Y_1}{X_1} = \frac{z^{-1}}{1 - (2 - g_1)z^{-1} + z^{-2}}. \qquad (7.24)$$

A similar result holds for the second resonator loop for which we can write similar equations to determine the z response.

The poles established by equation (7.24), or similarly for the second resonator, are on the unity circle at the angular frequencies

$$\omega_{1p} = \pm \frac{1}{T_s} \arccos\left(1 - \frac{g_1}{2}\right) \simeq \frac{\sqrt{g_1}}{T_s} \qquad (7.25)$$

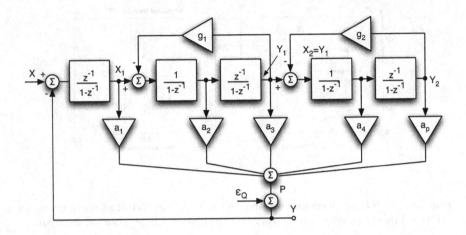

Figure 7.10. Modulator which uses local feedback to obtain complex conjugate zeros.

where the equation holds for $g_1 \ll 1$. Accordingly, the local feedback gains g_1 or g_2 control the position of poles.

Since the poles of the loop filter become zeros of the *NTF*, the gains of the local feedback actually determine the shift of zeros. Therefore, in order to place the zeros in a region around the signal band limit, it is necessary to use gains $g_i \simeq 1/OSR$.

7.2.5 Chain of Integrators with Distributed Feedback

A high-order architecture with multiple feedbacks is a generalization of the already studied second-order modulators that uses two feedbacks from the digital output to the inputs of the integrator used.

The scheme shown in Fig. 7.11 uses the cascade of a number of delayed integrators whose inputs are the previous integrator output minus an amplified (or attenuated) version of the quantized output.

The input P of the quantizer becomes

$$P = \frac{Xz^{-p}}{(1 - z^{-1})^p} - Y \sum_{1}^{p} a_{p-i+1} \frac{z^{-i}}{(1 - z^{-1})^i}. \tag{7.26}$$

The equation describing the modulator is

$$Y = P + \epsilon_Q. \tag{7.27}$$

Solving equations (7.26) and (7.27) leads to an *NTF* with p zeros at $z = 1$. Possible poles of the *NTF* and *STF* are generated by the polynomial

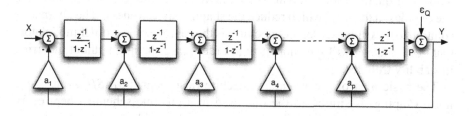

Figure 7.11. Chain of integrators with distributed feedback.

$$D(z) = (1 - z^{-1})^p + \sum_{1}^{p} a_{p-i+1} z^{-i} (1 - z^{-1})^{p-i}. \qquad (7.28)$$

A suitable set of coefficients a_i can move the poles to $z = 0$ giving $D(z) = 1$ and, therefore, giving rise to a STF equal to z^{-p}, and a *p-th* order high-pass noise shaping, $(1 - z^{-1})^p$.

Observe that the input of the generic integrator is always given by the subtraction of two terms

$$V_{in,i} = V_{out,i-1} - a_i V_{out} \qquad (7.29)$$

where V_{out} is the analog conversion of Y, limited within the $\pm V_{ref}$ range. Since at low frequency the input of an integrator must be very small (and zero at *dc*), $V_{out,i-1}$ is very close to $a_i V_{out}$. Therefore, the maximum amplitude of the *(i-1)-th* integrator output is, neglecting the noise contribution, a_i times the reference. Although this value can often be too high (as the feedback factors can be larger than 1), it is possible to adjust the dynamic range of the integrators by using scaling. As already studied an attenuation of the integrator gain followed by a matched amplification does not change the signal and noise transfer functions but optimizes the op-amp dynamic range.

The best procedure for scaling any design is to run simulations using the maximum input amplitude at a number of different frequencies thus determining the produced voltage swing of the internal nodes as a function of frequency. Knowing these maximum and minimum swings enables the optimum choice of the scaling factors whose values, as known, must set the op-amps swing below the saturation limits and obtain a minimum level with very low input amplitudes that is large with respect to the thermal noise.

7.2.6 Cascaded $\Sigma\Delta$ Modulator

The previous study outlined that stability of architectures whose order is higher than 2 is problematic especially for *1-bit* or, more in general, when number of quantization intervals is small. Moreover, the stability requires the use of poles in the *NTF* which reduce the shaping effectiveness. The alternative solution to these high-order architectures is the cascade of low order modulators that obtain, as we will see shortly, high-order noise shaping without incurring in stability troubles.

The basic idea of a cascade architecture (also named *MASH*, *m*ulti-st*a*ge noise *sha*ping) is similar to the one used in a pipeline scheme where every stage, in addition to the digital result, generates the residual to be processed by a subsequent stage. The *MASH* does the same with a $\Sigma\Delta$ modulator in the cells that, further to the digital output, provide the quantization noise as the input to the following stage, as shown in Fig. 7.12. The combination of the digital results is not as simple as the one used in a conventional pipeline, but is made more complex to facilitate the canceling of quantization error contributions.

Assume that the two stages of the scheme of Fig. 7.12 uses $\Sigma\Delta$ orders p_1 and p_2 ($NTF_1 = (1 - z^{-1})^{p_1}$ and $NTF_2 = (1 - z^{-1})^{p_2}$); moreover, the *STF*'s equal to z^{-r_1}, z^{-r_2} respectively. Therefore

$$Y_1 = X \cdot STF_1 + \epsilon_{Q1} NTF_1 = Xz^{-r_1} + \epsilon_{Q1}(1 - z^{-1})^{p_1} \qquad (7.30)$$

$$Y_2 = \epsilon_{Q1} STF_2 + \epsilon_{Q2} NTF_2 = \epsilon_{Q1} z^{-r_2} + \epsilon_{Q2}(1 - z^{-1})^{p_2} \qquad (7.31)$$

which enables the cancellation of ϵ_{Q1} by using the following processing in the digital domain

$$Y_{out} = Y_1 STF_2 - Y_2 NTF_1 \qquad (7.32)$$

and, using the given signal and noise transfer functions, becomes

$$Y_{out} = Y_1 z^{-r_2} - Y_2(1 - z^{-1})^{p_1} = Xz^{-(r_1+r_2)} - \epsilon_{Q2}(1 - z^{-1})^{p_1+p_2}. \quad (7.33)$$

The result shows that the *STF* is just a delay, while the noise is the quantization error of the second modulator, ϵ_{Q2}, shaped by an *NTF* of order $(p_1 + p_2)$.

The result does not depend on ϵ_{Q1} because its effect is cancelled out by the digital processing provided that the *NTF* of the first modulator exactly matches the multiplying term $(1 - z^{-1})^{p_1}$ used in (7.33). However, possible op-amp

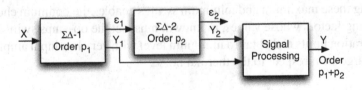

Figure 7.12. Multi-stage noise-shaping architecture.

non-idealities and component mismatches can make the actual *NTF* different from the ideal expected function, causing an incomplete cancellation of ϵ_{Q1}. The residual noise is

$$\epsilon_{n,out} = (NTF_{real} - NTF_{ideal})\epsilon_{Q1}. \qquad (7.34)$$

Assume that $\Sigma\Delta_1$ is a first order scheme and that non-idealities move the zero of the noise transfer function from $z = 1$ to $1 - \delta_I$. The *NTF* becomes $[1 - (1 - \delta_I)z^{-1}]$ yielding a residual noise

$$\epsilon_{n,out,1} = \delta_I z^{-1} \epsilon_{Q1}. \qquad (7.35)$$

The spectrum is white and only reduced by the oversamping ratio. Therefore, the *SNR* does not get worse if the in-band power of ϵ_{Q1} attenuated by δ_I^2 is lower than the one of ϵ_{Q2} shaped by a $(p_1 + p_2)$ order. If the number of bits of $\Sigma\Delta_1$ and $\Sigma\Delta_2$ are equal the spectra of the quantization noise are also equal, therefore

$$\frac{\delta_I^2}{OSR} < \frac{\pi^{2L}}{(2L+1)} \frac{1}{OSR^{2L+1}}; \quad L = p_1 + p_2. \qquad (7.36)$$

Since a finite op-amp gain causes a pole shift δ_I equal to $1/A_0$, equation (7.36) becomes a necessary condition for the required op-amp gain.

Consider now the use of a second order sigma delta in the first cell of the *MASH* and that the op-amp limits cause an equal shift by δ_I of the integrator poles. Since the actual *NTF* becomes $[1 - (1 - \delta_I)z^{-1}]^2$ the effect of ϵ_{Q1} is not totally cancelled but is just attenuated by

$$(1 - z^{-1})^2 - [1 - (1 - \delta_I)z^{-1}]^2$$
$$= \delta_I^2 z^{-2} + 2\delta_I z^{-1}(1 - z^{-1}) \qquad (7.37)$$

made by two terms: one with amplitude δ_I^2 (the square of the first order counterpart) and the other equal to $2\delta_I$ passed through a first order shaping function. Since this value of attenuation is significantly lower than its first order counterpart, it is advisable to use a second-order modulator (which is the maximum order that does not create stability problems) in the first cell of the *MASH*.

The inaccuracy of the generated quantization error is another possible drawback. Since ϵ_{Q1} is the analog difference between the input of ADC_1 and its *DAC* conversion, then both *DAC* and subtractor can give rise to inaccuracies. We don't consider the non-linearity of the *DAC* as its errors can be corrected by digital techniques; we only account for the gain errors of the DAC, δ_D, and the gain error of subtractor, δ_S, leading to an incorrect quantization error, $\epsilon'_{Q,1}$, given by

$$\epsilon'_{Q,1} = [(1 + \delta_D)Y_1 - P_1](1 - \delta_S) \qquad (7.38)$$

which, after simple mathematics, and neglecting $\delta_D \delta_S$, becomes

$$\epsilon'_{Q,1} = \epsilon_{Q,1} + \epsilon_{Q,1}\delta_S + Y_1 \delta_D (1 + \delta_S). \tag{7.39}$$

The use of $\epsilon'_{Q,1}$ in (7.31) and (7.32) gives rise to additional contributions in the output error. $\epsilon_{Q,1}$ is the term cancelled out. The multiplication by NTF of the first $\Sigma\Delta$ shapes $\epsilon_{Q,1}\delta_S$; moreover, remembering the relationship $Y_1 = X \cdot STF_1 + \epsilon_{Q,1} \cdot NTF_1$, the last term of (7.39) produces $X\delta_D$ shaped by NTF_1 and $\epsilon_{Q,1}\delta_D$ shaped by the square of NTF_1. Therefore, disregarding the delay caused by the STF as it is not relevant for this analysis, the gain error of the DAC and subtractor gives rise to

$$\epsilon_{out} \simeq X(\delta_D \cdot NTF_1) + \epsilon_{Q,1}(\delta_S \cdot NTF_1 + \delta_D \cdot NTF_1^2) \tag{7.40}$$

showing that the gain error of the subtractor δ_S is more critical than the gain error of the DAC. Moreover, since the NTF of the first modulator shapes the errors, it is again a convenient choice for the first cell of a second order type.

The $MASH$ architecture can be expanded to the cascade of many stages. A suitable digital processing enables canceling the quantization noise of all but the one of the last stage giving rise to an output noise equal to the spectrum of the last quantization noise shaped by an order equal to the addition of all the orders. The result is significant; however, as studied above, either the mismatch between the analog NTF and its digital estimation or the DAC gain error in generating the quantization errors restrict this benefit. Therefore, because of practical limitations, three is the maximum number of stages that can be profitably used in a $MASH$ architecture.

The order of the used modulators specifies the type of $MASH$ architecture. Thus, for example, we can have a 1-1-1 $MASH$ for the cascade of three first order modulators, a 2-2-2 $MASH$ for two second order followed by a first order modulator and very rarely the cascade of four modulators. A trade-off between shaping benefit and digital filter complexity gives a typical total order between 3 and 6.

Warning

The cascade of $\Sigma\Delta$ modulators relies on the exact knowledge of the noise transfer function and the exact generation of the quantization error to be cancelled. Errors disputing the assumptions can greatly reduce the achievable SNR!

The analysis of a three stages $MASH$ follows a methodology similar to the one developed for the two stages scheme. We perform the study by using the architecture of Fig. 7.13 outlining the block diagram of a *2–1–1 MASH*. The second order modulator uses a first integrator without delay and a second one with delay; analog subtracters obtain the quantization errors.

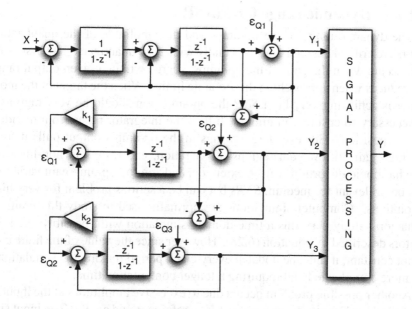

Figure 7.13. Schematic diagram of a 2-1-1 MASH $\Sigma\Delta$ modulator.

The three outputs are

$$\begin{aligned}
Y_1 &= Xz^{-1} + \epsilon_{Q,1}(1-z^{-1})^2 \\
Y_2 &= \epsilon_1 z^{-1} + \epsilon_{Q,2}(1-z^{-1}) \\
Y_3 &= \epsilon_2 z^{-1} + \epsilon_{Q,3}(1-z^{-1})
\end{aligned} \qquad (7.41)$$

which requires the following signal processing to cancel the first and second quantization noises for obtaining the output Y

$$Y = Y_1 z^{-2} - Y_2 z^{-1}(1-z^{-1})^2 + Y_3(1-z^{-1})^3. \qquad (7.42)$$

It can be easily verified that the above equation cancels the $\epsilon_{Q,1}$ and $\epsilon_{Q,2}$ contributions (as required), and yields

$$Y = Xz^{-3} + \epsilon_{Q,3}(1-z^{-1})^4 \qquad (7.43)$$

giving rise to noise shaping of the third quantization noise with order four: the addition of the $2+1+1$ orders used.

Again the cancellation of the quantization error of the first and the second modulator relies on the equality of the analog noise transfer functions and their estimations used in equation (7.42). Any possible mismatch gives rise to noise leakage.

7.2.7 Dynamic range for MASH

The dynamic range of the op-amps and the amplitude of the quantization errors are critical design issues especially with 1-*bit* quantizers. The first problem occurs when the gain of the op-amps is large within a given output range but reduces when approaching the dynamic limits. When the target of the modulator is achieving very high *SNR* the op-amp gain should be very high as it is necessary to ensure a minimum shift of the integrator poles from the ideal $z = 1$ position. However large swings in the op-amp's output pulls it from the high gain to low gain regions of operation, and results in non-linearity. The harmonic distortion, whose extent depends on the op-amp gain variation, can be irrelevant for medium *SNR*s but can be a serious problem for very high resolutions. Computer simulations are normally used to study this limit. A behavioral study performs a time-domain simulation with, possibly, the integrators described by equation (6.28). However, since the value of the finite gain is not constant, it must be updated every clock period making the simulation a bit more complicated and requiring a longer computation time.

Another possible problem occurs due to excessive amplitudes at the input of the following $\Sigma\Delta$ stages. A bipolar $\pm V_{ref}$ reference and a -6 dB_{FS} input sine wave results in a quantization error close to $\pm V_{ref}$ for a single order modulator and larger than $\pm 2.5 V_{ref}$ for a second order modulator, as shown by Fig. 7.14.

Having the modulator input well inside the $\pm V_{ref}$ range is possible but it would require at least two or three bits quantization which, in turn, imposes a very high linearity on the *DAC*s used in the feedback loop and for generating

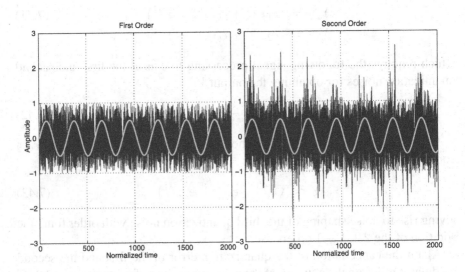

Figure 7.14.   Input and quantization error for a first and a second order 1-bit modulator.

the quantization error. As is known, the error of the *DAC* in the feedback loop gives rise to non-shaped noise, while the error of the *DAC* used to generate the quantization error is responsible for an incomplete noise cancellation. Both limits require using digital methods like the *DEM* studied shortly for ensuring an excellent linearity.

Observe that a signal at the input of a cascaded $\Sigma\Delta$ that is momentarily very close to, or exceeds the reference, can be admitted if the activity caused by the high-frequency terms leave enough room for the low-frequency components. Actually, what is really necessary is that the cascaded modulator makes a proper conversion of the in-band part of the quantization noise as possible inaccuracies of high frequency components are filtered out in the digital domain. That condition occurs when the amplitude of the signal to be converted is not too close to the full scale limits; therefore, the *MASH* start losing their effectiveness unless the input is somewhat below $0\ dB_{FS}$.

An obvious method used for reducing the amplitude of the processed quantization noise is, as shown in Fig. 7.13, by attenuating the input of the second and the third modulator. The method works but is not very practical: the *SNR* can actually reach its maximum at $0\ dB_{FS}$ but the peak value changes as it is necessary to amplify Y_2 by $1/k_1$ and Y_3 by $1(/k_1k_2)$ in the signal processing blocks. This leads to a loss in *SNR* by $20\ log_{10}(k_1 \cdot k_2)$ accompanied by an almost equal extension of the usable range.

> ### High Resolution MASH
>
> For a high resolution MASH it is recommended always using multi-bit modulators to limit the dynamic range of the op-amps and reducing the amplitude of quantization errors to be cancelled out.
> The linearity of the DACs must be enhanced by trimming or dynamic matching of elements.

Example 7.3

Study by computer simulations the operation of the 2-1-1 MASH 1-bit modulator of Fig. 7.13. Determine the plot of SNR as a function of the input amplitude for different conditions of operation. Estimate the effect of the finite gain of the op-amps used in the integrators and their possible clipping.

Solution

The Matlab file Ex7_3 controlled by the file Ex7_3Launch obtains the behavioral model of the 2-1-1 MASH as required by this example. The model enables the following options: setting gain of the four

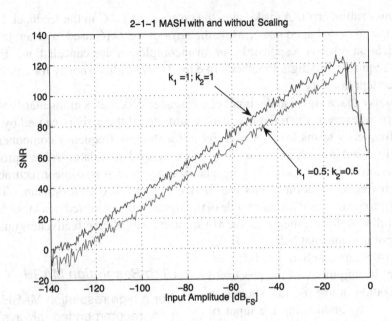

Figure 7.15.   SNR versus the input amplitude with ideal integrators.

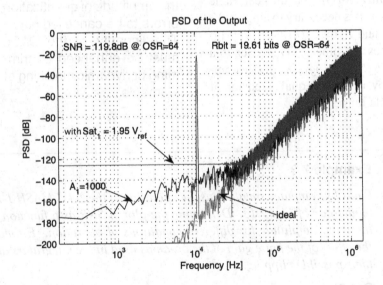

Figure 7.16.   Spectrum with finite gain or saturation in the first integrator of the first modulator.

op-amps; setting the hard saturation of the op-amps outputs; setting of the two scaling factors.

Simulations with scaling factors equal to 1 and 0.5 give the results of Fig. 7.15 for input amplitudes ranging from -140 dB_{FS} to -1 dB_{FS} (for these plots use version a of the launcher). The peak SNR is, with an oversampling equal to 64, around 120 dB, while the zero crossing (that marks the dynamic range) is approximately at the expected -143 dB. The use of the scaling factors moves the peak SNR to a higher input level but because of the less effective operation the plot is shifted down by 12 dB.

Simulations with different values of the design parameters permit the user to understand the limits. For example Fig. 7.16 compares the ideal spectrum with the ones corresponding to finite gain and saturation in the first op-amp. As expected the saturation causes an input referred white noise while the finite gain gives rise to a first order noise shaped term.

7.3   CONTINUOUS-TIME $\Sigma\Delta$ MODULATORS

A continuous-time (CT) modulator moves the interface between continuous-time and sampled-data inside the feedback loop as shown in Fig. 7.17. The discrete-time implementation assumes that the sampling is done before the modulator, giving rise to a processing that is entirely in the sampled-data domain. In contrast, the continuous-time circuit performs the sampling after the loop filter that, being continuous-time, uses an s domain transfer function. Moreover, the output of the *DAC* is assumed to be continuous time so its output is normally provided by a cascaded *S&H*.

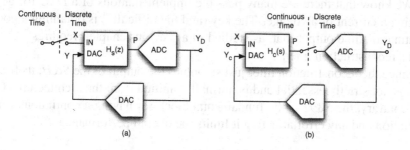

Figure 7.17.   (a) Sampled-data $\Sigma\Delta$ modulator. (b) Continuous-time $\Sigma\Delta$ modulator.

Since sampled-data circuits require the use of *MOS* switches, they are not conveniently implemented in pure bipolar technologies. Whereas the continuous-time modulators can be integrated in any technology, including *CMOS*, *BiCMOS*, and pure bipolar. Moreover, since the sampler of a *CT* modulator is inside the loop, its non-ideal operation is attenuated with the same effectiveness of the quantization noise reduction. On the contrary, the discrete-time counterpart samples the input outside the loop and any *S&H* non-idealities, including the non-linear on-resistance of the input switch, degrade the performance.

Another advantage of the *CT* solution concerns the slew-rate: the step transitions at the input of integrators used in sampled-data modulators require large slew-rates as both feedback and input signals are step functions. In contrast, the effect of the continuous-time input and the steps of the *DAC* is distributed over the entire clock period T.

Also, the supply voltage of a continuous-time modulator can be lower than its sampled-data counterpart since it does not require high voltages to properly switch-on sampled-data paths. As a matter of fact, an op-amp to be used in sampled-data circuit requires at least three overdrive voltages plus $V_{Th,n}$ + $V_{Th,p}$, while a transconductor used in continuous-time analog modulators can work with two overdrives plus one threshold.

Since the main problem of *CT* solutions is, as shown shortly, the limited linearity (which is a feature of most circuit implementations), the above advantages make the *CT* solution suitable for large signal bandwidth and low power consumption.

It was mentioned before that the output of the *DAC*s used for the digital-analog feedback path is continuous-time; therefore, while in a sampled-data circuit the *DAC* gives rise to samples of charge to be integrated in a switched capacitor integrator, the *DAC* of a *CT* scheme must provide a continuos current (or a voltage transformed into current by a resistance) to be injected into a capacitance for performing the integration function.

We know that there are many possible implementations of a *DAC* for generating a current or a voltage. The key (and the difficulty) is obtaining a good continuous-time output that, as studied in a previous chapter, requires a *S&H* and a reconstruction filter.

Since the reconstruction filter just smoothes the output of the *S&H*, its function is not strictly essential and is normally omitted from the architecture. On the contrary, the *S&H* is the fundamental block for having a continuous-time operation and any error affecting it limits the overall performance.

7.3.1   S&H Limitations

The jitter of the phase controlling the hold and the finite rise and fall-time of the generated waveforms limit the *S&H* linearity. To begin with, we will analyze the first limit using the diagram of Fig. 7.18 (a) that represents a typical *DAC* held output and its corresponding error at the *i-th* times, $\epsilon_{j,i}$.

The signal of Fig. 7.18 (a) assumes that because of the clock jitter the first $0 \rightarrow 1$ transition will happen before the expected time, and therefore results in a positive pulse error with amplitude V_{ref} lasting $\delta t_{j,1}$. The next $1 \rightarrow 0$ transition is anticipated by $\delta t_{j,2}$ and causes a negative pulse error and so forth. Therefore, at every $0 \rightarrow 1$ or $1 \rightarrow 0$ transition a positive or negative pulse with random duration $\delta t_{j,i}$ represents the jitter error. If the output of the *DAC* is bipolar $\pm V_{ref}$, the pulse amplitude is $2V_{ref}$.

Assuming that a given input gives rise to a fraction α_{tr} of $0 \rightarrow 1$ or $1 \rightarrow 0$ transitions and the variance of the clock jitter is σ_{ji}^2 with Gaussian distribution, then the noise power is

$$P_{n,DAC,j} = 4V_{ref}^2 \alpha_{tr} \frac{\sigma_{ji}^2}{8T_s^2}. \qquad (7.44)$$

Since the noise power in the signal band must be smaller than the power of the shaped quantization noise, then for an *L-th* order modulator with oversampling ratio *OSR* it is necessary to satisfy

$$\frac{\sigma_{ji}^2}{T_s^2} < \frac{\pi^{2L}}{6\alpha_{tr}(2L+1)} \cdot OSR^{-2L} \qquad (7.45)$$

which is a very demanding request at high resolution.

Assume, for instance, that $f_s = 40\,MHz$, $\alpha = 0.25\ L = 4$ and $OSR = 32$. The use of equation (7.45) gives rise to the condition $\sigma_{ji} < 0.63\,ps$, which is a fairly low figure.

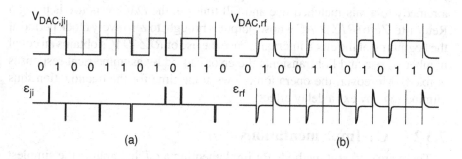

(a)   (b)

Figure 7.18. (a) Signal at the output of the DAC with jitter in the clock and its error. (b) Waveform with finite rise and fall time and its error.

Fig. 7.18 (b) shows the output of the DAC with finite rise or fall times and finite bandwidth. The error is a series of pulses synchronous with the bit transitions and with the same sign as the bit transitions. The shapes and areas of the positive and negative pulse errors are equal and opposite if the rise and fall times are equal; but the shape and area differ for asymmetric responses. These two possibilities cause quite different effects on the output spectrum. Consider, for the former case, to control the DAC with two consecutive ones. The generated signal is a pulse whose duration is $2T_s$ with a finite rise and falling edge. Midway on the output waveform, we can assume to have two equal and opposite pulse errors corresponding to a fictitious transition $1 \rightarrow 0$ superposed to another fictitious $0 \rightarrow 1$ transition. Therefore, we can assume that the DAC generates a non rectangular pulse under the control of the input data, and for this case, the pulse limited rise and fall times are equivalent to a a low-pass filter used after an ideal S&H cascading the DAC.

In contrast, if the pulse error caused by the $1 \rightarrow 0$ transition differs from the $0 \rightarrow 1$ it is necessary to account for the waveform mismatch at all times that, for instance, a $1 \rightarrow 0$ transition occurs causing extra noise proportional to the fall rise time difference and similar to one of the clock jitter.

Another limit is caused by the delay in the feedback signal due to the finite time response of the quantizer. This results in an altering of the NTF and, in turn, the SNR of CT modulators. A possible solution to the problem, and also a remedy to a mis-matched rise and fall times in the DAC response is using a Return-to-Zero (RTZ) DAC whose output is brought to zero every clock period at the beginning of the clock interval. Since the use of RTZ–DACs obtains an equal number of rise and fall transients the error caused by asymmetrical responses is fixed. Moreover, the operation allows some time for the quantization thus eliminating any extra delay effect.

7.3.2   CT Implementations

There are different methods for implementing a CT integrator: the simplest one is an active RC circuit using an op-amp or an OTA. Another possibility that enables tunability is given by employing a MOS transistor whose equivalent resistance is controlled by the gate voltage. The scheme can be made linear by suitable circuit techniques. Other implementations use transconductors for realizing $g_m C$ schemes.

Various combinations of the above approaches lead to the following design solutions:

- Use of all RC integrators to implement low-voltage low-pass modulators.

- Use of *Mosfet-C* integrators with on-line tuning capabilities.

- Use of $g_m C$ integrators with current steering *DAC*s for low-power, medium order modulators used in the audio band.

- Mixed use of RC and $g_m C$ integrators: the use of a first RC stage and the remaining $g_m C$ makes the architecture suitable for high-resolution audio band applications.

- Use of current-mirror based integrators for very low-power.

The above combinations mainly qualify the CT technique as suitable for low-voltage and low-power operation. Indeed, since for high-resolution the linearity of transconductors is not sufficient it is necessary to use an RC integrator in at least the first stage of the loop filter.

An important design request is to make the integrator time-constant proportional to the clock frequency. Obtaining this condition can be problematic as absolute accuracy, voltage and temperature coefficients of integrated capacitance and resistors or transconductors are poor and totally uncorrelated. The situation is different for switched capacitor schemes where the time constant of the integrators depends on capacitive matching and the clock frequency. Therefore, for high-linearity it may be necessary to foresee an on chip trimming of the resistance or using an external component.

The rest of this sub-section describes some of the possible schemes used, stepping from a simple RC active integrator which is well studied in basic circuit books. Fig. 7.19 shows a *Mosfet-C* integrator realized with a pseudo-fully differential op-amp and *MOS* transistors that replace the conventional resistance of an RC configuration. The voltages used to bias the gate of the four *MOS* control the value of the equivalent resistance thus making it possible to regulate the integrator time constant.

The control is typically done on-line by using a tuning circuit that incorporates a matched integrator whose time constant is locked to a precise time control: often the clock of the system or its divisions. The tuning *PLL* controls the gate voltages V_{C1} or V_{C2} of the integrator used in the tuning circuit and the integrators of the CT architecture.

The circuit of Fig. 7.19 uses transistors in the triode region and a cross-coupled scheme to obtain a resistance linearity that is adequate for medium *SFDR* requirements. The use of the op-amp or an *OTA* limits the frequency of operation but the consumed power is generally lower than the one of an *SC*

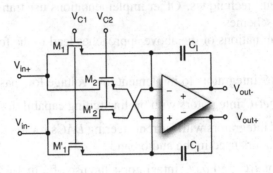

Figure 7.19. MOS-C integrator.

counterpart because of less demanding slew-rate specification. More details on the operation of the scheme can be found in specific literature.

Let us now consider the *MOS* transconductors whose design exploits the almost linear relationship between the differential voltage and the differential current in a *MOS* differential pair, or the almost linear voltage-current relationship of *MOS* transistors in the triode region of operation. Using the first feature leads to a tunable transconductance gain that is (for transistors in saturation) proportional to the square root of the current. However, since the extent of the linear *V-I* region of *MOS* differential pairs is less than twice the overdrive voltage, it is worth using source degeneration as shown in Fig. 7.20 (a) to increase the linearity region by about $R_s I_B$. The transconductance gain depends on both

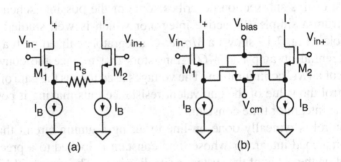

Figure 7.20. (a) Differential transconductor with source degeneration. (b) Source degeneration obtained with MOS in the linear region.

the transconductance of the differential pair and the degenerating resistance

$$G_m = \frac{g_m}{2 + R_s g_m}; \quad g_m = \sqrt{2\mu C_{ox}\frac{W}{L}I_B}. \tag{7.46}$$

The result shows a trade-off between linearity and tuning capability as it is necessary to obtain a relatively large $g_m R_s$ product to increase the linearity and a small $g_m R_s$ product for controlling the value of G_m by changing the bias current as shown by equation (7.46).

Since the linearity of integrated resistances is normally lower than capacitors the *SFDR* of a *CT* modulator with resistor based transconductors is typically lower than the one of its *SC* counterpart, unless the scheme uses thin film or external resistors that, in addition to a low voltage coefficient, enable tunability by trimming of the resistive value. The same feature (but obviously with a worse linearity) is permitted by using *MOS* transistors in the triode region as shown in Fig. 7.20 (b). The gate voltage of the two back-to-back elements and the choice of the common mode voltage regulate the value of the equivalent resistance.

The scheme of Fig. 7.21 (a) uses a transistor in the triode region as the feedback established by the amplifier controls the drain voltage of M_1 and the chosen value of V_d keeps the transisitor in the triode region. The result is a voltage-controlled resistance whose current is given by

$$I_{out\pm} = I_{out,q} \pm \mu C_{ox}\frac{W}{L}V_d V_{in} \tag{7.47}$$

where the input differential voltage is referred to the common mode value V_{cm} that determines the quiescent current $I_{out,q}$. The transconductance tuning is simply obtained by adjusting V_d which also changes the quiescent current level.

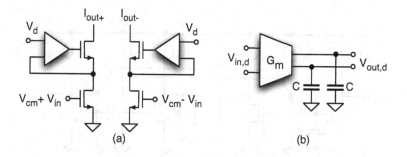

Figure 7.21. (a) Transconductor based on a MOS in the linear region. (b) Use of the transconductor to realize a $g_m C$ integrator.

The response of the scheme is almost linear if the transistors remains in the triode region. For this, it is therefore necessary to use a small value of V_d and a large common mode voltage, conditions that make the absolute value of G_m low and particularly sensitive to the noise on the drain voltage.

Fig. 7.21 (b) shows the use of a differential transconductor for realizing a $g_m C$ integrator. The transconductor output current (which is proportional to the input voltage) is integrated by the capacitance, giving rise to the output voltage. Observe that the result is proportional to the capacitance that receives the current; therefore, the response is also controlled by the top plate parasitic whose linearity can affect the accuracy. If the circuit does not use virtual grounds, the lack of feedback ensures a relatively high-speed operation.

The circuit of Fig. 7.22 is a fully differential current-mode integrator capable of allowing for very-low voltage and low power. The function is obtained thanks to the two cross-coupled current mirrors, equal transistors and equal bias currents. The input of the left mirror that generates I_{out-} is the addition of I_{in+} and a replica of I_{out+}. Symmetrically, the input of the right mirror is the addition of I_{in-} and a replica of I_{out+} and generates I_{out+}. The current injected into the diode connected transistor M_1 (or M_4) with transconductance g_m and capacitance C gives rise to a mirrored current multiplied by

$$H_I(s) = \frac{1}{1 + sC/g_m}. \tag{7.48}$$

Therefore, the differential currents $I_{d,in} = I_{in+} - I_{in-}$ and $I_{d,out} = I_{out+} - I_{out-}$ are related by

$$(I_{d,in} + I_{d,out})H_I(s) = -I_{d,out} \tag{7.49}$$

which, using (7.48), yields

$$I_{d,out} = \frac{I_{d,in}}{sC/g_m} \tag{7.50}$$

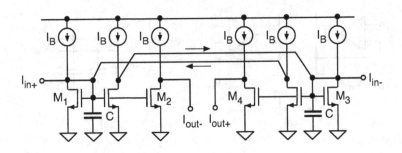

Figure 7.22. Current-mirror based *CT* integrator with currents as input and output.

which is the required continuous-time integration with time constant C/g_m.

The scheme enables tuning by changing the bias current that controls the transconductance of M_1 and M_4. Moreover, using the scheme for very low power results in a very low linearity because the transconductances change significantly when the signal current is a non-negligible fraction of the bias.

7.3.3 Design of CT from Sampled-Data Equivalent

The design of *CT* architectures is more critical than their sampled-data counterpart because the interface between continuous-time and sampled-data processing is inside the loop filter. The solution also depends on the type of waveform generated by the *DAC* that, as mentioned above, can be held for the entire clock cycle (non-return-to-zero, *NRTZ*), or brought to zero at a given point of the clock period (return-to-zero, *RTZ*).

The design of the *CT* modulator is normally done using an already designed sampled-data prototype capable of obtaining the desired features. The design task is therefore the identification of a corresponding *CT* architecture with an equal or very close noise transfer function.

Assume that the loop transfer function of the linear discrete-time model shown in Fig. 7.23 (a) is $H_s(z)$. Since the *CT* counterpart includes the response of the *DAC* (Fig. 7.23 (b)) its loop transfer function in the s-domain is

$$G_c = H_c(s) \cdot H_{DAC}(s). \qquad (7.51)$$

The equivalence of the sampled-data and the continuous time modulator requires

$$H_s(z) = \mathcal{Z}\{H_c(s) \cdot H_{DAC}(s)\} \qquad (7.52)$$

Neglecting speed limitations and a possible *ADC-DAC* delay, the impulse response of the quantizer is a rectangle with duration τ (which can be either a

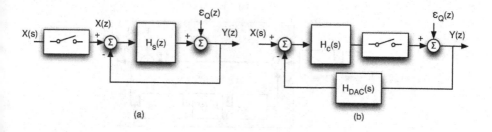

Figure 7.23. (a) Linear model of a sampled-data $\Sigma\Delta$ modulator. (b) Linear model of a continuous-time $\Sigma\Delta$ modulator.

fraction or the entire T_s depending on whether the *DAC* uses *RTZ* or *NRTZ*). The Laplace transform of the rectangle is

$$H_{DAC}(s) = \frac{1 - e^{-sT}}{s} \tag{7.53}$$

which used in (7.52) obtains the equivalence

$$H_s(z) = \mathcal{Z}\{G_c(s)\} = (1 - z^{-\tau/T_s})\mathcal{Z}\left\{\frac{H_c(s)}{s}\right\} \tag{7.54}$$

where $(1 - z^{-\tau/T_s})$ becomes $(1 - z^{-1})$ when using a *NRTZ DAC*.

The above equivalence or the more general relationship (7.52) determines the continuous-time response $H_c(s)$. Since the modulator operates in the time domain, the relationship (7.52) must be re-formulated as

$$\mathcal{Z}^{-1}[H_s(z)] = \mathcal{L}^{-1}\{H_c(s) \cdot H_{DAC}(s)\}. \tag{7.55}$$

Notice that the use of (7.55) requires complicated mathematics because of the time-domain convolutions. The use of optimization *MATLAB* functions simplifies the transformation from sampled-data to continous-time modulators. For a detailed study of the methods that determine $H_c(s)$ from (7.55) the reader should refer to the references given at the end of this chapter.

Let us now study how to estimate the signal and noise transfer functions using sampled-data relationships while accounting for the mixed continuous-

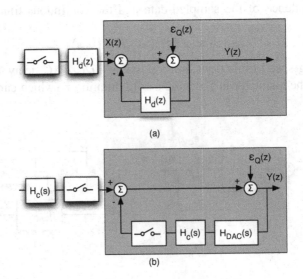

(a)

(b)

Figure 7.24. Linear model equivalents to the ones of Fig. 7.23. (a) Discrete-time. (b) Continuous-time.

time sampled-data processing: indeed, the output of the modulator is sampled-data but a large part of the processing is continuous-time. For this study we will transform the single-loop modulators of Fig. 7.23 into the ones of Fig. 7.24 (a) and Fig. 7.24 (b) by moving the blocks $H_d(z)$ and $H_c(z)$ into the lower part of the loop and compensating by using an equal block in the input branch. The obtained results are obviously equivalent to the generating diagrams of Fig. 7.23. Moreover, since the parts in the gray boxes that process the quantization error are sampled-data, they are suitable for calculating the noise transfer function in the sampled-data domain.

The *NTF* of the scheme of Fig. 7.24 (a) is obviously the one of the sampled-data modulator

$$NTF_d = \frac{1}{1 + H_d(z)} \tag{7.56}$$

while the *NTF* of the *CT* circuit is determined by the $\mathcal{Z}$-transform of the network that obtains in the s domain the response of the cascade of the $H_c(s)$ network and the $H_{DAC}(s)$ *S&H* response

$$H_{CT}^*(z) = \mathcal{Z}\left\{\mathcal{L}^{-1}\left[H_c(s)H_{DAC}(s)\right]\right\}. \tag{7.57}$$

Therefore, it results

$$NTF_c = \frac{1}{1 + H_{CT}^*(z)}. \tag{7.58}$$

Since the design method aims at obtaining the equivalence of the blocks of Fig. 7.24, the continuous-time *NTF* and the sampled-data *NTF* tend to be equal.

The estimation of the *STF* is done by applying the mapping $z \to e^{sT}$ to result given in the s-domain by the circuit of Fig. 7.24 (a). The result is therefore approximated by

$$STF_c(s) = \frac{H_c(s)}{1 + H_d(e^{sT})}. \tag{7.59}$$

7.4   BAND-PASS ΣΔ MODULATOR

If the loop filter has very high gain in a certain interval around a center frequency f_0, then the quantization noise is strongly attenuated in that bandwith and gives rise to a band-pass converter, provided that the digital filter used after the modulator rejects the noise outside the band interval around f_0. The conceptual scheme and the expected signal and noise transfer functions are given in Fig. 7.25.

A straightforward way to obtain a band-pass ΣΔ is by using resonators that move the poles of the loop transfer function $H(z)$ from $z = 1$ to complex

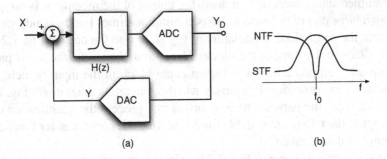

Figure 7.25. (a) Conceptual scheme of a band-pass modulator. (b) Typical STF and NTF.

conjugate positions on the z-unity circle which should be chosen to provide infinite loop gain at the required frequencies. Since

$$STF = \frac{H(z)}{1 + H(z)}; \quad NTF = \frac{1}{1 + H(z)} \tag{7.60}$$

then at the resonation frequency at which $H(z)$ becomes infinite the *NTF* is zero while the *STF* is 1. Far from the resonation frequency the module of $H(z)$ becomes small (and possibly lower than 1) thus causing an attenuation of the out-of-band input components and an amplification of the quantization noise.

The design of a band-pass modulator normally starts from a low-pass proto-type that is then transformed into band-pass by a suitable transformation. The low pass prototype can be a simple first or second order modulator, a cascade architecture or any single loop high-order scheme. The low-pass response is transformed into band-pass around the points $z_{bp} = \pm e^{j\Omega_0}$ by the transformation

$$z^{-1} \rightarrow \frac{z^{-1}cos\Omega_0 - z^{-2}}{1 - z^{-1}cos\Omega_0}. \tag{7.61}$$

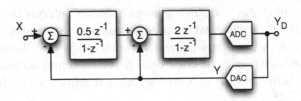

Figure 7.26. Second order modulator used as basis of the LP→ BP transformation.

If, for example, $\Omega_0 = \pm\pi/2$, then $z_{bp} = \pm j$ or $f_{bp} = f_s/4$, which is half of the Nyquist frequency. The corresponding transformation is $z^{-1} \rightarrow -z^{-2}$ and is not very difficult to attain. Another good choice of zero placement is at $\Omega_0 = \pm 2\pi/3$ which changes $(1 - z^{-1})$ into $(1 + z^{-1} + z^{-2})$. This and the previous transformation are achieved with relatively simple implementations using unity capacitances. In contrast, generic transformations of the low-pass prototype can give rise to a transfer function with non-integer coefficients making the circuit implementation difficult, as realizing accurate non-integer capacitances is problematic. The use of non-integer inaccurate elements alters the *NTF* and possibly moves the zeros out of the unity circle degrading the noise cancellation at the resonant frequency.

Suppose now that the transformation $z^{-1} \rightarrow -z^{-2}$ is applied to the low-pass modulator of Fig. 6.23, whose diagram is repeated in Fig. 7.26 for the reader's convenience. Fig. 7.27 (a) shows the same scheme with more details on the integrators implementations. Its transformed version is shown in Fig. 7.27 (b) where a double delay and a subtracter replace the single delay and the adder to obtain the $z^{-1} \rightarrow -z^{-2}$ transformation. The equation describing the transformed integrators is

$$(P - R)z^{-2} = R \quad \rightarrow \quad R = \frac{Pz^{-2}}{1 + z^{-2}} \qquad (7.62)$$

meeting the requirement of the $z^{-1} \rightarrow -z^{-2}$ transformation.

The minus sign in the forward direction is moved to the input of the first block and to the output of the second block. The result is that the sign of the

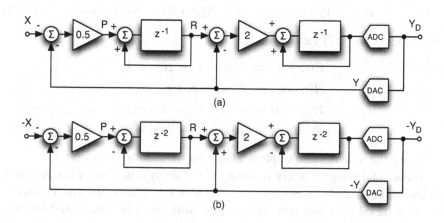

Figure 7.27. (a) Block diagram of a second-order ΣΔ modulator. b) Its fourth-order band-pass ΣΔ counterpart obtained with the $z^{-1} \rightarrow -z^{-2}$ transformation.

feedback injected in the first integrator does not change because the two minus signs cancel one another, while the feedback injected into the second block becomes positive. The input and the output are inverted to compensation for the two effects.

The circuit implementation of scheme of Fig. 7.27 faces two key problems. The first is a subtraction instead of an addition in the processing block: the plus sign means an accumulation that is simply implemented by a sampled-data integrator. In contrast the minus sign would require inverting the output before its use in an analog accumulator (Fig. 7.28 (a)). A convenient solution is to invert input and output every two clock periods as shown in Fig. 7.28 (b). The ± 1 modulation is straightforward in fully differential SC integrator as it is enough to switch the input and output differential connections for obtaining the required modulation.

The time-domain equations describing the circuit are

$$
\begin{aligned}
R'(n+1) &= P(n-1) + R'(n-1) \\
R'(n+2) &= P(n) + R'(n) \\
R'(n+3) &= P(n+1) + R'(n+1) \\
R'(n+4) &= P(n+2) + R'(n+2).
\end{aligned}
\tag{7.63}
$$

Observe that even the output R' is inverted every two clock periods to obtain R, therefore

$$
\begin{aligned}
R(n-1) &= -R'(n-1); \quad R(n) = -R'(n); \\
R(n+1) &= R'(n+1); \quad\;\; R(n+2) = R'(n+2); \\
R(n+3) &= -R'(n+3); \quad R(n+4) = -R'(n+4);
\end{aligned}
\tag{7.64}
$$

which used in (7.62) yields

$$
\begin{aligned}
R(n+1) &= P(n-1) - R(n-1) \\
R(n+2) &= P(n) - R(n) \\
R(n+3) &= P(n+1) - R(n+1) \\
R(n+4) &= P(n+2) - R(n+2)
\end{aligned}
\tag{7.65}
$$

showing the time domain implementation of the required function.

The second design concern is about the z^{-2} delay achievable with an analog delay line or an analog memory. The scheme of Fig. 7.28 (c) uses an alternative solution based on a two path architecture working in an interleaved fashion. The upper path processes the odd samples while the even samples pass through the lower path.

The switching between paths results from the zeros used in the modulation control. Therefore, each path works during the even or the odd periods. In

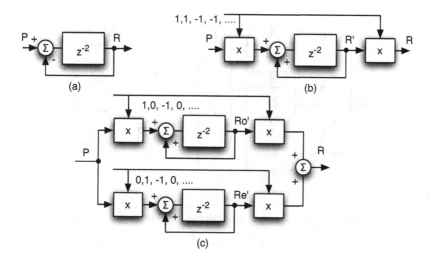

Figure 7.28. (a) Basic block of the band-pass modulator. (b) Implementation with a 2-delay integrator. (c) Time-multiplexed integrator for two-delay.

reality, the circuit implementation takes advantage of the inactivity period by using an operational path frequency of half the sampling rate. This operation reduces the power consumption because at half-clock the power of each half diminishes by approximately 4.

A possible offset mismatch between paths determines a tone at $f_s/2$ while a gain mismatch causes mixing of the input signal with $f_s/2$, as it happens for the time-interleaved Nyquist-rate schemes. Since this is a serious limit (as the input band is around $f_s/2$), then the offset and the gain of the paths must be carefully controlled.

7.4.1 Interleaved N-Path Architecture

The method used in Fig. 7.28 (c) can be viewed as an interleaved arrangement of ΣΔ converters. This interleaved scheme, also known as an N-path ΣΔ modulator, is a viable solution for band-pass converters by using two or more paths. The diagram of Fig. 7.29 is an interleaved 3-*path* ΣΔ architecture with each path working at $f_s/3$. Since the analog input of each path is every third input sample then there is a decimation by 3 before the ΣΔ processing that folds signals around $f_s/3$ down to the signal band at *dc*. The low-pass ΣΔs convert the three decimated signals giving rise to three $f_s/3$ bit streams. These are then combined in the output multiplexer to obtain an f_s data stream.

Realizing a bandpass response uses the same technique employed in sampled-data filters. An N-path filter is made using an interleaved array of low-pass

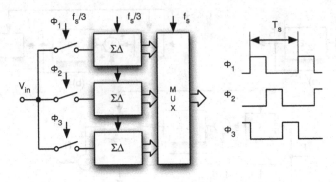

Figure 7.29.   Three-path $\Sigma\Delta$ modulator implementing $z^{-1} \to z^{-3}$.

filters each running at $1/N$ times the main frequency to obtain a $z^{-1} \to z^{-N}$ transformation. If the low-pass filter has a pole or a zero at $z_i = \rho_i e^{j\phi_1}$, then the transformation gives rise to poles or zeros at the N solutions of $z^N = z_i$,

$$z_k = \sqrt[N]{\rho_i}\, e^{(2\pi k + \phi_i)/N} \quad k = 0, \cdots, (N-1) \tag{7.66}$$

that are uniformly distributed around the unity circle and lie on the unity circumference if the initial poles or zeros are on the unity circumference, $\rho_i = 1$.

The study of an N-path $\Sigma\Delta$ modulator is a bit more complex than a simple N-path filter because a $\Sigma\Delta$ scheme processes two signals, the input and the quantization noise. Since the signal transfer function is just a delay, the signal response after the $z^{-1} \to z^{-N}$ transformation is again a delay. Furthermore, since the quantization noise is produced inside the modulator by the action of the quantizers, the system operates on N uncorrelated quantization noises which are shaped by the *NTF* modified by the $z^{-1} \to z^{-N}$ transformation.

Since the *MUX*'s interpolation by N, used to obtain the output at f_s, reduces the noise power of each path by N, then the total noise power of the N uncorrelated noise terms equals the noise power of a single path. Therefore, the total noise spectrum is the white spectrum $\Delta^2/6f_s$ of a single path shaped by the *NTF* of the whole N-path $\Sigma\Delta$.

Fig. 7.30 shows the *NTF* generated by applying a $z^{-1} \to z^{-3}$ transformation to an *NTF* with two zeros at $z = 1$. The resulting zeros are on the unity circumference at

$$z_1 = 1; \quad z_{2,3} = -\frac{1}{2} \pm j\frac{\sqrt{3}}{2}. \tag{7.67}$$

The 2 new pairs of zeros at $f_s/3$ and $2f_s/3$ modify the *NTF* by the addition of two noise shaping regions around $z = -1/2 \pm j\sqrt{3}/2$, in symmetrical positions of the upper and lower parts of the unity circumference. Consequently,

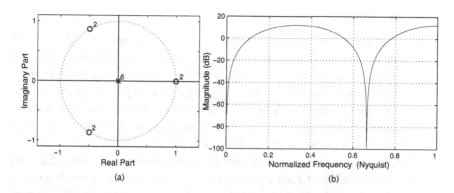

Figure 7.30. (a) Duplication of a doublet of zeros at $z = 1$ because of a $z^{-1} \rightarrow z^{-3}$ transformation and (b) transformed NTF.

the $z^{-1} \rightarrow z^{-3}$ transformation produces a modulator which can be be either low pass or band pass at $f_s/3 = 2f_N/3$ (Fig. 7.30 (b)).

The zeros out of the band of interest give rise to extra terms in the *NTF* whose amplitude at the used frequency f_{in} is

$$k_i = |z_{in} - z_i|; \quad z_{in} = e^{2\pi f_{in}T} \tag{7.68}$$

where z_{in} maps the input frequency and, for our purposes, we should account only for the z_i outside of the band of interest. Since k_i is the distance between points on the unity circumference, its module can be larger than 1 (as happens in the case illustrated in Fig. 7.30) resulting in an overall gain, $\prod_i k_i > 1$. This amplifies the shaped spectrum thus reducing the benefit of noise shaping.

Using an N-path sigma-delta is beneficial for reducing the power consumption as the clock frequency in each path diminishes by a factor N. Accordingly, the power consumption of each path diminishes by a factor of approximately N^2 so the entire N-paths consume about N times less than a single path running at the full clock speed.

Previously we studied that offset, gain mismatch and phase misalignment hamper the performance of interleaved schemes. Since the same limits affect an N-path $\Sigma\Delta$, the method is only suitable for medium resolution and for system specifications that tolerate some harmonic distortion.

Example 7.4

Simulate a three path sigma-delta modulator made by the interleaved connection of second order 1-bit modulators. Verify the SNR reduction caused by additional zeros out of the band of interest and estimate the effect of offset, gain mismatch and clock misalignment.

Solution

The scheme described by file Ex7_4 uses second order modulators with delay in the second integrator and single bit quantization. Control phases that become "on" one sampling period out of three enables the input of the three integrators. The control phases of the second and third paths are delayed versions of the first one by one and two clock periods respectively. The outputs of the three paths are then combined by the output multiplexer.

The used scheme compares the output spectrum of a normal $\Sigma\Delta$ and the one of the 3-path by operating the two manual switches used in the schematic. For this test, the input frequency must be close to zero to obtain a good response from the normal low pass $\Sigma\Delta$.

Since the distance between the extra zeros and the ones at $z = 0$ is $\sqrt{3}$, as shown by zeros diagram of Fig. 7.30 (a), the two pairs of extra zeros yield a total gain of 9. The spectra of Fig. 7.31 confirm this result: at low frequency the 3-path spectrum is about 19 dB higher than the normal case. Furthermore, for OSR = 50 the SNRs of the 1-path and 3-path schemes are 62.9 dB and 44.2 dB respectively. The 18.7 dB difference is almost equal to the expected value of 19.1 dB (corresponding to a gain of 9).

The study of various limitations is done by including an error in one of the paths: the first path can account for a gain error, the second path includes a possible offset and the third path simulates a possible

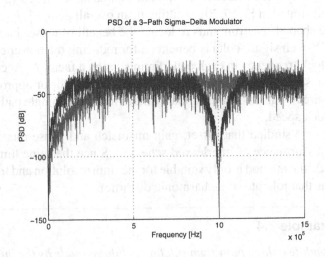

Figure 7.31. Comparison of the output spectrum of a single path and a 3-path $\Sigma\Delta$.

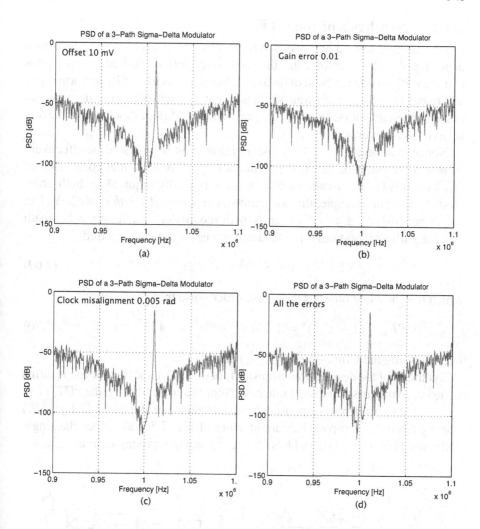

Figure 7.32. Details of the output spectrum with (a) an offset of 10 mV; (b) a gain error of 0.01; (c) a phase misalignment of 0.005 rad; and (d) all the errors together.

clock misalignment by using a replica of the input generator with a phase shift. The three controlling parameters can be specified by the launcher. The use of the following values: Gain error = 0.01, Offset =10 mV, Phase = 0.005 rad, gives rise to the plots of Fig. 7.32. The offset results in a tone at $f_s/3$ while the other errors produce an image of the input tone.

7.4.2   Synthesis of the NTF

Inband tones caused by mismatches are a limiting factor in the performance of an N-path $\Sigma\Delta$. To resolve this limit, the tones must be pushed to frequencies that are well outside the band of interest. This can be achieved by performing an additional $\{z^{-1} \rightarrow -z^{-1}\}$ transformation, but a better solution is the synthesis of a more suitable noise transfer function that, in addition, should avoid the non necessary extra zeros.

Notice that the $z^{-1} \rightarrow z^{-N}$ transformation of an N-path scheme effectively increases the order of the *NTF* polynomial by using a parallel connection of blocks instead of a cascade of blocks, as is typically required by high order modulators. For example, the noise transfer function of a *k-th* order $\Sigma\Delta$ $(1 \pm z^{-1})^k$ becomes $(1 \pm z^{-N})^k$ in an N-path modulator. Therefore, a two-path scheme with a second order modulator has a noise transfer function

$$NTF' = (1 - z^{-2})^2 = 1 - 2z^{-2} + z^{-4} \tag{7.69}$$

which has the same order as the fourth order noise transfer function:

$$NTF_4 = (1 - z^{-1})^4 = 1 - 4z^{-1} + 6z^{-2} - 4z^{-3} + z^{-4} \tag{7.70}$$

but is missing the terms $-4z^{-1}, 8z^{-2}$, and $-4z^{-3}$.

Synthesis of the desired noise transfer function must add these missing terms by using extra cross coupled branches. Suppose to synthesize the *NTF* $(1 + z^{-1} + z^{-2})$ (a band-pass response whose only zeros are at $z = -1/2 \pm j\sqrt{3}/2$) starting from the two-path $\Sigma\Delta$ architecture of Fig. 7.33 (a). Since the single path uses $1/(1 + z^{-2})$ the NTF is $(1 + z^{-2})$ and the missing term is z^{-1}.

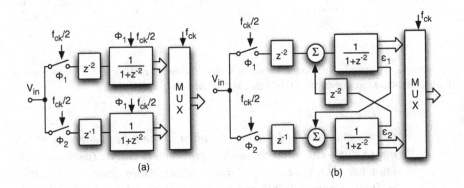

Figure 7.33. (a) Two-path sigma-delta modulator with $1/(1 + z^{-2})$ loop gain. (b) Addition of extra noise terms to obtain the $(1 + z^{-1} + z^{-2})$ band-pass NTF.

Observe that since there is a z^{-1} delay difference between the two paths, than it is possible to take the quantization error of one path, and process it on the other path with a z^{-1} delay. Thus the missing term can be synthesized by the feedback branches of Fig. 7.33 (b) that inject the noise of one path into the other path with a suitable delay. By inspection it results

$$Y_e = X_e z^{-2} + \epsilon_1(1 + z^{-2}) + \epsilon_2 z^{-2}$$
$$Y_o = X_o z^{-2} + \epsilon_2(1 + z^{-2}) + \epsilon_1. \tag{7.71}$$

When the even and odd outputs are combined together the result is

$$Y = Y_e + Y_o z^{-1} = X z^{-1} + \frac{1}{2}(\epsilon_1 + \epsilon_2)(1 + z^{-1} + z^{-2}) \tag{7.72}$$

where the fraction of $1/2$ is due to the interpolation of the two paths that diminishes the power of both quantization noises by 2.

The tones caused by mismatches are pushed far away from the signal band because any offset mismatch results in a tone at $f_N/2$ while the resonant frequency is at $2f_N/3$. Furthermore, a gain mismatch or a clock misalignment causes mirror images at dc and $4f_N/3$. Fig. 7.34 shows the output spectrum of a 1-1-1-*Mash* made by cascading three of the 2-path first order architectures shown in Fig. 7.33. The resonant frequency is at $40\,MHz$ with Nyquist at $60\,MHz$. The offset tone is at $30\,MHz$ and the other mismatch terms are at dc (not visible in the figure as they are below the noise level) and $80\,MHz$.

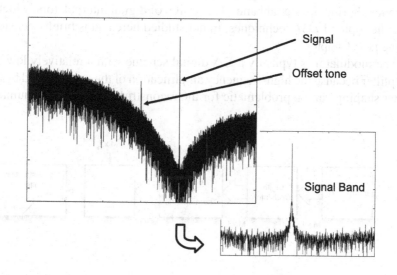

Figure 7.34. Spectrum of the output signal of a 1-1-1 MASH using the scheme of Fig. 7.33.

7.5   OVERSAMPLING DAC

The principle of operation of an oversampled *DAC* is very similar to the one of an oversampled *ADC* as it benefits from increased resolution when using high sampling rates and, for sigma-delta *DAC*s, improves *SNR* using noise shaping. Therefore, the earlier study of oversampling techniques and, in particular, the study of $\Sigma\Delta$ architectures is valid for both analog or digital domain use as the goal of the method is obtaining a suitably low in-band quantization noise. The difference between analog and digital oversampling converters is that for *A/D* the processing is performed in the analog domain and also the continuous-time input is somewhere transformed into sampled-data, while for *D/A* the processing is done in the digital domain to generate a digital result that is then transformed into a continuous-time analog signal by a *D/A* converter with a relaxed number of bits and a reconstruction filter.

Fig. 7.35 indicates the basic functions required by an oversampling *DAC*. The first block is an interpolator that increases the data-rate used for storing or transmitting digital signals to a much higher level: $2f_B \cdot OSR$. Then the digital modulator reduces the number of bits for a possible thermometric representation. The thermometric code is then used to control the low resolution *DAC* which precedes the reconstruction filter.

The design of the interpolator is critical because its specifications must account for the small margin f_B to $f_s - f_B$ normally used for storing or transmitting digital data. The same problem occurs for a Nyquist rate *DAC* but in that case the difficulty is in the design of the analog reconstruction filter. In both situations it is necessary to use a high order filter to reject very close images without altering the signal band. The design of digital interpolators, which is a specific topic of *DSP* techniques, in not studied here and is briefly considered in the next Chapter.

The modulator is typically a $\Sigma\Delta$ digital scheme with a relatively low order loopfilter because the high-frequency amplification of the noise caused by high-order shaping can be problematic for the reconstruction filter. The number of

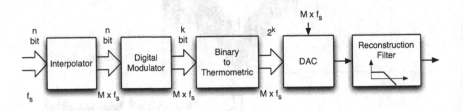

Figure 7.35.   Block diagram of an oversampling DAC.

bits used by a second or third order $\Sigma\Delta$ *DAC* depends on the resolution and the clock-frequency. A single-bit modulator achieves 100 *dB* bits of dynamic range with oversampling ratios ranging between 64 and 256 requiring, with digital audio bands, clock rates between 2.8 *MHz* and 11 *MHz*, which are easily realizable in present *CMOS* technologies. Larger bandwidths and 100 *dB* plus dynamic range would lead to excessive clock rates for the analog sections. Therefore, at a given point, the best design choice is a multi-bit *DAC* with the consequent requirement of matching between the unity elements used to implement the *DAC*.

Because of oversampling the reconstruction filter is not complex but is an essential part of the digital-to-analog conversion as it is necessary to remove not only the images of the signal band but also the high-frequency noise possibly caused by approximations in the digital processing. The reconstruction filter can be incorporated into the *DAC* circuitry or be partially obtained by the low-pass action of the transducer (as it is for audio applications).

7.5.1   1-bit DAC

Sigma-delta *D/A* converters that use 1-*bit* digital modulators must employ a 1-*bit DAC* followed by either a switched-capacitor and/or continuous-time analog filters. The noise specifications of the filter can be challenging for very high resolution because for a desired *SNR* the amount of noise power in the signal band must be $P_{max,in}/SNR$. For example, with a *SNR* of 110 *dB* and $2V_{FS}$ ($P_{max,in} = 0.707\,V^2$) the noise voltage caused by the reconstruction filter must be lower than $2.24\mu V$. For a bandwidth of 20 *kHz* the given figure corresponds to a noise spectrum of $15.84\,nV/\sqrt{Hz}$.

The noise power caused by a resistance or a pair of switched capacitors are

$$V_{n,R}^2 = 4kTRf_B, \quad V_{n,SC}^2 = \frac{2kT}{OSR \cdot C}; \tag{7.73}$$

remembering that $OSR = f_s/(2f_B)$ and that a switched capacitor is equivalent to a resistance $R_{eq} = 1/(Cf_s)$, the above equations becomes

$$V_{n,R}^2 = 4kTRf_B, \quad V_{n,SC}^2 = 4kTR_{eq}f_B. \tag{7.74}$$

Therefore, there is an upper limit to the value of the resistance or the equivalent resistance since $V_{n,R}^2$ (or $V_{n,SC}^2$) must be lower than $P_{max,in}/SNR$.

Suppose to consider only the thermal noise of an input resistance or the *kT/C* noise of two switches used to realize a low-pass time constant of $\tau = 1/(16\pi \cdot f_B)$ for an audio *D/A* converter with $f_B = 10\,kHz$, oversampling ratio 128 and *SNR* 100 *dB*. Since the maximum usable resistance is $7.5\,k\Omega$, the minimum switched capacitance is $51\,pF$ and the time constant is $2\,\mu s$, the *RC* filter should use a filtering capacitance as large as $266\,pF$. This is somewhat problematic because large capacitances occupies large silicon area (the specific capacitance of integrated capacitances is in the order of $4\,fF/\mu^2$). On the contrary the resistance is small and easily obtainable in integrated filters. However, the inaccurate time constant caused by the uncorrelated error or integrated resistors and capacitors requires a larger frequency margin or on-chip trimming of values.

The continuous-time output stages shown in Fig. 7.36 are 1-*bit DACs* that generate a voltage or a current which is fed into a single-pole active filter. There are also other variations that can be used to achieve high-order analog filtering using one or more op-amps. The scheme of Fig. 7.36 (a) also uses a passive filter made by R_1–C_1 that limits the voltage steps at the input of the op-amp by smoothing transients and avoiding any possible non-linear settling of the op-amp. However, the voltage coefficient or the integrated resistors can cause harmonic distortion. Therefore, it is necessary to study the problem and, if necessary, to use external components or integrated thin-film resistors whose voltage coefficient is very low.

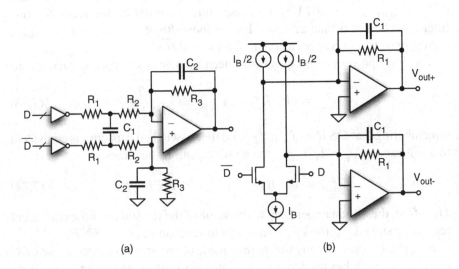

Figure 7.36.   (a) Continuous time 1-bit DAC and filter. (b) Pseudo-differential continuous time 1-bit current-DAC and filter.

The scheme of Fig. 7.36 (b) switches the bias current I_B between the two inputs of a pseudo-differential one-pole filter. The two generators $I_B/2$ compensate for the common mode term by having symmetrical outputs.

A problem with the schemes of Fig. 7.36 is their sensitivity to clock jitter. The point was already discussed for the *DAC* used in a continuous-time $\Sigma\Delta$ modulator; here the situation is similar as the clock edges contaminated by random jitter alter the area of the voltage or current pulse. Since the power of the representing noise is spread over the Nyquist interval the reconstruction filter removes part of the power. Therefore, the error caused by the jitter in the signal band can be estimated by

$$P_{n,ji} = V_{ref}^2 \alpha_{tr} \frac{\sigma_{ji}^2}{2 \cdot OSR \cdot T_s^2} \tag{7.75}$$

where, α_{tr} is the fraction of $1 \rightarrow 0$ and $0 \rightarrow 1$ transitions.

The condition that the noise power given by (7.75) must be lower than $P_{max,in}/SNR$ determines the maximum permitted jitter.

Example 7.5

The reference voltage of a 1-bit $\Sigma\Delta$ DAC is $\pm 1\,V$ and the signal band is 22 kHz with 128 OSR. Determine the clock jitter that ensures a SNR=110 dB with an yield of 99.9%. Assume $\alpha_{tr} = 0.25$.

Solution

The clock frequency is $f_s = 2 \cdot 128 \cdot 22\,kHz = 5.63\,MHz$ leading to $T_s = 0.18\,\mu s$.

Since the required jitter power must be 110 dB below the power of a full scale sine wave and the power of the jitter is proportional to the power of the input, then it is necessary to satisfy the condition

$$\alpha_{tr} \frac{\sigma_{ji}^2}{4 \cdot OSR \cdot T_s^2} = 10^{-11} \tag{7.76}$$

which results in the requirement of $\sigma_{ji} < 25.7\,ps$. Since the normal distribution obtains a yield of 0.999 at 3.3 σ, it is therefore necessary to ensure $\sigma_{ji} < 7.8\,ps$. The obtained figure is not so straightforward and requires a special care in the choice of the clock generator, in the clock regeneration inside the chip and its distribution.

Another problem affecting a 1-*bit* continuous-time *DAC* is its sensitivity to the difference in rise and fall times of a generated signal. This issue was also studied before and the conclusion was that if the rise and fall times are matched then the area error with any sequence of bits does not depend on the sequence. Conversely, if the rise and fall waveforms are different, $1 \rightarrow 0$ transitions uncorrelated with $0 \rightarrow 1$ transitions give rise to an error similar to the one caused by the jitter. The power of this error is given by the rise-time fall-time mismatch.

The use of a multi-bit $\Sigma\Delta$ modulator gives rise to an output voltage that drives the *DAC* up and down by one *LSB* around the expected output level. Since the error caused by the jitter is equal to the amplitude of the output step multiplied by the jitter error, the jitter drawback diminishes with a multi-bit $\Sigma\Delta$ modulator driving a multi-bit *DAC*.

In addition to a reduced jitter sensitivity, a multi level *DAC* allows for a lower *OSR* which simplifies the design of the last stage of the interpolation filter and also reduces the value of the *SC* capacitances used in the reconstruction filter. However, the need for high linearity requires a matching between unity elements that is normally not achieved by integrated capacitors, resistors or current sources. Therefore, regardless on the *DAC* scheme, it is necessary to use trimming, digital calibration or dynamic element matching. These digital assisted techniques, also used for the internal *DAC* of a multi-bit $\Sigma\Delta$ *ADC*, are studied in details the next Chapter.

The imperfect timing synchronization in the control of the unity elements of the *DAC* gives rise to delays between the voltage or current pulses generated by the element control causing glitches. The combination of the pulses generates can be problematic especially for video applications as they give rise to noise and high-frequency tones. The problem is addressed by a combination of synchronization block in the logic controls and careful layout with equalization of interconnections by the use of meanders to match path delays.

The oversampling *DAC* obviously requires a high-frequency clock synchronous with the input digital rate to drive the interpolation filter and the oversampled output *DAC*. In many application the source of digital data is remote, such as digital-audio radio broadcast, cable set-top boxes, and links over a network. In those cases the synchronous high frequency clock must be extracted and synthesized from the audio-bit stream. The function is normally obtained by a phase-locked loop (*PLL*) for clock recovery. Since the jitter cannot be satisfactory in some cases a second *PLL* and discrete time filters are used to preserve the *SNR*.

7.5.2   Double Return-to-zero DAC

The rise and the fall time mismatch in the pulse generated by the *DAC* is solved by the return-to-zero (*RTZ*) code, where each bit is turned off during half of the clock period. The method is illustrated in Fig. 7.37 using a typical 1-*bit* sequence at the input of the *DAC*. While the ideal non-return-to-zero

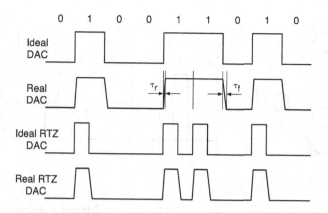

Figure 7.37. DAC voltages for ideal and real cases with normal and RTZ control.

DAC response switches instantaneously from 0 to V_{ref} at the codes transitions, the finite rise and fall responses, represented by constant slopes V_{ref}/τ_r and V_{ref}/τ_f, give rise to trapezoidal pulses. Therefore, a sequence of p consecutive ones gives rise to a pulse whose duration is $pT + \tau_f$ and area is

$$A_{DAC}(p) = V_{ref}(pT + \frac{\tau_f - \tau_r}{2}) = V_{ref} \, pT(1 + \epsilon_p) \qquad (7.77)$$

showing a non linear relative error equal to $(\tau_f - \tau_r)/(2pT)$.

On the contrary, the return to zero coding gives rise to disjoined pulses for any sequence of consecutive ones. The smaller area of the single pulse denotes a lower gain but there is no dependency on the symbol. Therefore, as expected, the method overcomes the rise-time fall-time mismatch. However, the area of the pulse depends on two jitter error one at the rise-time and the other at the fall time. Since the two jitter errors are uncorrelated, the associated noise voltage increases by 3 *dB*. A possible clock misalignment causes a gain error whose effect is normally inessential.

The increase of level transitions is not a problem if the *DAC* is used in a *CT A/D* modulator. However, when using the *DAC* to generate an analog output the augmented high frequency output components require a more complex reconstruction filter.

A solution to this problem is the dual *RTZ* scheme shown in Fig. 7.38 that uses two return-to-zero DACs controlled by the *RTZ* code and its 1/2 clock period delayed version. The two signals are added together to form a continuous output signal over the full clock period giving an output spectrum equal to the one of a conventional *NRTZ*.

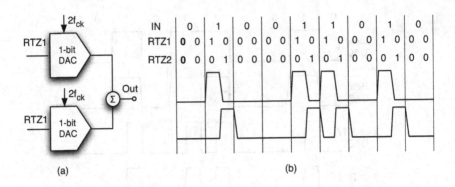

Figure 7.38. (a) Two return-to-zero DACs. (b) Input data and analog waveforms.

Since the spectrum of each *RTZ* signal is symbol independent, the superposition of the two *RTZ* signals is spectrally clean if the small glitches occurring at the rise fall instants give rise to a negligible effect. Indeed, different rise and fall time cause a glitch when the two return to zero pulses are summed up giving rise to a white spectrum and very-high frequency spectral terms. Furthermore, a possible jitter in the clock does not affect the dual *RTZ* scheme if the falling edge is at the same time as the rising edge. Actually the the rise fall time difference causes glitches that occurs all the time the *DAC* input is 1. The effect on the output spectrum is at high frequency and the impact on the *DAC* operation is negligible.

PROBLEMS

7.1 Simulate a $\Sigma\Delta$ second order modulator employing the SNR enhancement method illustrated in Fig. 7.1. Use n = 3 and a 2-level DAC. Estimate the number of levels required by the ADC and determine the SNR degradation caused by the use of op-amp with 40 dB gain. The OSR is 64.

7.2 Simulate the scheme of Fig. 7.2 without the error cancellation branch. Use a third order digital $\Sigma\Delta$ to reduce the number of bits of the DAC in the main feedback loop from 4 to 2. Remember that the digital $\Sigma\Delta$ must have a zero delay.

7.3 Add the the scheme studied in the previous problem the error cancellation branch. Use a second order digital $\Sigma\Delta$ and estimate the effect of a finite gain (60 dB) of the op-amp used.

7.4 Estimate the number of bits given by a fourth order 3-bit $\Sigma\Delta$ modulator (no poles in the NTF). The OSR is 32. Determine the loss caused by a pole in the NTF at $-3f_B$. How the loss changes with different oversampling ratios?

7.5 Design a second order modulator using the weighted feedback summation scheme. Both integrators are with delay. Determine the feedback coefficient that gives an $NTF = (1 - z^{-1})^2$. Simulate the scheme with a behavioral simulator and verify the stability.

7.6 Design a third order modulator using the weighted summation feedback scheme. Determine the position of the zeros and poles of NTF and STF as a function of the parameters used and plot the root locus with a gain block of gain k that replaces the quantizer.

7.7 Use scaling in the third order modulator with weighted feedback summation designed in Example 7.2 to optimize the dynamic range of the integrators. Assume first that the op-amp swing is equal to the input range, and then repeat for an opamp swing of twice the input range. The modulator uses 1-bit.

7.8 Design a resonator made by the cascade of two delayed integrators and a feedback path. The phase of the obtained conjugate zeros must be $\pi/12$.

7.9   Determine the best placement for the zeros of a fourth order low-pass modulator to maximize the SNR. Suppose that the NTF is just $\prod(z - z_i)$ (no denominator). Use OSR = 16. Determine the value of the optimum SNR with a 5-level DAC.

7.10   Design the architecture that obtains the NTF studied in the previous example. Use the configuration of Fig. 7.10 but omit the first integrator and the first feedback branch a_1. Plot the NTF in the 0, $0.1f_s$ interval and the SNR as a function of the input amplitude.

7.11   Use in the Example 7.3 2-bit quantization with and without clipping of the output to $\pm V_{ref}$. The gain used between modulators is 1. Plot the SNR versus the input amplitude with ideal integrator and with integrators made by an op-amp whose gain is 75 dB. The OSR is 32.

7.12   Repeat Example 7.3 but use a 2-2 MASH structure. Compare the results with the one of the 2-1-1 counterpart and try to find advantages and disadvantages of the two solutions. All the quantizers use 1-bit.

7.13   Design a 2-1-1-1-1 MASH architecture with 32 OSR. Estimate the SNR with ideal op-amps and with a finite gain of 60 dB. Compare the result with the SNR obtained with a 2-1-1-1 and a 2-1-1 counterpart.

7.14   For estimating the limit of the difference between rise and fall time of the DAC used in a continuous-time $\Sigma\Delta$ consider a sampled data second order scheme and add to the main DAC different and opposite terms at the $0 \rightarrow 1$ and the $1 \rightarrow 0$ transitions. The errors ranges between 0.01 and 0.001.

7.15   Use the sampled-data model of a second order sigma-delta to study the effect of the clock jitter on a CT architecture. A random number describes the jitter, whose value with a given variance changes the sampling time. Distinguish between the jitter on the DAC injection into the first or the second summing node.

7.16   Simulate at transistor level the Current-mirror based CT integrator with currents as input and output of Fig. 7.22. Use any transistor model available and use W/L = 20. The bias current is 10 μA and the capacitances are 0.2 pF.

7.17 Transform a second order low-pass $\Sigma\Delta$ modulator into a band-pass with resonant frequency at $0.12f_s$. Design the architecture that obtains the result.

7.18 Simulate the scheme of Fig. 7.27 (b) and determine the SNR for a signal band of $\pm f_N/64$ around the resonant frequency. Study the effect of the use of real op-amp including the effects of bandwidth and slew-rate.

7.19 Repeat Example 7.4 but add a $z \to -z$ transformation. The modification moves the NTF zeros at $z = 1$ into $z = -1$ giving a low-pass high-pass transformation. Verify the effect of the transformation on the spur tones given by offset and other mismatches.

7.20 Simulate with Spice the circuits of Fig. 7.36 (a). The supply voltage is 3.3V and the full scale output of the DAC is $\pm 1.5V$. The clock frequency is 64 MHz and the OSR is 128. The required SNR is 100 dB and the input referred noise of the op-amp is 12 nV/$\sqrt{Hz}$. Design the circuits for obtaining a corner frequency at 30 kHz and model the op-amp with finite gain and bandwidth.

7.21 A $\Sigma\Delta$ DAC with OSR 128 is made by a 1-bit second order modulator followed by a DAC without return-to-zero. The rise time of the DAC is 0.03T and the fall time is 0.043T. Estimate with computer simulation at the behavioral level the output spectrum. The input is a sine wave at 1/11 of the Nyquist frequency with -6 dB_{FS} amplitude.

7.22 Repeat Example 7.5 but use a multi-bit DAC. Estimate the jitter specification for 2, 3 and 4 bits.

REFERENCES

Books and Monographs

J. A. Cherry and W. M. Snelgrove: *Continuous-Time Delta-Sigma Modulators for High-Speed A/D Conversion: Theory, Practice and Fundamental Performance*, Kluwer Academic Publishers, Norwell, MA 2000.

A. Rodriguez-Vasquez, F. Medeiro, and E. Janssens: *CMOS Telecom Data Converters*, Kluwer Academic Publishers, Boston, Dordrecht, London, 2003.

Journals and Conference Proceedings

SNR Enhancement

T. Leslie and B. Singh: *An improved sigma-delta modulator architecture*, Proc. IEEE ICASSP, pp. 372–375, May 1990.

S. Kiaei, S. Abdennadher, G. C. Temes, and Y. Yang: *Adaptive digital correction for dual quantization $\Sigma\Delta$ modulators*, IEEE Int. Symp. on Circ. and Syst., vol. 2, pp. 1228–1230, 2003.

J. Yu and F. Maloberti: *A low-power multi-bit $\Sigma\Delta$ modulator in 90-nm digital CMOS without DEM*, IEEE Journal of Solid-State Circuits, vol. 40, pp. 2428–2436, 2005.

High Order

Y. Matsuya, K. Uchimura, A. Iwata, and T. Kaneko: *A 17-bit oversampling D-to-A conversion technology using multistage noise shaping*, IEEE Journal of Solid-State Circuits, vol. 24, pp. 969–975, 1989.

B. P. D. Signore, D. A. Kerth, N. S. Sooch, and E. J. Swanson: *A monolithic 20-b delta-sigma A/D converter*, IEEE Journal of Solid-State Circuits, vol. 25, pp. 131–1317, 1990.

P. Ferguson Jr., A. Ganesan, R. Adams, S. Vincelette, R. Libert, A. Volpe, D. Andreas, A. Charpentier, and J. Dattorro: *An 18b 20KHz dual $\Sigma\Delta$ A/D converter*, IEEE International Solid-State Circuits Conference, vol. XXXIV, pp. 68–69, February 1991.

D. B. Ribner, R. D. Baertsch, S. L. Garverick, D. T. McGrath, J. E. Krisciunas, and T. Fujii: *A third-order multistage sigma-delta modulator with reduced sensitivity to nonidealities*, IEEE Journal of Solid-State Circuits, vol. 26, pp. 1764–1774, 1991.

T. Ritoniemi, E. Pajarre, S. Ingalsuo, T. Husu, V. Eerola, and T. Saramaki: *A stereo audio sigma-delta A/D-converter*, IEEE Journal of Solid-State Circuits, vol. 29, pp. 1514–1523, 1994.

L. A. Williams III and B. A. Wooley: *A third-order sigma-delta modulator with extended dynamic range*, IEEE Journal of Solid-State Circuits, vol. 29, pp. 193–202, 1994.

B.-S. Song: *A fourth-order bandpass delta-sigma modulator with reduced number of op amps*, IEEE Journal of Solid-State Circuits, vol. 30, pp. 1309–1315, 1995.

J. Grilo, I. Galton, K. Wang, and R. G. Montemayor: *A 12-mW ADC deltasigma modulator with 80 dB of dynamic range integrated in a single-chip Bluetooth transceiver*, IEEE Journal of Solid-State Circuits, vol. 37, pp. 271–278, 2002.

J. Morizio, M. Hoke, T. Kocak, C. Geddie, C. Hughes, J. Perry, S. Madhavapeddi, M. H. Hood, G. Lynch, H. Kondoh, T. Kumamoto, T. Okuda, H. Noda, M. Ishiwaki, T. Miki, and M. Nakaya: *14-bit 2.2-MS/s sigma-delta ADC's*, IEEE Journal of Solid-State Circuits, vol. 35, pp. 968–976, 2000.

L. J. Breems, R. Rutten, and G. Wetzker: *A cascaded continuous-time $\Sigma\Delta$ modulator with 67-dB dynamic range in 10-MHz bandwidth*, IEEE Journal of Solid-State Circuits, vol. 39, pp. 2152–2160, 2004.

V. P. Petkov and B. E. Boser: *A fourth-order ΣΔ interface for micromachined inertial sensors*, IEEE Journal of Solid-State Circuits, vol. 40, pp. 1602–1609, 2005.

J. M. de la Rosa, S. Escalera, B. Perez-Verdu, F. Medeiro, O. Guerra, R. del Rio, and A. Rodriguez-Vazquez: *A CMOS 110-dB at 40-kS/s programmable-gain chopper-stabilized third-order 2-1 cascade sigma-delta modulator for low-power high-linearity automotive sensor ASICs*, IEEE Journal of Solid-State Circuits, vol. 40, pp. 2246–2264, 2005.

Continuous Time

J. A. Cherry and W. M. Snelgrove: *Excess loop delay in continuous-time deltasigma modulators*, IEEE Trans. Circuits and Systems-II, vol. 46, pp. 376–389, 1999.

J. A. Cherry and W. M. Snelgrove: *Clock jitter and quantizer metastability in continuous-time delta-sigma modulators*, IEEE Trans. Circuits and Systems-II, vol. 46, pp. 661–676, 1999.

J. A. E. P. van Engelen, R. J. van de Plassche, E. Stikvoort, and A.G. Venes: *A Sixth-Order Continuous-Time Bandpass Sigma-Delta Modulator for Digital Radio IF*. IEEE Journal of Solid-State Circuits, vol. 34 pp. 1753–1764, 1999.

R. H. M. van Veldhoven: *A triple-mode continuous-time ΣΔ modulator with switched-capacitor feedback DAC for a GSM-EDGE/CDMA2000/UMTS receiver*, IEEE Journal of Solid-State Circuits, vol. 38, pp. 2069–2076, 2003.

F. Gerfers, M. Ortmanns, and Y. Manoli: *A 1.5-V 12-bit power-efficient continuous-time third-order sigma-delta modulator*, IEEE Journal of Solid-State Circuits, vol. 38, pp. 1343–1352, August 2003.

S. Jan: *Base-band continuous-time ΣΔ Analog-to-Digital Conversion for ADSL applications*, Ph.D. Thesis, Texas A&M University, 2002.

H. Aboushady: *Design for reuse of current-mode continuous-time ΣΔ Analog-to-Digital Converters*, Ph.D. Thesis, University of Paris VI, 2003.

K. Philips, P. A. C. M. Nuijten, R. L. J. Roovers, A. H. M. van Roermund, F. M. Chavero, M. T. Pallares, and A. Torralba: *A continuous-time ΣΔ ADC with increased immunity to interferers*, IEEE Journal of Solid-State Circuits, vol. 39, pp. 2170–2178, 2004.

Band-pass

J. A. E. P. van Engelen, R. J. van de Plassche, E. Stikvoort, and A. G. Venes: *A sixth-order continuous-time bandpass sigma-delta modulator for digital radio IF*, IEEE Journal of Solid-State Circuits, vol. 34, pp. 1753–1764, 1999.

F. W. Singor and M. Snelgrove: *10.7MHz bandpass delta-sigma A/D modulators*, 1994 IEEE Custom Integrated Circuits Conference, pp. 163–166, 1994.

F. Francesconi, G. Caiulo, V. Liberali, and F. Maloberti: *A 30-mW 10.7-MHz pseudo-N-path sigma-delta band-pass modulator*, IEEE VLSI Symposium, Vol. 10, pp. 60–61, 1996.

Bang-Sup Song: *A fourth-order bandpass delta-sigma modulator with reduced numbers of op amps*, IEEE Journal of Solid-State Circuits, vol. 30, pp. 1309–1315, 1995.

R. Maurino and P. Mole: *A 200-MHz IF 11-bit fourth-order bandpass ΣΔ ADC in SiGe*, IEEE Journal of Solid-State Circuits, vol. 35, pp. 959–967, 2000.

T. O. Salo, S. J. Lindfors, T. M. Hollman, J. A. M. Jarvinen, and K. A. I. Halonen: *80-MHz bandpass sigma-delta modulators for multimode digital IF receivers*, IEEE Journal of Solid-State Circuits, vol. 38, pp. 464–474, 2003.

V. Colonna, G. Gandolfi, F. Stefani, and A. Baschirotto: *A 10.7-MHz self-calibrated switched-capacitor-based multibit second-order bandpass ΣΔ modulator with on-chip switched buffer*, IEEE Journal of Solid-State Circuits, vol. 39, pp. 1341–1346, 2004.

F. Ying and F. Maloberti: *A mirror image free two-path bandpass $\Sigma\Delta$ modulator with 72dB SNR and 86dB SFDR*, IEEE International Solid-State Circuits Conference, vol. XVII, pp. 84–85, 2004.

$\Sigma\Delta$ *DAC*

R. Adams and T. Kwan: *A stereo asynchronous digital sample-rate converter for digital audio*, IEEE Journal of Solid-State Circuits, vol. 29, pp. 481–488, 1994.

P. Ju, K. Suyama, P. F. Ferguson Jr., and Wai Lee: *A 22-kHz multibit switched-capacitor sigma-delta D/A converter with 92 dB dynamic range*, IEEE Journal of Solid-State Circuits, vol. 30, pp. 1316–1325, 1995.

T. Kwan, R. Adams, and R. Libert: *A stereo multibit $\Sigma\Delta$ DAC with asynchronous master-clock interface*, IEEE Journal of Solid-State Circuits, vol. 31, pp. 1881–1887, 1996.

S. Rabii and B. A. Wooley: *A 1.8-V digital-audio sigma-delta modulator in 0.8-μm CMOS*, IEEE Journal of Solid-State Circuits, vol. 32, pp. 783–796, 1997.

V. Peluso, P. Vancorenland, A. M. Marques, M. S. J. Steyaert, and W. Sansen: *A 900-mV low-power $\Sigma\Delta$ A/D converter with 77-dB dynamic range*, IEEE Journal of Solid-State Circuits, vol. 33, pp. 1887–1897, 1998.

I. Fujimori and T. Sugimoto: *A 1.5 V, 4.1 mW dual-channel audio delta-sigma D/A converter*, IEEE Journal of Solid-State Circuits, vol. 33, pp. 1863–1870, 1998.

R. Adams, K. Q. Nguyen, and K. Sweetland: *A 113-dB SNR oversampling DAC with segmented noise-shaped scrambling*, IEEE Journal of Solid-State Circuits, vol. 33, pp. 1871–1878, 1998.

M. Annovazzi, V. Colonna, G. Gandolfi, F. Stefani, and A. Baschirotto: *A low-power 98-dB multibit audio DAC in a standard 3.3-V 0.35-μm CMOS technology*, IEEE Journal of Solid-State Circuits, vol. 37, pp. 825–834, 2002.

Chapter 8

DIGITAL ENHANCEMENT TECHNIQUES

Proper accuracy and component matching in data converter implementations secure demanding features. Since the matching of integrated capacitors or resistors and the accuracy of the transistors parameters is in the order of 0.1%, it is necessary to calibrate the values or to correct the results using digital techniques. Therefore, various methods substantially assist the data converter designer in enhancing the expected performances. In this chapter we shall study approaches of enabling error measurement, thus allowing for their correction or calibration in the analog or digital domain. The methods used can be either on-line (where the converter continues to function), or off-line with a dedicated operation for the error measurement or calibration. We shall also study correction techniques that enhance the spectral performances by a dynamic averaging of elements. Finally it is worth noting that although these methods are suitable for Nyquist-rate schemes, they are more effective with sigma-delta architectures.

8.1   INTRODUCTION

The rapid advancement of digital technology motivates an increasing use of digital techniques that improve the *ADC* or the *DAC* design by the correction or calibration of static and possibly dynamic limitations. Since the use of digital enhancing techniques reduces the need for expensive technologies with special fabrication steps, a side advantage is that the cost of parts is reduced while maintaining good yield, reliability and long-term stability. Indeed, the extra cost of digital processing is normally affordable as the use of modern mixed signal technologies allows for efficient usage of silicon area even for relatively complex algorithms.

The methods studied in this chapter can be classified into the following categories:

- Trimming of elements.
- Foreground calibration.
- Background calibration.
- Dynamic matching.

8.2   ERROR MEASUREMENT

The basis of digital assistance methods is in the measurement of errors. Accurate error measurement allows for their correction by trimming, or by the storing of errors for a successive digital correction or analog calibration. When determining the digital representation of the error, the techniques can use an auxiliary *ADC* or can employ the same data converter used for the normal operation or part of it.

A typical requirement is estimating the difference between elements that are expected to be equal. Fig. 8.1 illustrates a two phase scheme for measuring the difference between C_1 and C_2. During Φ_s the capacitance C_1 samples the voltage V_B, and C_2 and the auxiliary capacitance C_3 are connected to ground. During Φ_e the switching of the roles of C_1 and C_2, and a *DAC* output of zero give rise to a voltage at node A of

$$V_A = V_B \frac{C_2 - C_1}{C_1 + C_2 + C_3} \tag{8.1}$$

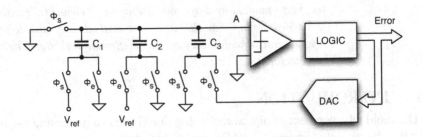

Figure 8.1.   Method for measuring the mismatch between C_1 and C_2.

which is proportional to the mismatch. Observe that the *DAC* and the logic form a successive approximation loop that through the action of C_3 can bring the error to zero. The *DAC* voltage at the end of N successive approximation cycles (the number of bits of the *DAC*) is

$$V_A = \frac{V_B(C_2 - C_1) + (V_{DAC} + \epsilon_Q)C_3}{C_1 + C_2 + C_3} \qquad (8.2)$$

where ϵ_Q is less than the *LSB* of the auxiliary *A/D* conversion.

If $\pm V_{ref}$ and $\Delta = V_{ref}/2^{(N-1)}$ are the references and the quantization step of the *DAC*, the range of measurement and its accuracy are

$$|(C_2 - C_1)|_{max} = \frac{V_{ref}}{V_B}C_3; \quad \Delta C_{mism} = \frac{\Delta}{V_B}C_3. \qquad (8.3)$$

Assuming that $V_B \simeq V_{ref}$ the value of C_3 must be larger than the expected mismatch and the number of bits of the *DAC* gives the accuracy of the measure with C_3 as full scale.

Often, the capacitance C_3 is a built in element of the DAC and is divided into binary weighted or unity elements. The logic switches the fractions of C_3 to $\pm V_{ref}$ or other intermediate values to obtain the redistribution of the charge proportional to the error.

Current based data converter architecture requires the use of another error measurement scheme to estimate the mismatch between a reference and the unity current sources forming, for instance, a current steering *DAC* as shown in

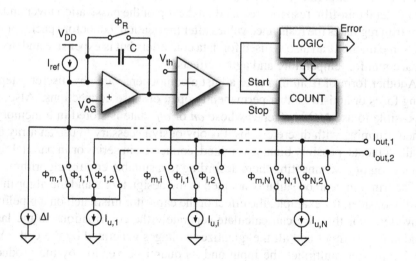

Figure 8.2. Method for measuring the error of unity current sources.

Fig. 8.2. The use of Φ_m sequentially selects one of the current sources which suspends its normal operation to be replaced by an extra unity current generator. The sink current under measure compensates for the reference current coming the positive supply voltage while the extra current ΔI ensures that, within the mismatch measure range, the current entering the virtual ground in always negative. Therefore, $I_{ref} - I_{u,i} - \Delta I$ integrated on the capacitance C gives rise to a ramp whose slope is measured by the V_{th} crossing and a digital counter that starts at the beginning of the measure and stops at T_{stop}

$$T_{stop} = k_{meas}T_{ck} = \frac{CV_{th}}{\Delta I + \delta I_{u,i}}; \quad \delta I_{u,i} = I_{ref} - I_{u,i} \qquad (8.4)$$

enabling the estimation of $\delta I_{u,i}$.

Although this scheme has many possible variants, the basic concept is that an extra element permits a cyclic substitution of one element of the array for its calibration on a dedicated circuit. The digital representation of the error is updated after every measure and used for digital correction of performances.

8.3 TRIMMING OF ELEMENTS

The previous method aimed at obtaining a digital measure of an error that, with some methods, is to be accounted for in a digital correction section. In contrast, the trimming technique slightly changes the value of one or more elements to ensure the required analog accuracy. The trimming method is widely used in integrated filters where it is necessary to regulate the time constant of integrators or other parameters for ensuring an accurate frequency response. The use of metal thin-film resistors realized on the top of the passivation layer and a laser trim regulates the resistance values after fabrication, but before packaging. The same method can also be used for data converters but is expensive and does not account for temperature and aging effects.

Another form of trimming is to adjust a component value by discrete steps using fuses or anti-fuses that permanently open or close connections. Also, it is possible to use *MOS* switches whose *on* or *off* state is stored in a memory. When trimming with discrete steps it is obviously necessary to use an array of small elements, possibly binary weighted, connected in series or in parallel for connecting or disconnecting them according to a suitable control algorithm.

The trimming can be suitably adapted to the design target and the algorithm used. Consider, for example, the effect of the capacitor mismatch on a pipeline converter. For this, we can calculate separately the contributions to the last residue by the input and all the quantized voltages generated by the cell. An ideal converter multiplies the input and its quantized version by the product of all the inter-stage gains for obtaining almost equal values. Any error in the inter-stage gains just causes a fixed gain error in the input term but, more

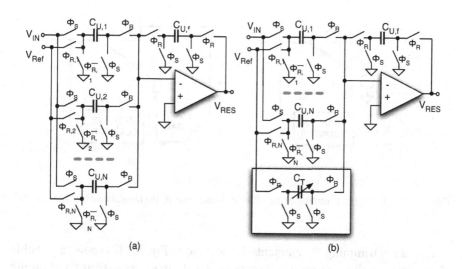

Figure 8.3. Capacitor mismatch correction by equivalent trimming.

critically, gives rise to a non-linearity in the quantized terms. Therefore, what is important is ensuring a precise gain in the paths from the *DAC*s to the output. Since the required accuracy diminishes while moving along the pipeline the above request is particularly stringent for the stages that establish the first few *MSB*s.

Suppose now that the first cell of a pipeline architecture has a sufficient number of bits, n, and that the residual, as shown in Fig. 8.3, is generated by an *M-DAC* scheme with $N = 2^n$ unity capacitances that inject their charge into a single unity feedback capacitance to give a nominal gain of N. Accounting for the real value of used capacitances the residual voltage is

$$V_{res} = \frac{V_{in} \sum_1^N C_{U,i} - V_{Ref} \sum_1^N C_{U,i} b_i}{C_{U,f}} \tag{8.5}$$

where b_i are 1 or 0 according to the *DAC* control.

Assuming $C_{U,i} = C_{U,f}(1 + \epsilon_i)$, equation (8.5) becomes

$$V_{res} = V_{in} \left[N + \sum_1^N \epsilon_i \right] - V_{DAC} \left[k + \sum_1^N \epsilon_i b_i \right] \tag{8.6}$$

Therefore, while a fixed error on the input signal can be tolerated, the error caused by the second term depends on the code and cannot exceed given limits often requiring correction.

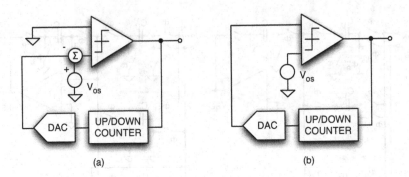

Figure 8.4.   (a) Digital trimming of the offset of a comparator. (b) Digital measure of the offset.

Instead of trimming the elements the scheme of Fig. 8.3 (b) shows a possible alternative solution. An array of binary weighed capacitances forms a trimming element that injects into the virtual ground a negative charge proportional to the reference voltage (a positive charge is also possible by using an inverting configuration). The extra injected charge cancels the error in the second term of equation (8.6) thus eliminating the non-linearity caused by the capacitance mismatch. The sensitivity of the digital correction depends on the capacitances of the array and can be increased by using an attenuated version of the reference voltage.

Suppose now that it is necessary to compensate for the offset of a comparator used in converter architecture. The use of a *DAC* and a digital logic enables the trimming and the measure of the offset as shown by the schemes of Fig. 8.4. The first one, outlining the need for a shift of the lower terminal to cancel its offset, establishes a loop around the comparator with an up/down counter and the *DAC*. After a transient the lower input of the comparator changes between positive and negative giving a correction accuracy lower than 1 *LSB* of the *DAC*.

The scheme of Fig. 8.4 (b) obtains a digital measure of the offset for a possible digital storing and calibration. The circuit is a slight modification of the one used for offset trimming, as it is not necessary to use the addition required for cancellation. Both schemes of Fig. 8.4 are slow *ADC* converters that obtain the result using a minimum number of extra elements.

8.4   FOREGROUND CALIBRATION

The digital trimming of elements is a special case of foreground calibration for which the normal operation of the converter is interrupted for performing the trimming of elements or the mismatch measurement by a dedicated calibration cycle. Because of this operation the technique is named *foreground.*

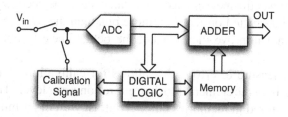

Figure 8.5. Conceptual scheme of a foreground calibrated ADC.

The calibration cycle is normally performed at power-on or during periods of inactivity of the converter, but any miscalibration or sudden environmental changes such as power supply or temperature may make the measured errors invalid. Therefore, for devices that operate for long periods it is necessary to have periodic extra calibration cycles.

Since every trimming or measurement requires one or more clock cycles the calibration of the entire set of elements or parameters can take a significant number of clock cycles. If the measurement cycles are performed just after power is turned on, there is a start-up time that can last fractions of a second.

Fig. 8.5 shows the conceptual scheme of a foreground calibrated *ADC* and Fig. 8.6 of a *DAC*. For the *ADC*, during the measure phase the input connection is disabled and a suitable input is applied to the *ADC*. It can be an analog linear ramp, a sine wave or pseudo-white noise, possibly generated by a slow and accurate on-chip *DAC* driven by an on-chip digital signal. In this case, since the signal is generated by the system itself, the method is called self-calibration.

Various types of inputs can be used to obtain the input-output characteristics and, in turn, the *INL* error. How the differential inputs are generated, and how they are processed to produce the the *INL* will not be discussed in detail here as the next chapter will specifically study the testing procedures. For now it is sufficient to know that the calibration provides the *INL* to be stored in a memory. The input switch restores the data-converter to normal operational after the mismatch measurement and every conversion period the logic uses the output of the *ADC* to properly address the memory that contains the correction quantity. In order to optimize the memory size the stored data should be the minimum word-length which, obviously, depends on technology accuracy and expected *A/D* linearity. As

Observe

The digital addition of the calibrating term to the digital conversion result can give rise to an output digital word with an equal or an increased number of bits.

outlined in the inset the measure accuracy can exceed the accuracy of the *ADC* giving rise to a corrected output with more bits than the *ADC*. Note however, that this does not improve the *SNR* but the better linearity obtains a higher *SFDR*.

The digital measure of errors, that allows for calibration by digital signal processing, can be at the element, block or entire converter level. The calibration parameters are stored in memories but, in contrast with the trimming case, the content of the memories is frequently used, as they are the input of the digital processor. If, for instance, it is required to correct the interstage gain error, ϵ_k, of a two-step converter, the mismatch data is used every clock period to perform the calculation

$$Y_{out} = \frac{Y_C 2^{n_c}(1 + \epsilon_k) + Y_F}{(1 + \epsilon_k)} \tag{8.7}$$

where Y_C and Y_F are the coarse and fine outputs of the two flashes, and n_c is the number of bits of the coarse flash. The processing required by (8.7) involves multiplications and divisions. Often the division can be avoided if the inverse of the denominator is estimated once after the gain measure, or if the system can tolerate a gain error. Since the processing effort is significant anyway, it is necessary to consider carefully the trade-off between digital trimming and a mismatch measurement followed by digital computation.

When the error measure is done at converter level the computation requirements are less demanding as the measure provides all the information necessary to create a look-up table for the *INL*, provided that a monotonic reponse is ensured. Therefore, after every *A/D* conversion, the correction logic must just add the stored *INL* error. Another possibility is to use the look-up table of the input-output characteristics. The memory size is larger than the one containing the *INL* but does not require an adder which operates at the converter speed.

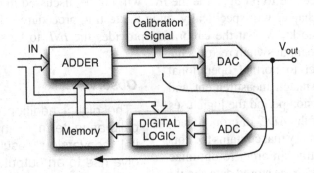

Figure 8.6.   Conceptual scheme of a foreground calibrated DAC.

Fig. 8.6, that describes the operation of a foreground calibrated *DAC*, enables the measure of the response using the blocks of the indicated path. The normal input is disabled to allow for a suitably generated calibration input. An on-board or external high-accuracy *ADC* provides static data necessary for storing the *INL* or the input-output response in the memory. Since the calibration aims at a static result, the converters and the input signal can be very slow as to permit an accurate analog to digital conversion with an on-chip *ADC*.

The need for a specific time-slot for the foreground method can be avoided using circuit redundancy. This enables calibration or error mismatch of a section of the converter while a redundant section replaces the part under calibration and completes the architecture. For the calibration of unity elements, described in Fig. 8.2, only one extra element is required. The concept can be applied to equal stages of a pipeline architecture or, as we will study shortly, to converters of an interleaved scheme.

8.5   BACKGROUND CALIBRATION

Methods using background calibration work during the normal operation of the converter by using extra circuitry that functions all the time synchronously with the converter function. This makes the measurement hidden in the background and does not interfere with the normal converter operation.

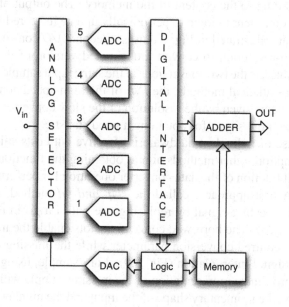

Figure 8.7.   4-path interleaved ADC with an extra-path for background calibration.

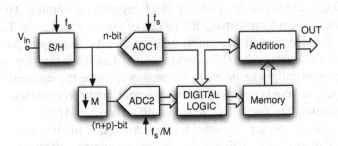

Figure 8.8. Calibration with a slow high-resolution converter.

These circuits use hardware redundancy to perform a background calibration on the fraction of the architecture that is not temporarily used. For example, the interleaved scheme of Fig. 8.7 is a 4-*path* architecture with 5 *ADC*s. While 4 converters obtain the required operation the fifth converter is in calibration mode. The input selector receives two analog signals, one is the main input which is distributed among the paths according to the interleaved algorithm and the other is a test input generated by a *DAC* under the control of the logic. The second of these is used to calibrate the converter.

The digital interface sends the 4 regular outputs to the adder where they are corrected according to the content of the memory. The output of the element under calibration is sent to logic to periodically update its stored *INL*.

The technique illustrated in Fig. 8.8 uses a second *A/D* converter with low-speed and high-resolution to convert a decimated version of the input. The digital logic subtracts the two conversions of the same input samples and obtains, possibly using statistical methods, the *INL* of the high-speed converter with a maximum accuracy given by the resolution of the slow converter. The method relies on a busy input signal for obtaining a good statistic.

Since the use of redundant hardware is effective but costs silicon area and power consumption, other methods aim at obtaining the functionality by borrowing a small fraction of the data converter operation for performing the self-calibration. A first approach, called the *skip and fill* method, exercises the converter with a testing signal by replacing one out of a given number of input samples (say *p*). The borrowed conversion slots enable the measuring and storing of the required conversion parameter while the missing conversion is recovered by digital interpolation (Fig. 8.9). For example, using an *FIR* filter obtains an approximate representation of the missing sample with an accuracy that depends on the momentary shape of the input and the number of taps of the *FIR* filter. Moreover, by delaying the digital output (causing a latency time in

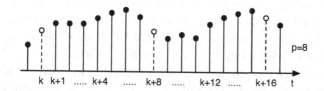

Figure 8.9. Interpolation of a skipped sample replaced by an input used for testing.

the result) the samples both before and after the skipped sample are available, thus allowing an effective non causal filter.

Suppose that $V_{ref}\epsilon_{sk,i}$ is the error in estimating the skipped samples. Since the error power $V_{ref}^2 \overline{\epsilon_{sk}^2}/p$ must be less than the quantization power, it is required to make sure that

$$\overline{\epsilon_{sk}^2} < \frac{p}{12 \cdot 2^{2n}}; \tag{8.8}$$

which, for $p = 16$ and 12-*bit* ($n = 12$), requires an error as small as $\overline{\epsilon_{sk}} < 0.03\%$.

The method shown in Fig. 8.10 is also based on the concept of borrowing time-slots for calibration. To obtain the necessary room the conversion rate is higher than the sampling rate. Assume, as shown in Fig. 8.10 (a) that the sampling and conversion frequency are such that $f_{conv} = 7f_s/6$. Therefore, the lower sampling frequency accommodates for 7 conversions in 6 sampling

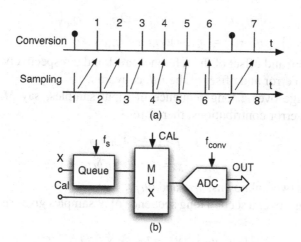

Figure 8.10. (a) Timing scheme of a queue based background calibration. (b) Block diagram of a possible implementation.

periods, thus making the converter available one extra slot every 7. The shift between sampling and conversion times is managed by a queue as shown in Fig. 8.10. The queue consists of one or more $S\&H$s used for storing the input sample between the sampling instant and the time of its conversion. The queue management is such that, for instance, the sample 1 of Fig. 8.10 (a) is held until the next conversion time 1. The same is for the next samples until at the 7-*th* the queue is empty and the slot can be used for calibration.

A simple way to realize the queue is using two interleaved $S\&H$s, however, the mismatch between cells can be problematic. The problem is fixed by a complex clock scheme enabling the use of only one $S\&H$.

8.5.1   Gain and Offset in Interleaved Converters

As known, interleaved architectures are very sensitive to offset and gain channel mismatch because they cause tones at the path frequency, and signal replicas at the frequencies that correspond to the mixing of input and path frequencies. Therefore, even if the applications can tolerate gain or offset errors and the converter used has a good linearity it is necessary to perform calibration to obtaining good matching of offset and gain.

The use of just an extra-path as shown in Fig. 8.7 and a suitable algorithm obtains the result. The extra path, used as a reference, is periodically placed in parallel with one of the interleaved converters. Since the inputs are equal, the difference between the outputs gives the mismatches that, considering only offset and gain errors, is

$$y_{ref}(i) = x_{in}(i) + \epsilon_{Q,r}(i)$$

$$y_p(i) = x_{in}(i)\delta_{G,i} + O_{G,i} + \epsilon_{Q,p}(i)$$

$$y_p(i) - y_{ref}(i) = x_{in}(i)\delta_{G,i} + O_{G,i} + \epsilon_{Q,p}(i) - \epsilon_{Q,r}(i) \qquad (8.9)$$

where the gain and offset of the reference are 1 and 0 respectively and $\delta_{G,i}$ and $O_{G,i}$ are gain error and offset of the *i-th* converter.

The average over a long sequence of input samples, say M, cancels the quantization error contributions; therefore

$$<y_{ref}>_M = <x_{in}>_M$$

$$<y_p - y_{ref}>_M = <x_{in}>_M \, \delta_{G,i} + O_{G,i} \qquad (8.10)$$

that contains two unknowns, $\delta_{G,i}$ and $O_{G,i}$.

The average over a second long sequence of N samples gives rise to a second equation

$$<y_p - y_{ref}>_N = <y_{ref}>_N \, \delta_{G,i} + O_{G,i} \qquad (8.11)$$

that together with (8.10) obtains the gain error and offset of the *p-th* converter.

The process is then repeated by cyclicly connecting the reference converter in parallel to another converter of the interleaved architecture.

8.5.2   Offset Calibration without Redundancy

Amplifiers for instrumentation use the chopper stabilization technique for removing the offset. The input is multiplied by a square wave (± 1) before the amplifier to move the signal spectrum across the modulating frequency (Fig. 8.11 (a)). The spectrum at the output of the amplifier includes the offset, but a second synchronous modulation brings the signal spectrum back to the original position and moves the offset to a high frequency. A low pass filter removes the high-frequency components without disturbing the signal band.

The same method can be used for data converters that, granting some over-sampling, make available a frequency interval (close to the Nyquist frequency) where the signal band can temporarily moved.

Another possibility is chopping the input with a pseudo-random sequence of ± 1 whose spectrum is almost white (Fig. 8.11 (b)). The second chopper control should be the same as the first one but delayed by the *ADC* latency. The input spectrum is transformed into a spread white spectrum and the second synchronous modulator reconstructs the *A/D* conversion of the input. The offset is transformed into white noise whose power in the signal band is further reduced if using oversampling. Any kind of ± 1 modulation is straightforward when using a fully-differential signal as it just necessary to cross couple the input connections after the *S&H*.

Since at the *ADC* output the signal has little power around *dc* it is possible to estimate the offset (and the $1/f$ noise) by filtering for removing the corre-

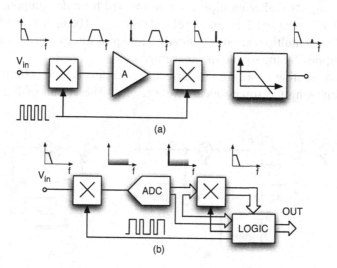

Figure 8.11.   (a) Block diagram of a chopper stabilized amplifier. (b) Modified use of the technique for the ADC offset reduction.

sponding contribution from the digital output thus giving rise to a reduced noise floor caused by offset.

The spectrum of the bit-stream is a measure of the effectiveness of the method. Since tones in the bit-stream mixed with an input signal component can give rise to *dc* terms equivalent to the offset, the accuracy of offset measure is limited. For this reason the method is suitable for interleaved converters with not more than 10-12-*bit*.

8.5.2.1   Gain Calibration in Interleaved Converters

Using an extra channel in the interleaved converters for calibration is costly for architectures with only few channels. The methods discussed in this subsection are more hardware effective but require the use of more complex digital signal processing.

Consider the scheme of Fig. 8.12 that adds a test signal T to the input and subtracts the same signal, possibly delayed, after the *A/D* conversion. With a unity *ADC* gain the subtraction cancels out the added term and the only effect is a reduction in the dynamic range of the converter, which is an acceptable cost with small T amplitude. In contrast, a gain error δ_G would leave a fraction of T in the output, given by

$$Y = X(1 + \delta_G) + T\delta_G + \epsilon_Q \qquad (8.12)$$

$T\delta_G$ can be measured by digital processing as shown in Fig. 8.12 (a). A digital multiplication of Y by T obtains $T^2\delta_G$ that contains a *dc* term equal to $<T^2> \delta_G$. No other multiplication terms will have *dc* components if the average of T is zero and T is not correlated with V_{in}. The operation of the *DSP* after the digital multiplication is basically a low-pass filtering to obtain the gain error multiplied by the square mean average of T.

It is also possible to secure the corrected gain directly thus eliminating the need for knowing the square mean average of T. The circuit of Fig. 8.12 (b)

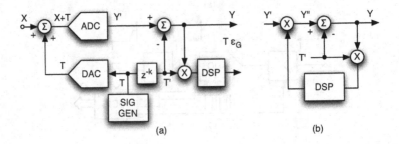

(a)   (b)

Figure 8.12.   (a) Measurement of the gain error of the ADC. (b) Feedback loop that makes the measure independent of the amplitude of T.

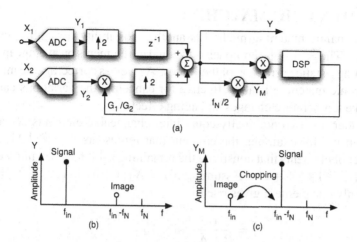

Figure 8.13. (a) Gain calibration scheme. (b) Signal and image caused by gain mismatch. (c) Swapping of signal and image frequency with chopping modulation.

is a modification of the digital processing used in Fig. 8.12 (a) that employs a second digital multiplier for adjusting in feedback the amplitude of Y' with the goal of obtaining zero average at the input of the *DSP*.

Observe that actually the gain measure includes the gain of the *DAC* of Fig. 8.12 (a). Therefore, the absolute accuracy of the result can be impaired by a possible *DAC* gain error. However, for interleaved schemes the use of the same DAC for calibrating the gain of various paths ensures that the gain matching is not affected by the *DAC* gain error.

The method of Fig. 8.13 shows the digital gain mismatch correction for a two-channel time-interleave. The digital output of the first converter is unchanged whereas the second channel is multiplied by the gain correcting factor.

Since the gain mismatch is equivalent to the multiplication of the input by a square-wave, the output spectrum Y contains an image of the input at $f_N - f_{in}$ as shown for a sine wave in Fig. 8.13 (b). The amplitude of the tone is $\delta_G A_{in}/2$.

The digital processing uses two multipliers, one for modulating Y at $f_N/2$ and the other for multiplying Y and its modulated version. Since the modulation by $f_N/2$ that obtains Y_M flips the spectrum around $f_N/2$, the modulated signal has the input tone at the spur position and vice versa. Accordingly, the second multiplication of Y by Y_M determines a *dc* component whose amplitude is proportional to the square of the input amplitude and the gain mismatch.

The role of the *DSP* after the two multipliers is extracting the *dc* signal and controlling in feedback the gain correcting multiplier of the second path for trimming the gain until the *dc* signal is zero. This condition corresponds to perfect gain matching between the two channels.

8.6 DYNAMIC MATCHING

The dynamic matching method is another possibility for calibrating unity elements. The goal of the approach is to equal the elements on average instead of performing a static correction of the values. As well as correcting the mismatch, the dynamic matching is also effective for canceling the variations caused by drifts like temperature changes and aging effects.

Consider just two nominally equal inter-changeable elements X_1 and X_2. They can be, for example, the element that represents the 2^{k+1} bit and the entire set of elements that constitute the remaining k-bit plus one unity element $(1 + 1 + 2 + 4 + \cdots + 2^k)$. Assuming $X_2 = X_1(1 + \delta)$ and $X_1 + X_2 = 1$, the use of only one element gives rise to

$$Y_1 = \frac{1}{2 + \delta} \quad or \quad Y_2 = \frac{1 + \delta}{2 + \delta} \tag{8.13}$$

depending whether the element X_1 or X_2 is used. The resulting errors are equal and opposite

$$\epsilon_{1,2} = \mp \frac{\delta}{2 + \delta} \tag{8.14}$$

The dynamic averaging uses one or the other element under the control of a random bit-stream. Accordingly, the error is randomly positive or negative with spectrum given by the bit-stream control. The total power is approximated by $\delta^2/[(2 + \delta)^2 \cdot 12]$ and is almost uniformly spread over the Nyquist interval.

A possible example of the method is illustrated by the scheme of Fig. 8.14 (a) where the current source I_{ref} is splitted into two equal parts by the transistor

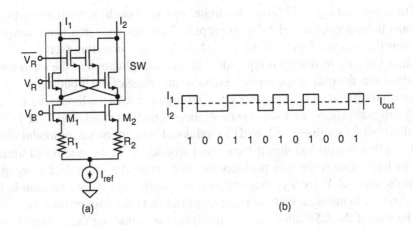

(a)   (b)

Figure 8.14.   (a) Current splitting and dynamic matching of the outputs. (b) Bit-stream control and I_1 current.

pair $M_1 - M_2$. The degeneration resistances R_1 and R_2 improve the matching by attenuating the possible limit given by the thresholds mismatch. However, since the current divider accuracy is also limited by the resistor mismatch, the scheme is not usable for high resolution without calibration or by using the method discussed here: the dynamic matching of the two elements.

The four switches on the top of the current divider obtain the dynamic matching. They use a randomizing signal V_R similar to the one of Fig. 8.14 (b). To make the average of the output currents I_1 and I_2 equal it is required that the number of 1 equals the number of 0.

The method, illustrated for two elements, can be extended to many elements by using a proper algorithm for randomizing the selection pursuing the goal of transforming the mismatch error into a noise-like term. If only a fraction of the Nyquist band is used a possible shaping of the noise-like spectrum further reduces the mismatch limit.

Example 8.1

A 7-bit DAC employs binary weighted elements obtained by the successive division by two of a reference current. Circuits similar to the one of Fig. 8.14 (a) split the currents but they experience a random error caused by the resistor mismatch. Determine the output spectrum with a full-scale sine wave and design a digital correction scheme that randomly exchanges the two divided paths. Since the LSB of this 7-bit DAC is the input of a cascaded stage, then to obtain an overall 10-bit converter the tones must be below -60 dB_{FS}.

Solution

The core part of the dynamic element matching is the random bit stream. A possible generation circuit is the one of Fig. 8.15 that uses an n-bit delay line and simple processing in a closed loop. If the

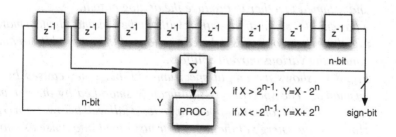

Figure 8.15. Possible generator of a random bit stream.

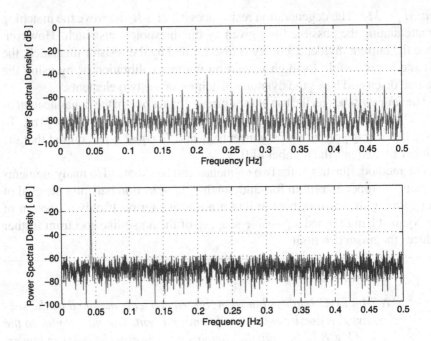

Figure 8.16. Output spectrum with error (top) and with dynamic correction of all the splitting coefficients (bottom).

initial content of the registers is pseudo-random the MSB of the output data is a suitable control for the current splitter. The file Ex7_1 and its launcher provide the behavioral model for this study. A set of random numbers whose variance is 0.1 (a large value for outlining the effects) generates the error in the current splitters. The bits of the digital sine wave are multiplied by a corresponding weighted element and summed together to produce the analog output.

A flag enables simulations with or without the dynamic element matching, that when "on" uses delayed versions of the output bit-stream to control the various current splitters.

Fig. 8.16 shows the output spectrum with big tones caused by the mismatch. The spectrum is significantly smoothed by the dynamic matching as the tones disappear or, at least fall well below -60 dB_{FS}. However, the energy of the tones is transformed into noise as shown by an increase of the noise floor.

8.6.1 Butterfly Randomization

The control of the dynamic matching of many elements as used in *DAC*s with a thermometric selection of unity elements can be problematic. Typically the equal resistors, capacitors or current sources, are summed up using a random representation of the thermometric input as shown in Fig. 8.17. The selection of the unity element is defined by the randomizer that receives N thermometric ones out of M input lines and generates a scrambled set of M controls, N of which are one and the others are zero.

Notice that the number of possible scrambled outputs is $M!$ which is a very large number even for a relatively small number of levels: it is 5040 for $M = 7$, 40320 for $M = 8$ and more than 3.6 million for $M = 10$. Such a large number of connections is difficult to code but also are not strictly necessary for obtaining a random result. What is actually necessary is avoiding frequent repetitions of the same or similar code which would produce tones and not noise-like spectra.

It is typically enough to randomize the elements with a subset of possible connections. A very simple solution uses an M-port barrel shifter which rotates one increment after each clock. A more effective way is the butterfly randomizer shown in Fig. 8.18 consisting of a series of butterfly switches coupling inputs to outputs. The use of $log_2 M$ butterfly stages, as done by the scheme of Fig. 8.18, ensures that any input can be connected to any output; adding more butterfly stages increases the number of possible connections. The control of the butterfly switches can be done using $log_2 M$ bits of a k-*bit RNG* (random number generator) $(k > log_2 M)$. More simply the control can be done by

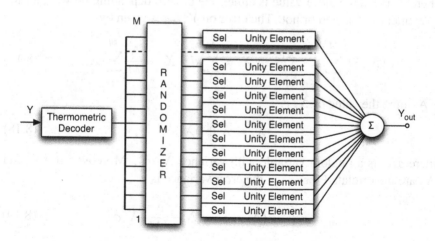

Figure 8.17. Randomization of the selection of unity elements of an M elements (M+1 levels) DAC with thermometric control.

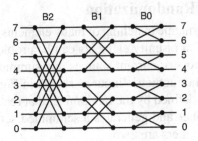

Figure 8.18.   Three stage butterfly randomizer of 8 elements with three bit control.

successive divisions by 2 of the clock. In this case the method is called *clocked averaging.*

If the value of the M elements of the set is X_i, $i = 1, \cdots, M$, then their average is

$$\overline{X} = \frac{1}{M} \sum_1^M X_i; \tag{8.15}$$

the addition of N randomly selected elements gives rise to

$$Y(N) = \sum_1^M d_i \cdot X_i \tag{8.16}$$

where d_i is a flag whose value is either one or zero depending on whether the X_i element is selected or not. The error on $Y(N)$ is given by

$$\epsilon_Y(N) = \sum_1^M d_i \cdot X_i - N\overline{X} = \sum_1^M d_i \cdot X_i - \frac{N}{M} \sum_1^M X_i. \tag{8.17}$$

Assume the value of X_i is

$$X_i = \overline{X} + \delta X_i \tag{8.18}$$

where δX_i is a random variable with variance $\overline{X}^2 \sigma_X^2$. Moreover, if δX_i and δX_j are uncorrelated for $i \neq j$, the error variance is

$$\sigma_y^2 = E\left\{ [\epsilon_Y(N)]^2 \right\} = (N - \frac{N^2}{M})\overline{X}^2 \sigma_X^2 \tag{8.19}$$

which is dependent on the input amplitude, is zero for zero or M elements, and has its maximum at $N = M/2$.

In summary, the element randomization is equivalent to a transformation of mismatch in space into mismatch in time that, assuming the randomizer works properly, eliminates tones in the output spectrum at the expense of an additional noise. Indeed, a busy input with all amplitudes equally probable causes an average mismatch noise power equal to $M\,\overline{X}^2\sigma_X^2/6$.

Since the peak-to-peak output is $M \cdot \overline{X}$, the power of a full scale sine wave is $M^2 \cdot \overline{X}^2/2$. Therefore, the *SNR* determined by just the mismatch error and an *OSR* oversampling results

$$SNR = \frac{3M}{OSR \cdot \sigma_X^2} \qquad (8.20)$$

With $M = 8$, $OSR = 1$ (Nyquist-rate converter), and $\delta = 0.002$, the *SNR* is 65 dB. If $OSR = 32$ the *SNR* improves to 80 dB. What matters most is whether the butterfly randomization is effective in transforming the elements mismatch into noise. With a limited number b of butterfly stages the clocked averaging repeats the same randomized pattern every 2^b clock periods causing tones at $f_s/2^b$, that is problematic especially for small input sine wave components. The use of a pseudo-random generator would require more hardware but is more effective in removing tones especially with small numbers of unity elements.

> **Observe that**
>
> An effective dynamic matching should transform the unity elements mismatch into noise for any input waveform and amplitude.

Example 8.2

A second order sigma-delta modulator with a 3-bit quantizer and OSR = 20 utilizes, in the main feedback path, a DAC with butterfly randomization using three clocked averaging stages. Determine the spectrum of the output and the one of the error caused by a 0.5% random mismatch in the eight elements of the DAC. Compare the results with and without butterfly randomization.

Solution

The second order modulator with an ideal 3-bit DAC and OSR=20 obtains a peak SNR of about 69 dB at −2 dB_FS, no tones and the expected shaping of the spectrum. The results can be verified by using the Simulink file Ex7_2 and its launcher by setting the unity element mismatch to zero (mism = 0).

The three stage, 8-level Butterfly randomizer can be inserted or disabled by a suitable flag. Moreover, although the ADC is ideal, the DAC uses unity elements whose value is affected by a random mismatch. Disabling the randomizer leads to the spectra of the error and

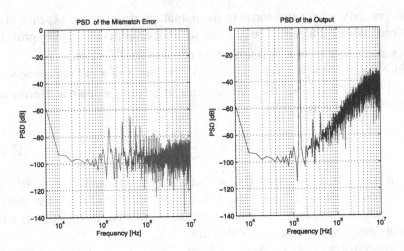

Figure 8.19. Spectrum of the DAC error (left) and spectrum of the output (right) with 0.005 random mismatch.

the modulator output shown in Fig. 8.19. Observe that the DAC error has tones at the input frequency and its multiples. Since the error is injected at the input of the modulator no shaping reduces them as shown by the right diagram of Fig. 8.19 where the second and the third harmonics emerge well visible above the shaped quantization noise.

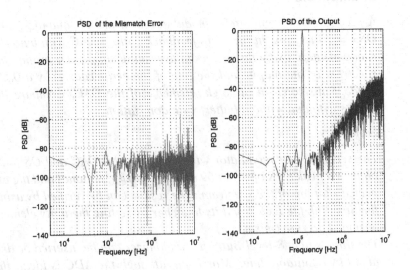

Figure 8.20. DAC error and output spectra with butterfly randomization.

Fig. 8.20 shows that the 0.005 mismatch error gives rise to a more noisy spectrum especially at low frequency; the spectrum has tones at high frequencies that are out of the band of interest. The high frequency tones are not visible in the output spectrum because they become lower than the shaped noise. To observe these tones it would be possibly necessary to use a very long sequence for the FFT.

The simulation results depend on the specific set of random values used; however, the SNDR without randomization is always around 50 dB while with randomization is around 60 dB. The loss with respect to the ideal case, even with randomization, is because at low frequency the spectrum does not go down to zero but becomes flat leaving, possibly without distortion tones, some mismatch power in the signal band.

The randomization of elements transforms the mismatch error that produces tones in the output spectrum into a pseudo noise. Since the result smoothes the tones but does not achieve a reduction in the total error power, then for Nyquist rate data converters the *SNDR* remains almost unchanged while the *SFDR* improves. In contrast, for an oversampling converter, since only a fraction of the Nyquist band is used, the *SNR* improves but with a white mismatch noise spectrum the benefit is just 3 *dB* per octave as it happens for a plain oversampling architecture. Instead, for $\Sigma\Delta$ converters it would be profitable to have, in addition to the quantization noise shaping, a shaping of the mismatch error for pushing part of its power to higher frequencies where it is removed by the digital filter. This is the goal of the methods discussed in the following two sub-sections.

All the methods aim at using all the elements of the array in fast cycles because when all the elements are used the total obtained error is just a gain error. Since a fast cycle gives rise to high frequency noise terms, the spectrum will show more power at high frequency.

8.6.2 Individual Level Averaging

The individual level averaging (*ILA*) approach aims at exercising each unity element with equal probability for each digital input code. The algorithm uses a register of indexes, one index $I_k(i)$ for each possible input code k. The elements that are used when code k is applied are $I_k(i), I_k(i) + 1, \cdots, I_k(i) + k - 1$, where i represents the time-index. When the index exceeds the number of elements, then the selection wraps around and the first element is used.

The result is that successive occurrences of the same code leads to the use of all the unity elements of the array. As mentioned, the method requires using

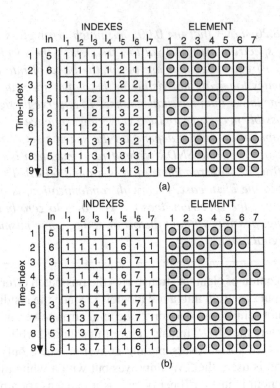

(a)

Time-index	In	I_1	I_2	I_3	I_4	I_5	I_6	I_7	1	2	3	4	5	6	7
1	5	1	1	1	1	1	1	1	○	○	○	○	○		
2	6	1	1	1	1	6	1	1	○	○	○	○	○	○	
3	3	1	1	1	1	6	7	1	○	○	○				
4	5	1	1	4	1	6	7	1	○	○	○			○	○
5	2	1	1	4	1	4	7	1	○	○					
6	3	1	3	4	1	4	7	1			○	○	○		
7	6	1	3	7	1	4	7	1	○	○	○	○	○		○
8	5	1	3	7	1	4	6	1	○			○	○	○	○
9	5	1	3	7	1	1	6	1		○	○	○	○	○	

(b)

Figure 8.21. ILA indexes and usage of 7-elements for a given sequence of input data. (a) Rotation approach. (b) Addition approach.

indexes, as shown in Fig. 8.21, for remembering the used set of elements for converting a given code so a different set is used at the next occurrence of that input code. In this way, after a few conversions of the same input, all the elements of the *DAC* are employed, and the mismatch is averaged out.

There are two methods for selecting elements: the rotation approach and the addition approach. The rotation approach increase the index of the code k, I_k, by 1 each time the code k occurs. The addition method increases by k the index I_k, modulo M, all the times an input whose value is k occurs.

In order to better understand the method consider the cases illustrated in Fig. 8.21 that uses an input sequence made by $\{5\,6\,3\,5\,2\,3\,6\,5\,5\}$ and 7 elements. Initially all the indexes are equal to 1 and the first input data determines the use of the first 5 elements. Then, the index $I_5(2)$ becomes 2 for the rotation approach and 6 for the addition approach. Since the input at the time indexes 2 and 3 the input is not 5, $I_5(3)$ and $I_5(4)$ remain unchanged. The next time the input is 5 is at the time-index 4 and $I_5(4)$ determines the use of different elements (from 2 to 6 for the rotational and $\{6\ 7\ 1\ 2\ 3\}$ for the additional

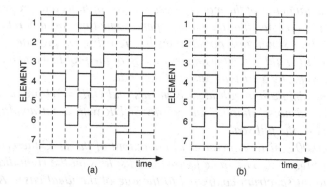

Figure 8.22. 7-elements waveforms for the input sequence of Fig. 8.21. (a) Rotation approach. (b) Addition approach.

method). Then, the index $I_5(5)$ becomes 3 for the rotation approach and 4 for the addition approach.

The *ILA* method obtains a shaping of the mismatch error if every element frequently switches on and off. Consider again the example of Fig. 8.21, every element gives rise to the waveforms of Fig.8.22 showing busy signals for both the rotation and the addition method. Only the elements 2 and 7 are switched once for the rotational approach. All the other elements switch at least twice in the 9 clock periods sequence.

The results are good but a more solid effectiveness of the method should be verified by computer simulations of a $\Sigma\Delta$ multi-bit modulator.

Example 8.3

A second order $\Sigma\Delta$ modulator uses an 8-level ADC with an 8-elements DAC. The modulator uses the ILA for the dynamic matching of elements. Determine the effect of the method with a mismatch between the DAC elements equal to 0.2%.

Solution

This example and the next one are solved by the help of the behavioral models included in the folder SD_DEM. The used scheme, described in SD2_DEM.mdl is a second order modulator whose second integrator output is processed by the subsystem DEM made by two blocks: the ADC and the DEM+DAC. The output of the ADC is the output signal while the second output is used in the mail feedback path. Since the second feedback path is assumed ideal the diagram uses the output of the ADC.

The launcher of the model foresees many options that are set by flags. It is suggested that the reader spend some time for understanding the various simulation possibilities before using the provided files. This would enable a better understanding of the problem and a more general use by a customization of the code.

The simulation with ideal elements and OSR = 64 produces the expected noise shaping and 95 dB SNDR with input amplitude 6 dB below the full scale.

A 0.002 mismatch causes at the output of the DAC the error shown in Fig. 8.23. The same figure also displays on the right diagram the output spectrum compared to the one of the ideal case. As known, since the tones of the error are not shaped, they appear unchanged on the output spectrum.

Observe that, for making sure that the spectra use the same scales, a marker at Nyquist is added to the signals before processing the data for obtaining spectra normalized with respect to the marker amplitude.

The flag "DEM_sel" selects the type of dynamic matching while the flag "nochange" enables to keep unchanged the mismatch for comparing the results of different algorithms with the same conditions. The simulation results for the two possible ILA approaches are shown in Fig. 8.24 and Fig. 8.25. Both methods remove the tones and, also, show an attenuation of the spectrum error at low frequency. However, the addition approach pushes more error power at high frequency as it is verified by the obtained SNDR that is 84 dB for the rotation and 87 dB for the addition method. The difference is not visible

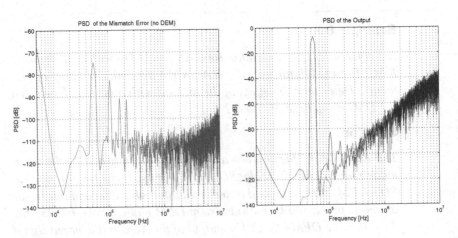

Figure 8.23. Spectrum of the DAC error (0.002 mismatch) and output spectra with and without mismatch.

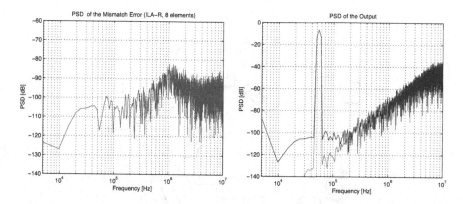

Figure 8.24. Spectrum of the DAC error with the ILA rotation method and output spectra with and without mismatch.

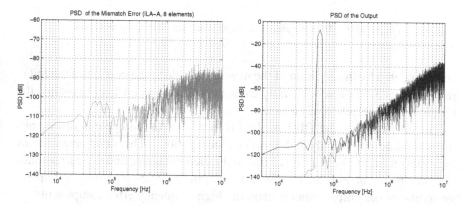

Figure 8.25. Spectrum of the DAC error with the ILA addition method and output spectra with and without mismatch.

> *from the plotted spectra because of the used scales. Both results are less than the ideal case but much better than the SNDR without dynamic matching that is, for the same conditions, as low as 75 dB and depending on the random set of mismatches can go from 65 to maximum 80 dB.*

8.6.3 Data Weighted Averaging

The second approach, early studied for the use with $\Sigma\Delta$ before the *ILA* by the son of the author in his master thesis, is now referred to as data weighted averaging (*DWA*). The method uses just one index, in common with all the input

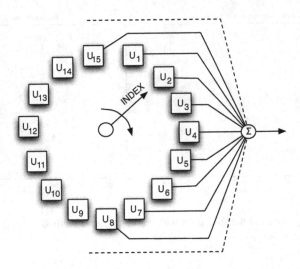

Figure 8.26. Data weighted algorithm selection of 15 unity elements.

codes updated by the addition of the new input code to the content of the index register. Fig. 8.26 depicts 15 unity elements arranges as a wheel to outline the rotation in the selection of elements.

The advantage of the *DWA* method is that the rotation cycle is fast thanks to the update of the only index every clock period. The same sequence of input data used in Fig. 8.21 gives rise to the element usage and the corresponding waveforms of Fig. 8.27. The fast calibration cycles give rise to frequent switching of the unity elements denoting high frequency error components.

The comparison of the *ILA* and *DWA* methods done by computer simulations shows that both give rise to shaping of the mismatch error. The *ILA* seems to be more effective for a small number of elements while the *DWA* method works well for 7 or more elements.

It is relatively easy to show that *DWA* method determines a first order shaping of the mismatch error. For this, suppose that $X_i = \overline{X} + \delta X_i$ is the value of the *i-th* of the M elements. Neglecting a possible gain error, the mismatches verify the condition

$$\sum_{1}^{M} \delta X_i = 0 \tag{8.21}$$

The set of random variables δX_i, $i = 1 \cdots M$, generates another random variable $\Delta_i(k)$

$$\Delta_i(k) = \sum_{i}^{i+k-1} \delta X_k \quad for \; i + k - 1 < M$$

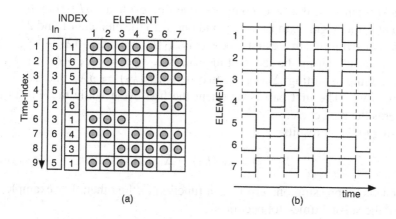

Figure 8.27. (a) DWA index and usage of 7-elements for a given sequence of input data. (b) Elements waveforms.

$$\Delta_i(k) = \sum_{i}^{M} \delta X_k + \sum_{1}^{i+k-1-M} \delta X_k \ \ for \ i+k-1 > M \qquad (8.22)$$

that is the total error given by of the elements used for converting an input k with the *DWA* index pointing at i as first element.

$\Delta_i(k)$ is a set of M^2 random numbers determined by the combination of M values δX_i. One of them is the error of the *DAC* that converts k, as shown in Fig. 8.28 (a). The example of Fig. 8.28 (b) ($M = 8$) depicts the *DWA* operation during two consecutive cycles. The first cycle starts from the first element and uses during the clock period nT the first 3 elements. The index moves to 4 and since the new input equals 4 the *DAC* uses 4 5 6 7. The new index is 7 and

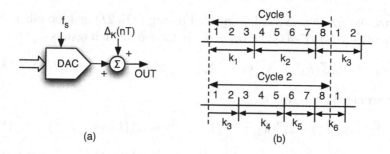

Figure 8.28. (a) Modeling the DWA with an injection of error. (b) Possible two cycles use of elements.

the next input is 3; the *DAC* completes the first cycle at *(nT+2)* by using the element 8 and begins part of the second one with the elements 1 and 2.

The noise injected in the first time-slot is $\Delta_1(k_1) = \delta X_1 + \delta X_2 + \delta X_3$, while the one caused during the second time-slot is $\Delta_2(k_2) = \delta X_4 + \delta X_5 + \delta X_6 + \delta X_7$. Assume splitting $\Delta_3(k_3)$ in two parts $\Delta_3'(k_3) = \delta X_8$ and $\Delta_3''(k_3) = \delta X_1 + \delta X_2$. The first part is pertinent the first cycle while the second part concerns the next cycle.

Using (8.21) it results

$$\Delta_1(k_1) + \Delta_2(k_2) + \Delta_3'(k_3) = 0 \qquad (8.23)$$

that enable expressing one errors as a function of the other. For example, the one of the second time-slot becomes

$$\Delta_2(k_2) = - \left[\Delta_1(k_1) + \Delta_3'(k_3) \right]. \qquad (8.24)$$

The noise of the time-slot nT, $(nT+T)$ and the fraction $\Delta_3'(k_3)$ injected during $(nT+2T)$ in the z-domain give rise to

$$\Delta_1(k_1) - \left[\Delta_1(k_1) + \Delta'(k_3) \right] z + \Delta'(k_3) z^2 \qquad (8.25)$$

that rearranged yields

$$-z^{-1}\Delta_1(k_1)(1 - z^{-1}) + \Delta_3'(k_3)(1 - z^{-1}) \qquad (8.26)$$

showing a first order shaping of the sample injected during nT and the fraction injected during $(nT+2T)$.

Similar considerations for the next clock periods obtain $\Delta_4(k_4)$, $\Delta_5(k_5)$ and the fraction of $\Delta_6(k_6)$ pertinent the second time slot.

The use again of (8.21) obtains

$$\Delta_4(k_4) = - \left[\Delta_3''(k_3) + \Delta_5(k_5) + \Delta_6'(k_6) \right] \qquad (8.27)$$

Since the fraction $\Delta_3''(k_3)$ is injected during $(nT+2T)$ and the other terms during $(nT+3T)$, $(nT+4T)$ and $(nT+5T)$, in the z-domain it results

$$\Delta_4(k_4)z^2 - \left[\Delta_3''(k_3) + \Delta_5(k_5) + \Delta_6'(k_6) \right] z^3 + \Delta_5(k_5)z^4 + \Delta_6'(k_6)z^5 \qquad (8.28)$$

that rearranged gives rise to

$$\left[z^{-1}\Delta_5(k_5) - \Delta''(k_3)z^{-3} \right] (1 - z^{-1}) + \Delta_6'(k_6)(1 - z^{-2}) \qquad (8.29)$$

therefore, two terms are filtered by $(1 - z^{-1})$ while $\Delta_6'(k_6)$ passes through $(1 - z^{-2})$. The latter shaping is because $\Delta_6'(k_6)$ is injected with two period delay of with respect to $\Delta_4(k_4)$ that, by turns, is due to the four clock periods (or fractions of clock periods) necessary to complete the second cycle.

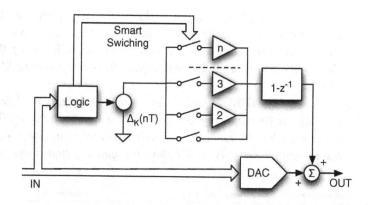

Figure 8.29. Model of the processing performed by the DWA method on the mismatch error samples.

If the input is small, the algorithm completes a cycle in many clock periods. If the input is large, it is worth making the study using the complementary error $\overline{\Delta_i(k)} = \sum_i^M \delta X_k - \Delta_i(k)$ for obtaining similar conclusions.

The above study shows that the mismatch error (or its complement) passes through shaping function $(1 - z^{-d})$, where d is the distance in clock-periods between the sample and the one almost halfway the considered cycle.

Since at low frequency the shaping function becomes

$$(1 - z^{-d}) \simeq d \cdot (1 - z^{-1}); \quad z \to 1, \tag{8.30}$$

the mismatch error is amplified by d and passed through $(1 - z^{-1})$, as shown in the scheme of Fig. 8.29 where a smart switch directs the error to the proper amplification block before the $(1 - z^{-1})$ filter.

Example 8.4

A second order sigma-delta modulator with 8 unity elements in the DAC and OSR = 64 utilizes the butterfly or the DWA randomization in the DAC of the main feedback path. Compare the spectra of output and error caused by a 0.004 random mismatch of elements.

Solution

This example uses the same simulation environment utilized for the previous example as it enables different kind of dynamic matching of elements by changing a flag. As known the butterfly randomization transforms the tones caused by the elements mismatch into noise. This is actually what results by looking at the spectrum of error of Fig. 8.30 that, being added to the input of the modulator, causes the flat until about 400 kHz. Since the signal band is 156 kHz the flat is well above the shaped noise and causes a significant degradation of the SNR. The simulation results shown in Fig. 8.30 corresponds to SNR = 70 dB.

The use of the DWA algorithm gives rise to the expected first order shaping, as shown in Fig. 8.31. The error spectrum has a 20 dB/dec slope until low frequencies. Thanks to the shaping the spectrum of the mismatch error is well below the quantization noise from few

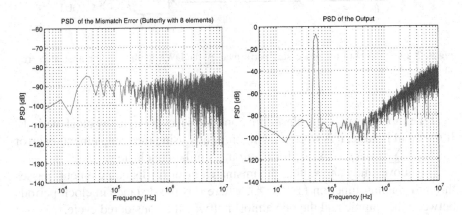

Figure 8.30.   Spectrum of the mismatch error and output spectrum for the Butterfly dynamic matching method.

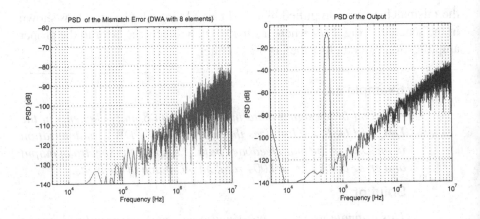

Figure 8.31.   Spectrum of the mismatch error (top) and output spectrum for the DWA dynamic matching method.

kHz. Therefore, since the signal bandwidth is 156 kHz, there is no degradation of the SNR with respect to ideal case even with a 0.4% mismatch error.

The input amplitude equal to −3 dB exercises all the codes at the output of the ADC. On the contrary, a small input exercises only codes around half scale that conclude the cycle in two clock periods. This will cause high frequency tones in the error spectrum.

8.7 DECIMATION AND INTERPOLATION

The digital signal at the output of a $\Sigma\Delta$ modulator is processed in the digital domain for performing two functions: removing the out-of-band shaped noise and reducing the clock-rate by decimation. The two features are one related to the other as decimation means sampling at lower frequency of a high data-rate signal that, as known, requires anti-aliasing for removing the out-of-band spurs.

The interpolation is an opposite operation used for increasing the sampling frequency and obtaining oversampling as required by a $\Sigma\Delta$ *DAC*. The interpolation is performed before the digital $\Sigma\Delta$ modulator as the sampling frequency of digital transmitted or stored data is close to the Nyquist limit. The result is a high resolution oversampled signal that, exploiting the quantization noise shaping, is transformed into a low resolution oversampled *DAC* digital control.

8.7.1   Decimation

As known shaping leaves little noise in the signal band while augmenting the noise near the Nyquist limit. Since in order to preserve the *SNR* it is necessary to have a total aliased noise power much smaller than the in-band noise, high-order modulators require a very high in-band filtering effectiveness.

Consider, for example, an *L-th* order $\Sigma\Delta$ modulator with oversampling *OSR* and k quantization intervals, granting a given *SNR*. Because of the *NTF* the power of the quantization noise in the band near Nyquist is $2^{2L}\Delta^2/(12\cdot OSR)$. On the contrary the noise power in the signal band is very low being the power of a full scale sine wave divided by *SNR*. For avoiding the *NTF* degradation caused by aliasing the filter H_{dec} before decimation must verify the condition

$$\frac{V_{FS}^2}{8}10^{-SNR/10} \gg H_{dec}^2(f_s/2)\frac{2^{2L}V_{FS}^2}{12\cdot k^2\cdot OSR} \qquad (8.31)$$

that, with a modulator that achieves $SNR = 104\,dB$ using, for instance, $L = 3$, $OSR = 32$ and $k = 8$ requires an stop-band gain much lower than $-87\,dB$ (say, at least $OSR/2 \rightarrow 12dB$).

Often, in addition to the amplitude response the system requires a flat group delay in the signal band, feature that is obtained by an all pass filter that adjust the phase or, better, using *FIR* filters.

With large decimation factors it is not convenient using a single block but is more effective performing the operation in successive steps for reducing the complexity of the high frequency signal processing sections. If the decimation factor K_D $(K_D = 2^{k_d})$ is divided into the product

$$K_D = 2^{k_{d_1}} 2^{k_{d_2}} \cdots 2^{k_{d_p}}, \qquad (8.32)$$

the decimation scheme can be like the one of Fig. 8.32 that uses the cascade of four blocks. In the cascaded decimator the clock of the cells goes down step by step giving rise to successive aliasing that requires for every step an in-band low residual noise.

Since the first filter operates at the maximum speed it is necessary using for it simple architectures like a *sinc FIR* filter. If, for example, the first decimation is by 4 out of a 64 total, the critical frequency intervals are $(31 \rightarrow 32)/64 f_N$, $(32 \rightarrow 33)/64 f_N$, and $(63 \rightarrow 64)/64 f_N$. Therefore the first filter should just take care of those intervals. Assuming now that the spur from those regions critical are properly removed the sampling frequency of the second stage can reduce by M_1 for a further filtering before the second decimation and, possibly a third decimation. The scheme of the second and the third filter is often half-band, meaning that the coefficients of every second tap (except the central one) are equal to zero.

The advantage of the cascade solution is, as mentioned, a reduced hardware complexity but, also, reduced power consumption because a lower clock frequency diminishes the power usage.

Decimator Architecture

The architecture of a decimation must obtain minimum hardware complexity and power consumption. Performing decimation in successive steps is the best choice.

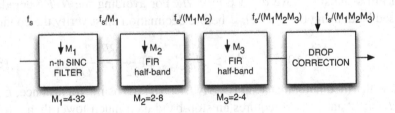

Figure 8.32. Cascade of decimation cells for realizing the decimator.

A *sinc* filter estimates the running average of N consecutive input samples whose output is

$$Y_{sinc}(n) = \frac{1}{N} \sum_{0}^{N-1} X(n-i). \qquad (8.33)$$

A signal passed through the *sinc* filter obtains the signal transfer function

$$H_s(z) = \frac{1}{N} \sum_{0}^{N-1} z^{-i} = \frac{1}{N} \frac{1 - z^{-N}}{1 - z^{-1}}; \qquad (8.34)$$

that has $(N-1)$ zeros at $z = \sqrt[N]{-1}$ excluded $z = 1$.

Using a *sinc* filter is efficient because it attenuates the input amplitude only at the required frequencies but the obtained attenuation cannot be enough for ensuring the desired noise rejection. Actually, for an *L-th* order modulator it is required using a $sinc^{L+1}$ filter because, as shown in Fig. 8.33 for $N = 4$, a high order *sinc* well improves the noise rejection at the critical frequencies.

Let us verify that the simple *sinc* filter is not an adequate anti-aliasing filter for a first order $\Sigma\Delta$ that diminishes its oversampling from *OSR* to *OSR'=OSR/N* with a decimation by N.

The spectrum of the quantization noise, shaped by $(1 - z^{-1})^2$ and filtered by $H_n^2(z)$, is

$$v_{n,out}^2(z) = v_{n,Q}^2 (1 - z^{-1})^2 \frac{1}{N^2} \left[\frac{1 - z^{-N}}{1 - z^{-1}} \right]^2 = \frac{v_{n,Q}^2}{N^2} (1 - z^{-N})^2. \qquad (8.35)$$

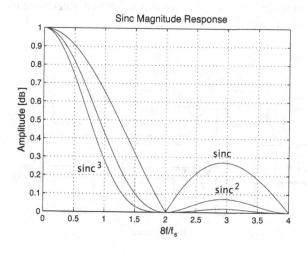

Figure 8.33. Frequency response of *sinc*, $sinc^2$ and $sinc^3$ filters.

The use of the $z \to e^{j\omega T}$ transformation obtains

$$v_{n,out}^2(\omega) = v_{n,Q}^2 \left[2\frac{sin(N\omega T/2)}{N}\right]^2 \tag{8.36}$$

Using the approximation $sin(N\omega T/2) \simeq \pm N\omega T/2$ in the frequency intervals $f = k/(NT) \pm f_B$; $k = 0, \cdots (N-1)$ that are the band base and the regions folded by aliasing in the band base, it results

$$v_{n,out}^2(2\pi f) = v_{n,Q}^2 \left[2\pi f T\right]^2 \tag{8.37}$$

that shows a same transfer function for the band base and in the aliased intervals. Accordingly, the superposition caused by aliasing will amplify by N the in-band noise causing an SNR worsening by $10 \cdot log_2 N$.

The same study repeated for a $sinc^L$ filter used after an $L - th$ order $\Sigma\Delta$ modulator leads to equal results. Therefore, it is necessary to enforce the filtering action by a using cascaded additional filter or, better, by increasing by 1 the order of the $sinc$.

The $sinc^{L+1}$ filter is a transversal structure that requires delays multipliers by coefficients and accumulators. The used coefficients are integer with a limited value if the order and the interpolation factors are limited. For example for $L = 3$ and $N = 4$ the filter is

$$H_s(z) = (1 + z^{-1} + z^{-2} + z^{-3})^3 \tag{8.38}$$

$$= 1 + 3z^{-1} + 6z^{-2} + 10z^{-3} + 12z^{-4} + 12z^{-5} + 10z^{-6} + 6z^{-7} + 3z^{-8} + z^{-9}$$

The coefficients can be stored in a memory or generated step by step during the filtering operation.

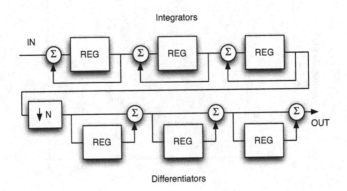

Figure 8.34.   Implementation of a $sinc^3$ filter with simple hardware.

For large values of N the number of coefficients and the number of taps of the transversal filter can become impractical. In those cases a possible implementation is the one shown in Fig. 8.34 for $L = 2$ made by the cascade of three digital integrators followed by a decimator by N and three differentiators. The implementation avoids the need for multipliers or coefficient storage or generation. Moreover, the amount of hardware running at the high sampling rate is kept to a minimum for a minimum use of power.

The next stages of the chain of filters operate at a decimated frequency but must satisfy more complex requirements because of the lower oversampling ratio. As mentioned, a good solution is using half band filters that have all but one of the odd coefficients equal to zero. By choosing symmetric impulse responses number of coefficients is minimum and the computational complexity is reduced by nearly 50% as compared to general direct-form filter architecture.

8.7.2 Interpolation

The digital interpolation increases the oversampling rate of digital data before a $\Sigma\Delta$ *DAC* and suppresses the images replica at multiples of the input data-rate. The interpolator specifications foresee a moderate suppression of high order images because the reconstruction filter used to remove the high frequency components of the shaped quantization noise attenuates them. On the contrary, the first images must be carefully suppressed as they are typically inside the

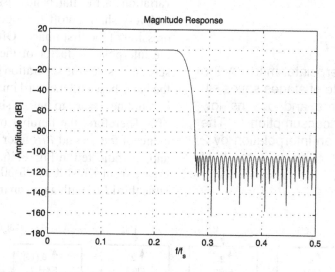

Figure 8.35. Frequency response of a FIR filter with 125 taps.

pass band of the analog reconstruction filter. Therefore, if the signal band and the Nyquist limit are close the pass-band stop-band transition is sharp.

The interpolation factor is normally large because a $\Sigma\Delta$ *DAC* aims at a significant reduction of the number of bit (at the limit it should go down to 1-*bit*. If the digital input is, for example, 16-18-*bit* corresponding to an *SNR* of 98-110 *dB*, it is necessary using a third-order *1-bit* modulator with 64 *OSR* to secure the 16-*bit* and *OSR*= 128 for ensuring the 18-*bit* resolution. Since using higher order modulations gives rise to an eccessive amplification of the quantization noise at high frequency, lower *OSRs* would require multi-bit schemes.

The interpolator can be realized with a single stage or by the cascade of many stages. The latter is the preferred method, as a single stage solution would require operations at the over-sampled frequency, causing excessive power consumption. Only for relaxed transitions between pass-band and stop band it is acceptable using a single stage interpolator. For example, a transition region $0.45 - 0.55 \ f_s$ with more than 100 *dB* of attenuation in the stop band and a ripple of about 0.0001 *dB* in the pass banda is obtained by a 125 taps *FIR* filter. The possible response is shown in Fig. 8.35.

On the contrary, a cascade solution can use an increasing clock in the stages, thereby saving power. The interpolation factor of the first stage is normally low: 2 or 4. The cascade of Fig. 8.36 uses a first interpolation by 2 as a *FIR* filter with not so many taps (around 100-140 compared with 380 of a single stage counterpart) can obtain a frequency response with very small gain variation in the flat band-pass region and very sharp cutoff for a good suppression of the first image. Often even the interpolation factor of the second stage is 2; since the separation between the band-base and the third image is f_s it is not necessary to have a sharp cutoff. Therefore, the number of coefficients of the second *FIR* filter is lower and, as indicated in Fig. 8.36, can decrease from 100-140 to 25-50. A possible third *FIR* with again an interpola-

> **Remind**
>
> An interpolator made by the cascade of stages saves silicon area and obtains low power consumption. The use of an interpolation by 2 in the first stage is the best choice.

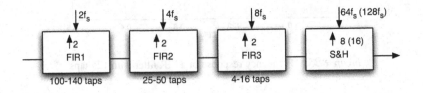

Figure 8.36.   Cascade of interpolation sections to obtain interpolation.

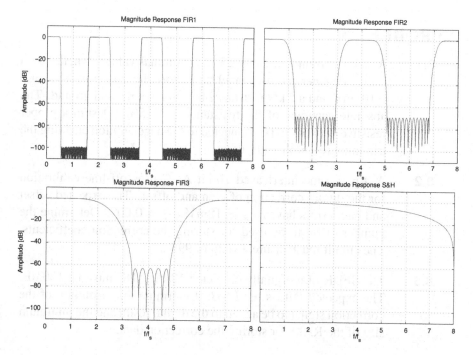

Figure 8.37. Frequency responses of the various sections of an interpolation scheme.

tion factor equal to 2 uses a further reduced number of taps (4-16). After that, since the obtained interpolation by 8 is fairly large, the remaining increase of the sampling-rate can be obtained with a simple *S&H* as the given attenuation in the signal band is acceptable.

Fig. 8.37 shows the possible responses of the four sections of the interpolator described in Fig. 8.36 with a displayed frequency ranging from 0 to $8f_s$. The Nyquist frequency of the first *FIR* is f_s and the response replicates itself around $2f_s, 4f_s, 6f_s, \cdots$. The replicas at $2f_s, 6f_s, \cdots$ and higher bands are suppressed by the second *FIR* whose cutoff border are not very sharp. The third *FIR* cancels the replicas at $4f_s, 12f_s, \cdots$, while the *sinc* takes care of the multiples of $8f_s$. The attenuation of the second and third Nyquist zone equal to 100 *dB* ensures a good suppression of those images. For the images from 4 to 8 an attenuation by 70 *dB* is likely acceptable.

PROBLEMS

8.1   A unity current source of 25 μA (nominal value) is made by a source degenerated n-channel transistor whose overdrive is (300 ± 10) mV and a 5 kΩ resistance between source and ground. The absolute accuracy of the resistance is $\pm 15\%$. Design a proper resistive network to be connected to the degeneration enabling the trimming to 1o-bit with fuses.

8.2   A 4-path time interleaved 10-bit ADC uses off-line calibration for the digital correction of gain and offset. The expected offset and gain errors have $\sigma_{Os} = 12mV$ $\sigma_G = 0.002$. Determine the number of bit to be used for storing the correction coefficients. The circuit must ensure a yield of 99%.

8.3   The SFDR of a digitally corrected 12-bit ADC must be 100 dB. The expected INL is $\pm 4\, LSB$. Determine the resolution of the correction signal to be added to the digital result and the minimum size of the RAM for storing the correction data.

8.4   Repeat Example 8.1 to determine the maximum error in splitting the currents that enables 90 dB SFDR. Use for the simulation a suitable number of points for the FFT sequence.

8.5   Use butterfly randomization in a 6 bit MDAC that uses 32 nominally equal unity capacitances. Estimate the SFDR for 0.005 random error.

8.6   Repeat Example 8.2 but use a 4 bit quantizer. The random mismatch between unity elements is 0.3%. Use a clocked averaging randomizer with 5-bit control.

8.7   A 2-1 Mash $\Sigma\Delta$ modulator uses 3-bit DACs (7 elements). The mismatch between unity elements of 0.0015 is averaged with the ILA rotation method. Determine the SNR as a function of the oversampling ratio. Assume ideal the response of the op-amps.

8.8   Repeat the previous problem but use the DWA method. Use 7-element and a 15 element DACs. Compare the results with the one of the previous problem.

8.9   Determine the in-band attenuation of $sinc^3$ and a $sinc^4$ filters used in a converter with OSR = 32.

REFERENCES

Books and Monographs

J. McClellan, T. Parks, and L. Rabiner: *FIR linear phase filter design program, Programs for Digital Signal Processing*, IEEE Press, Piscataway, NJ, ch. 5.1, 1979.

R. Crochiere and L. Rabiner: *Multirate Digital Signal Processing*, Prentice-Hall, Englewood Cliffs, NJ, 1983.

A. P. Chandrakasan and R. W. Brodersen: *Low Power Digital CMOS Design*, Kluwer, Norwell, MA, 1995.

Journals and Conference Proceedings

Error Measurement and Self Calibration

P. Li, M. J. Chin, P. R. Gray, and R. Castello: *A ratio-independent algorithmic analog-to-digital conversion technique*, IEEE Journal of Solid-State Circuits, vol. 19, pp. 828–836, 1984.

C. Shih and P. R. Gray: *Reference refreshing cyclic analog-to-digital and digital-to-analog converters*, IEEE Journal of Solid-State Circuits, vol. 21, pp. 544–554, 1986.

H. Ohara, H. X. Ngo, M. J. Armstrong, C. F. Rahim, and P. R. Gray: *A CMOS programmable self-calibrating 13-bit eight-channel data acquisition peripheral*, IEEE Journal of Solid-State Circuits, vol. 22, pp. 930–938, 1987.

B. Song, M. F. Tompsett, and K. R. Lakshmikumar: *A 12-bit 1-Msample/s capacitor error-averaging pipelined A/D converter*, IEEE Journal of Solid-State Circuits, vol. 23, pp. 1324–1333, 1988.

M. Mitsuishi, H. Yoshida, M. Sugawara, Y. Kunisaki, S. Nakamura, S. Nakaigawa, and H. Suzuki: *A sub-binary-weighted current calibration technique for a 2.5V 100MS/s 8bit ADC*, Proceedings of the 24th European Solid-State Circuits Conference, pp. 420–423, 1998.

H. Lee: *A 12-b 600 ks/s digitally self-calibrated pipelined algorithmic ADC*, IEEE Journal of Solid-State Circuits, vol. 29, pp. 509–515, 1994.

J. M. Ingino and B. A. Wooley: *A continuously calibrated 12-b, 10-MS/s, 3.3-V A/D converter*, IEEE Journal of Solid-State Circuits, vol. 33, pp. 1920–1931, 1998.

I. Mehr and L. Singer: *A 55-mW, 10-bit, 40-Msample/s Nyquist-rate CMOS ADC*, IEEE Journal of Solid-State Circuits, vol. 35, pp. 318–325, 2000.

E. B. Blecker, T. M. McDonald, O. E. Erdogan, P. J. Hurst, and S. H. Lewis: *Digital background calibration of an algorithmic analog-to-digital converter using a simplified queue*, IEEE Journal of Solid-State Circuits, vol. 38, pp. 1059–1062, 2003.

C. R. Grace, P. J. Hurst, and S. H. Lewis: *A 12-bit 80-Msample/s pipelined ADC with bootstrapped digital calibration*, IEEE Journal of Solid-State Circuits, vol. 40, pp. 1038–1046, 2005.

Y. Chiu, B. Nikolic, and P. R. Gray: *Scaling of analog-to-digital converters into ultra-deep-submicron CMOS*, 2005 IEEE Custom Integrated Circuits Conf., pp. 375–382, 2005.

Foreground and Background Calibration Method

D. Fu, K. C. Dyer, S. H. Lewis, and P. J. Hurst: *A digital background calibration technique for time-interleaved analog-to-digital converters*, IEEE Journal of Solid-State Circuits, vol. 33, pp. 1904–1911 1998.

K. C. Dyer, D. Fu, S. H. Lewis, and P. J. Hurst: *An analog background calibration technique for time-interleaved analog-to-digital converters*, IEEE Journal of Solid-State Circuits, vol. 33, pp. 1912–1919, 1998.

U. Gatti, G. Gazzoli and F. Maloberti: *Improving the Linearity in High-Speed Analog-to Digital Converters*, Proceedings IEEE Int. Simp. on CAS, vol. 1, pp. 17–20, 1998.

O. E. Erdogan, P. J. Hurst, and S. H. Lewis: *A 12-b digital-background-calibrated algorithmic ADC with -90-dB THD*, IEEE Journal of Solid-State Circuits, vol. 34, pp. 1812–1820, 1999.

V. Ferragina, A. Fornasari, U. Gatti, P. Malcovati and F. Maloberti: *Gain and Offset Mismatch Calibration in Time-Interleaved Multipath A/D Sigma-Delta Modulators*, IEEE Trans. on Circuits and Systems I, vol. 51, pp. 2365–2373, 2004.

Dynamic Matching

R. J. van de Plassche: *Dynamic element matching for high accuracy monolithic D/A converters*, IEEE J. Solid-State Circuits, vol. SC-11, pp. 795–800, 1976.

L. R. Carley: *A noise-shaping coder topology for 15+ bit converters*, IEEE Journal of Solid-State Circuits, vol. 24, pp. 267–273, 1989.

A. Maloberti: *Convertitore digitale analogico sigma-delta multilivello con matching dinamico degli elementi*, Tesi di Laurea, Universita degli Studi di Pavia, 1990–1991.

B. H. Leung and Sehat Sutarja: *Multibit Sigma-Delta A/D converter incorporating a novel class of dynamic element matching techniques*, IEEE Trans. Circuit Syst. II, vol. CAS-39, pp. 35–51, 1992.

F. Chen and B. H. Leung: *A high resolution multibit sigma-delta modulator with individual level averaging*, IEEE Journal of Solid-State Circuits, vol. 30, pp. 453–460, 1995

R. T. Baird and T. S. Fiez: *Linearity enhancement of multibit A/D and D/A converters using data weighted averaging*, IEEE Trans. Circuits Syst. II, vol. 42, pp. 753–762, 1995.

E. Fogleman, I. Galton, W. Huff, and H. Jensen: *A 3.3-V single-poly CMOS audio ADC delta-sigma modulator with 98-dB peak SINAD and 105-dB peak SFDR*, IEEE Journal of Solid-State Circuits, vol. 35, pp. 297–307, 2000.

E. Fogleman, J. Welz, and I. Galton: *An audio ADC Delta-Sigma modulator with 100-dB peak SINAD and 102-dB DR using a second-order mismatch-shaping DAC*, IEEE Journal of Solid-State Circuits, vol. 36, pp. 339–348, 2001.

K. Vleugels, S. Rabii, and B. A. Wooley: *A 2.5-V Sigma-Delta modulator for broadband communications applications*, IEEE J. of Solid-State Circuits, vol. 36, pp. 1887–1899, 2001.

Decimation and Interpolation

E. Hogenauer: *An economical class of digital filters for decimation and interpolation*, IEEE Trans. Acoustics, Speech, and Signal Processing, vol. 29, pp. 155–162, 1981.

J. C. Candy: *Decimation for Sigma Delta Modulation, IEEE Trans. Communications*, vol. COM-34, pp. 72–76, 1986.

T. Ritoniemi, E. Pajarre, S. Ingalsuo, T. Husu, V. Eerola, and T. Saramaki: *A stereo audio sigma-delta A/D-converter*, IEEE Journal of Solid-State Circuits, vol. 29, pp. 1514–1523, 1994.

B. P. Brandt and B. A. Wooley: *A Low-Power, Area-Efficient Digital Filter for Decimation and Interpolation*, IEEE Journal of Solid State Circuits, vol. 29, pp. 679–687, 1994.

J. T. Ludwig, S. H. Nawab, and A. P. Chandrakasan: *Low-power digital filtering using aproximate processing*, IEEE Journal of Solid-State Circuits, vol. 31, pp. 395–400, 1996.

C. J. Pan: *A stereo audio chip using approximate processing for decimation and interpolation filters*, IEEE Journal of Solid-State Circuits, vol. 35, pp. 45–55, 2000.

R. Amirtharajah and A. P. Chandrakasan: *A micropower programmable DSP using approximate signal processing based on distributed arithmetic*, IEEE Journal of Solid-State Circuits, vol. 39, pp. 337–347, 2004.

Chapter 9

TESTING OF D/A AND A/D CONVERTERS

This Chapter deals with the methods used for testing and characterizing data converters. We shall start with the static method for testing DNL and INL. Following this, we shall consider the testing of DACs dynamic performances, namely settling time, glitch and distortion. Static ADC testing will also be considered. Subsequently, we shall study the histogram method with different types of input. Distortion and intermodulation tests are also discussed. The use of sine wave and FFT in extracting part specifications is also considered.

9.1 INTRODUCTION

After a data-converter is designed and manufactured, then before it is delivered to customers, a set of measurements verifies that the performance corresponds to the expectation. At the very beginning of the product lifetime these measurements enable the debugging of the circuit thus improving future yields. This phase, named characterization, is usually performed manually by engineers using bench testing equipment like power supplies, signal generators, oscilloscopes etc.. Full characterization is very important as it helps in identifying problems in the early stage of production, possibly involving silicon changes that the designer adjusts according to the measurement results. In addition, characterization verifies the safe operating limits.

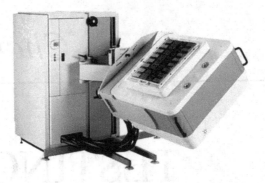

Figure 9.1.   Typical ATE platform.

After the design is validated there is routine production testing that, in the modern *IC* industry, is performed on an *ATE* (Automatic Test Equipment) platform. *ATE* provides a wide range of testing resources and handles different tests from simple ones, such as a continuity check, to fairly complex ones, such as the *SNR* test for static and dynamic linearity. The *ATE* can test a large quantity of *IC*s very efficiently performing hundreds of tests in a few seconds on more than one hundred pin-count devices as its computer controlled test and measurement equipment is arranged in such a way as to be able to test a unit with minimal human interaction.

A complete *ATE* system includes both hardware, and software (testing operating system). The hardware generally consists of a tester head, mainframe, and workstation used to access the measurement resource. Probers interface to the wafer dies through customer designed probe cards, which have probe tips

Figure 9.2.   Probes for wafer level testing (left) and screened chip (right).

that touch the die pads. Testers control the probers movement and exchange analog and digital electrical signals with probes through *PIB* (probe interface board). The wafer level testing screens out the bad dies (Fig. 9.2). They are scraped and not packaged. Once the good dies have been packaged, final tests are carried out to prove their functionality after the damage due to packaging process stress. The advantage of *ATE* is the repeatability of procedures and cost efficiency of high volumes. The chief disadvantages are the upfront costs: programming and setup.

The above described procedure is the same for any mixed-signal *IC*. The specific testing of data-converters is obtained using dedicated software and measurement resources.

9.2 TEST BOARD

The characterization and the test of packaged parts uses a board, like the one shown in Fig. 9.3, which houses the *IC*, brings the signal and supplies into the circuit and conveys the conversion results to the processing unit. The design of this interface board is often critical as it must be able to preserve the quality of the parameters being measured. More specifically, *pcb*'s used for testing high resolution, high speed circuits require special care when measuring total harmonic distortion (*THD*), spurious-free dynamic range (*SFDR*), intermodulation distortion (*IMD*) and other dynamic features at conversion rates of hundreds of *MS/s*.

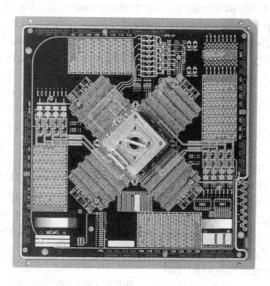

Figure 9.3. Typical pcb for ATE interface.

The design of the board for testing should follow design recommendations similar to the ones given in Chapter 2 for qualifying data converters. The voltage drops on the ground traces must be minimum. The transient switching currents generated by digital circuits flowing through the power path trace and the ground-return must be well separated from the analog paths. When large device currents cause the power-supply trace to suffer a voltage drop then the board must be able to to measure the error with a Kelvin connection and, if necessary, enable an adjustment of the power supply.

For testing *DAC*s running at speeds around and over *1 GS/s* the board uses good power-supply decoupling, and its digital traces need to be well separated from analog traces to avoid interfering signals.

Since many *DAC*s provide differential analog outputs, which minimize errors caused by common mode fluctuations, it is necessary to perform on the test board a differential to single ended conversion for testing purposes that, for very high speed, is conveniently achieved using a transformer rather than an amplifier. The solution normally simplifies the board design as the transformer converts differential to single-ended outputs with a linearity that is difficult to obtain with an amplifier output stage. Often, since transformers have their own optimum range of operation, it is necessary to use more than one transformer to perform the testing over the entire *DAC* frequency range.

The clock separation from analog sections must be verified in advance before using the board. This is done by powering down the analog part of the *DAC* and measuring possible spurs caused by the powered digital section.

The control of the instruments like voltage and current generators, pattern generators and spectrum analyzers can be done by connecting the various instruments to an *IEEE-488* bus (also called *GPIB* or *HPIB* bus) and using software. Another possibility for low-budget and medium-scale testing is using the serial and parallel ports of a *pc*. For example, the standard parallel port provides *12* logic outputs and five inputs, which can be connected directly to *TTL/CMOS* circuits. The parallel port often drives 2-wire *I2C* interfaces and can be used to transmit or receive data to and from a device under test (*DUT*).

Warning!

The board used for validating and, more important, for testing data converters must be carefully designed to avoid false errors caused by the board instead of the converter. Also, ensure accuracy and calibration of the used instrument.

Since for the testing of dynamic performances it is often necessary to use source generators or spectrum analyzers with optimum accuracy in different frequency ranges, then the board must foresee multiple input and outputs and enable the switching between them. The switching can be controlled by software that also regulates the measuring set-up and defines the signal processing performed during or after (off line) the data acquisition.

The testing of *ADC*s and *DAC*s involves the handling of digital input and outputs that, depending on the conversion speed, require low voltage differential signaling (*LVDS*) or single-ended *CMOS* digital drivers. Typically, 350 *MS/s* is the switchover point from *CMOS* to *LVDS*.

The design of boards for high-speed data converters requires a special skill that typically follows these practical recommendations:

- All the bypass capacitors must be very close to the device, preferably on the same side as the converter, using surface-mount components to achieve minimum trace length, low inductance, and low parasitic capacitance.

- The analog and digital supplies, references, and common-mode inputs must be bypassed to ground with ceramic capacitors effective at high frequency in parallel with large electrolytic capacitors effective at low frequency.

- Use multilayer boards with separate ground and power planes to ensure the highest level of signal integrity.

- Use a split ground plane arranged to match the physical location of analog and digital ground pins on the converter's package and keep the impedance of the two ground planes as low as possible.

- When using split grounds join them at a single point (typically in the gap between analog and digital ground) with a low-value surface-mount resistor of 1 to 5 Ω, a ferrite bead, or a direct short. This ensures that noisy digital ground currents do not interfere with the analog ground plane.

- Route high-speed digital signal traces well away from all the sensitive analog traces.

- Keep all the signal lines short and avoid 90° turns. Use 45° or rounded turns instead.

- Always treat the clock input as an analog signal but route it away from both the actual analog signals and other digital lines.

9.3 QUALITY AND RELIABILITY TEST

Quality and reliability of integrated circuits are important features that must obviously be ensured for data converters. The required quality is determined by the acceptable quality level (*AQL*), which is the planned minimum fraction of shipped units that should be defect-free. It is necessary to obtain a process yield that is higher than the required *AQL*, otherwise each part must be tested so that defective units can be identified and removed before or after packaging.

The quality of any product is assured by consistent actions in all the steps from product planning to development, designing, trial production, mass production

Be Aware

Quality and reliability is not just a metter of process but also design. Circuit schemes that are less sensitive to parameter's variation voltage changes and mismatches and avoid hot carrier damage do ensure quality!

and shipping. In general quality and reliability are obtained by ensuring the optimum work environment and the use of reliable technologies. For product design, it is necessary to strictly target the product specifications by performing careful design reviews before starting mass production. In mass production, product quality and reliability is assured by controlling all the production steps starting from raw materials, manufacturing equipment, and the manufacturing environment.

Since high quality means reliable circuits, stress testing and accelerate aging tests are done to determine how the products will perform under extreme field conditions and how long the performance will remain unchanged.

The most common stress tests are High Temperature Operating Life (*HTOL*), Autoclave (*PCT*, or pressure cooker test), Electrostatic discharge (*ESD*), and Latch-up (*LU*). The circuit under stress test is measured before and after each stress test to look for possible parameter shifts.

The *HTOL* test is used to determine the products performance after many years of operation. Since the operation of a circuit at high temperature speeds up the aging effects, parts are operated in an oven (usually at 125°C) for several thousand hours to simulate the performance over decades of field operation proving the long-term reliability.

Autoclave uses extreme pressure, humidity, and temperature conditions to test the package and die's protective capabilities. This kind of stress test verifies a lack of die contamination, contact corrosion, and other phenomena resulting from any packaging weaknesses.

As is known, electrostatic discharge can permanently damage an *IC* that is not properly protected and handled. The internal gate electrodes of a *CMOS IC* chip are insulated from the substrate by thin oxide films. Where these electrodes are connected to input pins then, in order to prevent oxide breakdown, input protection structures are used. The *ESD* test verifies the effectiveness of the these input protections against large voltages caused by three possible mechanisms: human handling, machine overvoltage generation, and, during fabrication, electrostatic device charging.

Latch-up is another *IC* problem affecting reliability. It causes a part to draw too much current which permanently alters its performances. The latch-up test applies current to the pins of the part to look for this problem. If the part is not designed correctly, the test can triggers latch-ua and an internal electrical short-circuit occurs.

9.4 DATA PROCESSING

Suitable data processing of measured data enables the estimation of key specifications of a data converter. The fitting technique which examines the output resulting from an input ramp or sine wave obtains gain, offset and harmonic coefficients by a linear, polynomial, or sinusoidal fitting.

Further processing provides the histogram of the output (for either a ramp or sine wave input) which obtains the *DNL* and, consequently, the *INL*.

9.4.1 Best-fit-line

The best-fit-line of the *ADC* input-output characteristics uses a sequence of n digital data $Y_i, i = 1, \cdots, n$ generated by linear input signals covering the entire dynamic range. For a *DAC* the used data is the *ADC* conversion of the analog output obtained with linear digital inputs that, as for the *ADC* case, cover the entire dynamic range. The fitting line is

$$\hat{Y}(i) = G \cdot i + Y_{os} \qquad (9.1)$$

where G is the gain and Y_{os} is the offset of the data converter.

The least squares method minimizes the summed square of residuals that for the *i-th* data point is the difference between the measured response value Y_i and the fitted response value $\hat{Y}_i$

$$r_i = Y_i - \hat{Y}_i. \qquad (9.2)$$

The summed square of residuals is

$$S = \sum_1^n r_i^2 = \sum_1^n (Y_i - \hat{Y}_i)^2 = \sum_1^n [Y_i - (G \cdot i + Y_{os})]^2. \qquad (9.3)$$

The minimum of S in the coefficients space requires that the partial derivatives of S equal to zero

$$\frac{\partial S}{\partial G} = -2 \sum_1^n i \cdot [Y_i - (G \cdot i + Y_{os})] = 0 \qquad (9.4)$$

$$\frac{\partial S}{\partial Y_{os}} = -2 \sum_1^n [Y_i - (G \cdot i + Y_{os})] = 0 \qquad (9.5)$$

that, using the values

$$S_1 = 2 \sum_1^n i; \quad S_2 = 2 \sum_1^n Y_i; \quad S_3 = 2 \sum_1^n i^2; \quad S_2 = 2 \sum_1^n iY_i \qquad (9.6)$$

as intermediate variables, yields

$$G = \frac{nS_4 - S_1 S_2}{nS_3 - S_1^2}; \quad Y_{os} = \frac{S_2}{n} - G\frac{S_1}{n} \qquad (9.7)$$

which are the gain and offset that fit the best straight line response.

The method can be extended to a higher degree polynomial to determine the fitting coefficients used for harmonic distortion specifications. The analytical study is based on similar computation methods described in specialized books or embedded in computation or system simulation packages.

The least square fitting assumes that the precision quality of all the used data is equal. However, it may happen that some data points have reduced accuracy as they represent inputs in critical regions or where the input was occasionally corrupted by a spur. Since the poor quality of a fraction of the used data set influences the fit, the poor data must be identified and removed from the sequence. Another possibility is to weight each residual term of equation (9.3) before its use in the fitting process. Weighted least squares fitting minimizes the expression

$$S' = \sum_1^n w_i (Y_i - \hat{Y}_i)^2. \tag{9.8}$$

Obviously data weighting requires knowledge that the accuracy of the conversion varies in different regions of the explored range.

9.4.2   Sine Wave Fitting

Some *ADC* test methods use a precise analog sine wave generator for producing a digital sine wave at the output. Generally, possible limits of the *ADC* give rise to a result that is not an exact sine wave and so it is necessary to extract the best sine wave approximation (normally with offset) from a large measured set. The processing methods used are the three-parameters least-squares fit (if the frequency is known) or the four-parameters least-squares fit when the frequency must also be determined. The three parameters fit is used when the data set does not exactly represent an integer number of cycles, as in this case the *DFT* can be more straightforward.

The fit method determines the values of A_0, B_0, Y_{os} (and ω_0) that minimize the sum of squared differences

$$\sum_{i=1}^M [y_i - A_0 cos(\omega_0 iT) - B_0 sin(\omega_0 iT) - Y_{os}]^2 \tag{9.9}$$

where $y_1, y_2, \cdots, y_M$ is a sequence of M input samples taken at successive sample times.

If the frequency is known we can define the matrixes

$$D_0 = \begin{bmatrix} cos((\omega_0 T) & sin(\omega_0 iT) & 1 \\ cos((2\omega_0 T) & sin(2\omega_0 iT) & 1 \\ \cdots & \cdots & \cdots \\ cos((M\omega_0 T) & sin(M\omega_0 iT) & 1 \end{bmatrix}$$

$$y = \begin{bmatrix} y_1 \\ y_2 \\ \cdots \\ y_M \end{bmatrix} \quad x_0 = \begin{bmatrix} A_0 \\ B_0 \\ Y_{os} \end{bmatrix}$$

that give rise to the matrix notation of (9.9)

$$(y - D_0 x_0)^T (y - D_0 x_0) \tag{9.10}$$

where T designates the transpose.

Since the minimum of the above equations determines the result, the testing method is split into the following steps: apply a sine wave with specified parameters to the input of the *A/D* converter, take a record of the output data, fit a sine wave to the sequence of samples by estimating phase, amplitude, dc value, and (if needed) frequency by minimizing the sum of the squared difference expressed by (9.10).

If the input frequency is unknown then the fitting method uses an estimated frequency value in the calculation and repeats the determination several times to obtain the value of ω_0.

9.4.3 Histogram Method

The histogram method is a statistical study of a sequence of output samples obtained with an *ADC* input whose magnitude distribution (or probability density function), $p_{in}(x)$ is known. The occurrence probability P_i of a certain output code V_i is, for an ideal *ADC*, the integral of the probability of having the input in the range of V_i. Therefore, an ideal *ADC* with N equal quantization intervals and dynamic range V_{FS} gives rise to

$$P_i = \int_{(i-1)\Delta}^{i\Delta} p_{in}(x)dx; \quad i = 1 \cdots N; \quad \Delta = V_{FS}/(N-1). \tag{9.11}$$

If the converter is not ideal the integral that defines the occurrence probability of the output code V_i must extend between the actual code transitions limits

$$P_{i,r} = \int_{V_{L,i}}^{V_{U,i}} p_{in}(x)dx; \tag{9.12}$$

where lower and upper code transitions are

$$V_{L,i} = \sum_{j=1}^{i-1} \Delta_j; \quad V_{U,i} = V_{L,i} + \Delta_i. \tag{9.13}$$

Assuming that the number of samples M is large, P_i and $P_{i,r}$ are approximately the number of samples M_i and $M_{i,r}$ that give rise to the code V_i in the ideal and real case divided by M

$$P_i = \frac{M_i}{M}; \quad P_{i,r} = \frac{M_{i,r}}{M}. \tag{9.14}$$

The number of quantization intervals and/or the probability density function are typically such that $p_{in}(x)$ can be assumed constant within the *i-th* quantization interval and equal to $p_{in}(V_{L,i})$. Therefore, using equation (9.12) we obtain

$$P_i = p_{in}(V_{L,i})\Delta; \quad P_{i,r} = p_{in}(V_{L,i})\Delta_i \tag{9.15}$$

which combined with (9.14) yields

$$\Delta_i = \frac{M_i}{M \cdot p_{in}(V_{L,i})}. \tag{9.16}$$

Moreover, if the inputs have constant probability over the entire dynamic range (like a linear ramp or a saw tooth extended over the entire analog range) then $p_{in}(x) = 1/V_{FS}$, where the integral of $p_{in}(x)$ over the $0 \cdots V_{FS}$ range is 1. Therefore,

$$\frac{\Delta}{V_{FS}} = \frac{M_i}{M} = \frac{1}{N}; \quad \frac{\Delta_i}{V_{FS}} = \frac{M_{i,r}}{M} \tag{9.17}$$

which determines the *DNL* of the *i-th* channel

$$DNL(i) = \frac{\Delta_i - \Delta}{\Delta} = \frac{M_{i,r} - M_i}{M_i} \tag{9.18}$$

also expressed by

$$DNL(i) = \frac{N \cdot M_{i,r}}{M} - 1. \tag{9.19}$$

Since the accumulation of the *DNL* obtains the *INL*, the use of the histogram method determines both *DNL* and *INL* with an accuracy that is inversely proportional to the number of samples stored in each channel. If, for instance, a bin that should ideally contain 10 samples counts only 9 samples, the *DNL* is -0.1 *LSB* with 10% accuracy. In contrast having 1000 ideal samples and 898 counts denotes an almost equal *DNL=-0.102 LSB* while the measurement has 0.1% accuracy. Notice that the improved accuracy is costly as to build up the second histogram requires $(1000 \cdot N)$ instead than $(10 \cdot N)$ samples.

In addition, the accuracy of the *DNL* also depends on the uncertainty in the measure caused by noise and non-linearities affecting the test signal. If, for example, the test signal is a perfect slow ramp $V_{in}(t) = tf_s V_{FS}/M$ that crosses an ideal channel $\Delta = V_{FS}/N$ for M/N clock periods ($T = 1/f_s$) then the resulting sample count in each channel should be equal to $floor(M/N)$ or $ceil(M/N)$. This will not be the case however, as noise and distortion alter the samples that are close to the channel borders. The noise blurs both sides of the

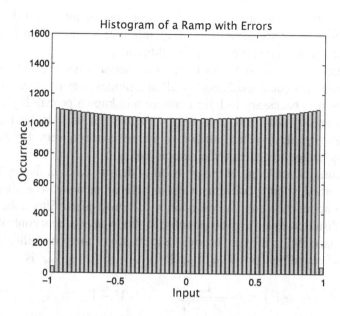

Figure 9.4. Histogram of an input ramp with errors expressed by equation (9.21).

channel causing a *DNL* inaccuracy that depends on the noise variance $\sigma_{n,in}$ and the number $M_{i,r}$ of samples in the bin

$$\epsilon_{DNL}^2 \simeq \frac{1}{M_{i,r}^2} + \frac{2\sigma_{n,in}^2}{M_{i,r}\Delta^2} \tag{9.20}$$

showing that an accurate *DNL* measurement requires the use of a low noise test signal or a large data set.

The nonlinearity of the ramp influences the *INL* measure as can happen, for example, with a non-linear ramp affected by a gain error and a third order nonlinearity (plus a little noise)

$$X(t) = 0.99kt - 0.02(kt)^3 + x_n(t); \quad -1/k < t < 1/k. \tag{9.21}$$

The histogram obtained for such a situation looks like the one of Fig. 9.4 showing that the errors move samples out from the first and the last bin, while the harmonic term causes a curvature in the histogram; the samples in the bins slightly fluctuate because of the noise. Observe that the histogram curvature does not have much effect on the *DNL*, however the accumulation of errors can cause an unacceptable *INL* giving rise to linearity requirements that are quite stringent for high resolutions.

The capability of generating an on chip linear ramp enables the self-testing of the data converter by a suitable reconfiguration of the architecture and the storing of the result for a successive self-calibration.

An alternative solution to a slow ramp or a triangular test signal is using a random signal with equal likelihood of all amplitudes over the measurement range. Since it is necessary to have constant and known probability density function, an analog white noise cannot be the best choice. Instead a more convenient test signal is a pseudo-random digital sequence converted into analog and low-pass filtered to generate the random voltage.

Since accurate knowledge of the probability density function is key for the histogram method, another convenient test signal is a sine wave. The filtering of harmonic components is an easy operation with high Q quartz filters; therefore, even with a distorted sine wave it is possible to obtain an excellent control of the sinusoidal shape and in turn, an accurate estimation of the probability density function that, including a possible offset $(V = A \cdot sin(x) - V_{os})$ is

$$p(V) = \frac{1}{\pi\sqrt{A^2 - x^2}}; \quad x = V - V_{os}. \tag{9.22}$$

The histogram with amplitude 1 and offset 0.01, depicted in Fig. 9.5, is not symmetrical because some samples of the first left bin are moved out to the next bin; some of the second bin goes in the third and so forth.

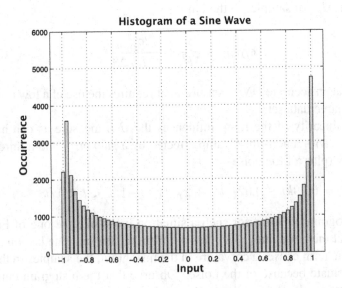

Figure 9.5.   Histogram of a sinewave with full amplitude 1 and offset 0.01.

The integration of the density function from $V_{L,i}$ to $V_{U,i}$ is

$$P(V_{L,i}, V_{U,i}) = \frac{1}{A\pi} \left\{ \arcsin\left[\frac{V_{U,i} - V_{os}}{A}\right] - \arcsin\left[\frac{V_{L,i} - V_{os}}{A}\right] \right\} \quad (9.23)$$

which used in (9.16) determines the value of each Δ of the transfer characteristics. Observe that since the value of amplitude A and offset are unknown it is necessary to estimate them before using (9.23). These estimations are normally done by the three parameter sine wave best-fitting using the digital data record.

The noise and interferences do not influence the histogram test to a great extent if the spur signal has zero mean. Indeed, the spurs occasionally move one sample from bin M to bin $M+1$, but there is equal probability that another spur will move a different sample back into bin M. Since the process is zero mean, the error just depends on the difference between the probability density functions of bins within the spur range. Therefore, the main effect of noise is increasing the uncertainty (standard deviation) on *DNL* and *INL* estimation.

If the analog input is obtained with different instruments for securing optimum performances over different but partially overlapped ranges, the resulting histograms must be combined by equalizing the values. Assume that the overlap region is X_A to X_B, $\Delta X = X_B - X_A$, with a fraction of samples of the two sets M_1 and M_2 in the ΔX interval given by

$$\alpha_1 = \frac{\sum_{X_A}^{X_B} O_1(i)}{M_1}; \quad \alpha_2 = \frac{\sum_{X_A}^{X_B} O_2(i)}{M_2} \quad (9.24)$$

where O_1 and O_2 are the occurrences of the two histograms.

The two histograms are combined together by multiplying, for example, the occurrences of the second histogram by α_1/α_2 and using for the overlapped region the first, the equalized second, or a weighted average of the two occurrences.

9.5 STATIC DAC TESTING

The basic test setup of Fig. 9.6 obtains most of a *DAC*s static specifications as it allows for the application of any set of digital codes to the input and employs a precise digital voltmeter for measuring the output. The measured output is sent to the computer for the required processing. The same setup can be used for both characterization and testing, the only difference being that for characterization the control is manual while with for testing the *ATE* controls all the measurement steps by executing programmed test sequences and data-logging the results.

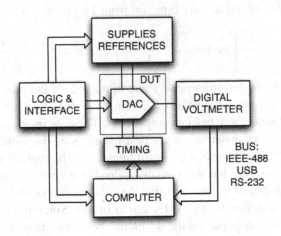

Figure 9.6. Basic setup for the static testing of DACs.

9.5.1   Transfer Curve Test

As is known, the *DAC* transfer curve is the analog output in response to all the possible input codes. To measure it the test setup uses a slow digital ramp possibly repeated a number of times to average the noise and the inaccuracies of the measurement.

The transfer curve is then the basis for estimating all the static specifications of the *DAC*. For example, the slope through the endpoint codes gives a rough measure of the gain; the offset is the voltage generated at the code that should generate zero. However, for accurate estimations it is better to use the best-fit line method for measuring gain and offset.

Since the difference between two successive codes is the quantization step, its variation with respect to the ideal step value is the *DNL*. Integrating the *DNL* leads to the *INL* according to the definition of Chapter 2. Other static specifications like the analog input range, monotonicity, hysteresis and others are easily derived.

9.5.2   Superposition of Errors

Testing high resolution *DAC*s requires a significant testing time. For instance, since the output levels of a 16-*bit DAC* are $2^{16} = 65,536$, measuring each level is very time-consuming even when using automatic methods. However, knowing the architecture and assuming a linear behavior helps in reducing the testing time by using the superposition of errors technique.

A property of a linear system is that the response to the addition of two inputs is the addition of the outputs with the two inputs applied separately

$$Y(X) = Y(X_1) + Y(X_2); \quad X = X_1 + X_2. \tag{9.25}$$

Therefore, an output voltage determined by the superposition of two terms causes an error equal to the addition of the two individual errors

$$\epsilon_{1+2} = \hat{Y}(X) - Y(X) = \epsilon_1 + \epsilon_2 \tag{9.26}$$

where $\hat{Y}$ denotes the output of the ideal *DAC*.

There are many *DAC*s whose output is the sum of individual parts such as, for example, the binary weighted scheme. Since the error of each part is independent of the other, the error is the sum of the errors of each binary part used. Accordingly, the test must only determine the error of the parts that are combined together to obtain the outputs; the error for a given input code is then calculated with software. Therefore, the superposition method (also called major carrier testing) greatly speeds up the process as the binary weighted architecture requires only n measurements instead of 2^n.

Critical points of the transfer curve of a binary weighted *DAC* are half, $1/4$ and $3/4$ of full scale and successive intermediate points. The error of the binary elements gives an immediate measure of the *DNL* at critical points. For example, at half scale the code switches between all but the *MSB* being 1 to all but the *MSB* being 0. Therefore

$$DNL(2^{n-1}) = \epsilon_n - \sum_1^{n-1} \epsilon_i. \tag{9.27}$$

The elements used before $1/4$ of the full scale are the first $(n-2)$ while just before and after the $3/4$ the *n-th* is also used. Accordingly

$$DNL(2^{n-2}) = DNL(2^{n-1} + 2^{n-2}) = \epsilon_{n-1} - \sum_1^{n-2} \epsilon_i, \tag{9.28}$$

and so forth for the other intermediate points.

The above equation shows that the *DNL* plot is, as expected, symmetrical with respect to the mid-point.

Even if the *DAC* architecture does not use superposition of elements knowing the converter architecture simplifies the testing. For example, in a resistive divider the error between two successive taps is not a normal concern; instead, it is the accumulation of errors that cause a large *INL*. It was studied that a linear string of elements or a folded string cause a bow-type or an s-type transfer curve. Therefore, instead of testing all the possible outputs it can be enough to test the output where the maximum *INL* is expected (a quarter, half or three quarter of the full scale). If the architecture is a mix of resistive divider and binary weighted capacitors, the testing can combine the superposition method with the measurement of critical points.

9.5.3   Non-linearity Errors

The transfer characteristics $Y(i)$, $i = 0 \cdots (2^n - 1)$ give rise to the non-linearity errors specified by the *INL* and the *DNL*. Indeed, the *DNL* in *LSB* is the discrete derivative of the *DAC* transfer curve, given by

$$DNL(i) = \frac{Y(i+1) - Y(i) - \Delta}{\Delta} \quad i = 0 \cdots (2^n - 2) \qquad (9.29)$$

where Δ is the ideal step size or, better, the average step size, obtained by the slope of the best-fit line.

The *INL* measures the transfer curve deviation from a reference line that can be the best-fit line, the end-point line or the ideal *DAC* line

$$INL(i) = \frac{Y(i) - Y_{ref}(i)}{\Delta} \quad LSB. \qquad (9.30)$$

The most suitable reference line for estimating the harmonic distortion is the best-fit line: accounting for gain error and offset the endpoints of the *INL* curve go to zero.

When the best-fit line calculation is not required, the endpoint line can be used to estimate the *INL*. In this case the curve corresponds to the running sum of the *DNL*

$$INL(k) = \sum_{i=1}^{k} DNL(i). \qquad (9.31)$$

9.6   DYNAMIC DAC TESTING

The digital signal used for the dynamic testing of *DAC*s is often a single sine wave or a combination of sine waves. The typical set-up shown in Fig. 9.7 uses a number of instruments to capture the analog performances after an analog conditioning block, typically a filter. The setup is suitable for measuring many relevant specifications like the *SNR*, *SNDR*, *SFDR*, and other spectral quality parameters. The measured or provided quantities are

- Sampling frequency.
- Input sine wave(s) frequency and amplitude(s).
- Amplitude of the fundamental output.
- Harmonics (and *IMD* products) levels.
- *FFT* plot.

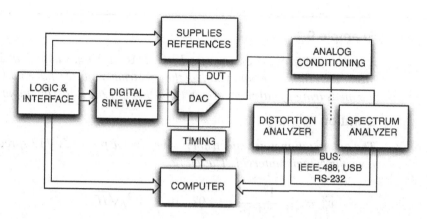

Figure 9.7. Basic setup for the dynamic testing of DACs.

9.6.1 Spectral Features

The output of a spectrum analyzer depends on the resolution bandwidth used, Δ_{BW} as it performs a sweep of the frequency in the measurement interval and determines the power of the input spectrum falling within $\pm\Delta_{BW}/2$ around the swept frequency. If the measured signal is made up of noise and a sine wave at f_x with peak amplitude A_x, then the amplitude displayed by the spectrum analyzer at f_x is

$$S_{out}(f_x) = v_n^2 \Delta_{BW} + \frac{A_x^2}{2}. \tag{9.32}$$

If the output contains harmonic distortion terms then the spectrum analyzer will represent a tone at frequency f_k with amplitude A_k as

$$S_{out}(f_k) = v_n^2 \Delta_{BW} + \frac{A_k^2}{2}; \tag{9.33}$$

therefore, if the tone is to be visible above the white noise contribution it is necessary to have

$$\Delta_{BW} << \frac{A_k^2}{2v_n^2} \tag{9.34}$$

which defines, similar to the condition established by the *FFT*, a maximum spectrum analyzer bandwidth. The condition, in turn, establishes a minimum measurement time as the speed of the sweep is lower and lower for low resolution bandwidths. Therefore, the cost of testing, mainly dominated by the cost per second due to the use of the *ATE*, can become significant.

Example 9.1

A tester measures the output of a 12-bit DAC, 1 V_{FS}, sampled at 20 MS/s. Determine the bandwidth of the spectrum analyzer that enables the measuring of harmonic tones 115 dB below the full scale input.

Solution

The quantization noise spectrum is given by the power $\Delta^2/12$ spread over the Nyquist interval. Therefore

$$v_n^2 = \frac{1^2}{12 \cdot 2^{24} 10^7} = 4.97 \cdot 10^{-16} \, V^2/\sqrt{Hz}.$$

The bandwidth that reveals tones -115 dB below the full scale sine wave ($3.95 \, 10^{-13} V^2$) is therefore 795 Hz.

The spectrum displayed by the spectrum analyzer provides various parameters like the *SFDR* resulting from the possible spectrum of Fig. 9.8. The noise floor at $-117 \, dB_{FS}$ makes the first 10 harmonic terms easily visible. The highest one that gives rise to the *SFDR* is the third harmonic whose amplitude is $-80 dB_{FS}$.

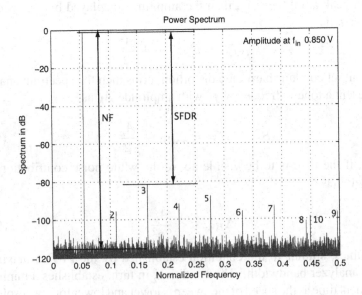

Figure 9.8. Possible spectrum of a spectrum analyzer.

Another specification derivable from the spectrum analyzer output is the *THD* which is the ratio of the signal power to the total power of the harmonics of the fundamental signal. The number of harmonics accounted for depends on the manufacturers choice, and normally follows the *IEEE* standard *1241-2000* recommendation: to use up to the tenth harmonic. Therefore,

$$THD = A_1 - 10 \cdot log_{10} \left[\sum_{i=2}^{10} 10^{A_i/10} \right]. \qquad (9.35)$$

After calculating the *THD* the *SINAD* can be estimated by

$$SINAD = 10 \cdot log_{10} \left[10^{SNR/10} + 10^{THD/10} \right]. \qquad (9.36)$$

9.6.2   Conversion Time

The parameter that measures the speed of the *DAC* is the *conversion time*, which is the time required by the output to stabilize within a given percentage of the final level after a step change of the digital input is applied.

The conversion time depends on the speed of the analog section and the maximum change of the *DAC* control. If the input is a sine wave at the Nyquist limit the *DAC* control can swing from the negative full-scale to the positive full-scale. If, in contrast, the *DAC* uses oversampling to ease the design of the reconstruction filter, then the maximum control step is only a fraction of the full scale. Since the maximum slope of a sine wave $A sin(\omega t)$ is $A\omega$ the largest change between two successive samples of that sine wave is $\pi V_{FS}/(2 \cdot OSR)$. Therefore, with $OSR = 4$ the output swing is about $0.4 V_{FS}$.

The measurement of the conversion time is normally done with a fast oscilloscope. A typical waveform looks like the diagram of Fig. 9.9 indicating three possible regions. First is the dead-time during which the circuit does not react because of the finite propagation delay of the control. The second period is the slewing transient limited by the maximum current available for charging capac-

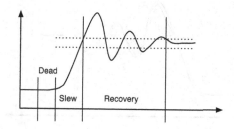

Figure 9.9.   Typical conversion-time of a DAC.

itances (including parasitic). Since the slew-rate normally causes an overshoot the third period is the time required for recovery and reaching the final value.

More in general, the conversion time measure determines the following parameters:

- Settling-time (slew plus recovery).
- Rise and fall-time (settling in the positive and negative direction with a full scale input step).
- DAC-to-DAC skew.
- Glitch energy.

The *DAC-to-DAC* skew features in circuts with multiple *DACs* and corresponds to the time mismatch between the rise of two *DACs*. This parameter is used when many *DACs* must work synchronously like, for example, for color video applications as possible skew between the time responses gives rise to blur pictures or ghost effect.

9.6.3   Glitch Energy

The glitch energy is in reality the area of the glitch impulse and is estimated from the settling response with a mid-scale transition.

The glitch is typically a very fast pulse or sequence of pulses; its measurement using an oscilloscope identifies the peak values and the transition times. Since it is normally acceptable to measure the glitch energy with an error that can be relatively large, the estimation can approximate the various part of the transient response by triangles. The Fig. 9.10 identifies four areas two positive and two negative. The value of the glitch impulse area, commonly measured in pV-sec (that is not units of energy), is

$$Glitch = A_1 - A_2 + A_3 - A_4 \qquad (9.37)$$

Or for some specifications suitable for video systems

$$Glitch = A_1 \qquad (9.38)$$

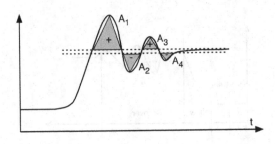

Figure 9.10.   Measurement of the glitch impulse area.

Obviously, an accurate measurement requires the use of oscilloscopes with fast-settling and wide-bandwidth response. Measuring the rise or the fall-time of the *DAC* or the glitch response requires a bandwidth of the oscilloscope amplifier equal to $\alpha\, 0.35/\tau_{DAC}$ (α is a suitable margin whose value is at least 1.5). Therefore, for $\tau_{DAC} = 1\,ns$ the bandwidth must be at least $524\,MHz$.

If the bandwidth (or settling time) of the op-amp is not sufficient it is possible to estimate the *DAC* transient time by recalling that the settling are composed quadratically; therefore

$$\tau_{DAC} = \sqrt{\tau_{meas}^2 - \tau_{OpAmp}^2} \qquad (9.39)$$

9.7 STATIC ADC TESTING

The main target of any static *ADC* testing is measuring the quantization intervals, i.e., the analog ranges that give rise to the same digital code. Accordingly, the test must primarily identify the code edges for obtaining, after some processing, all the static specifications.

The basic setup shown in Fig. 9.11 uses a precise voltage or current source whose value is controlled by computer. The digital output received by the interface and delivered to the computer enables processing or statistical studies.

The source generator must provide voltages or currents with an accuracy better than the *LSB* and with very low noise as any random fluctuation adds uncertainty to the code edge measurement. The input referred noise of the *ADC* (typically the one of the *S&H* and the first analog stage) can also add uncertainty making it necessary to use statistical methods.

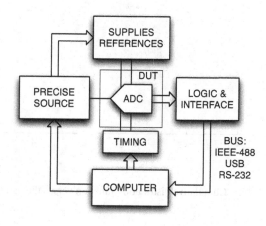

Figure 9.11. Basic setup for the static testing of *ADC*s.

If the noise is white its probability distribution function is normal (or Gaussian), described by

$$p(V_n) = f(V_n, \sigma) = \frac{1}{\sigma\sqrt{2\pi}} e^{-V_n^2/(2\sigma^2)} \tag{9.40}$$

where σ is the variance of the random noise.

Since the noise adds uncertainty to the edge measurement, then it is not possible to state that if the input is within the thresholds of a quantization interval that the code is also in that interval but, instead, it is necessary to give the code probability. If the difference between the edge and the nominal source level is much larger than σ, that is also a small fraction of Δ, the code probability almost equals 0 before the channel and goes to 1 after crossing the code edge as the cumulative distribution function depicted in Fig. 9.12, that is the integral of the Normal distribution. Therefore, the code edge for a noisy ADC or a noisy source generator is defined as the input voltage for which the probability of generating a given code is $1/2$.

Since uncertainty affects the edge measurement, the static test normally aims at measuring the center of the code that is the middle of the region with probability of having that code almost equal to 1.

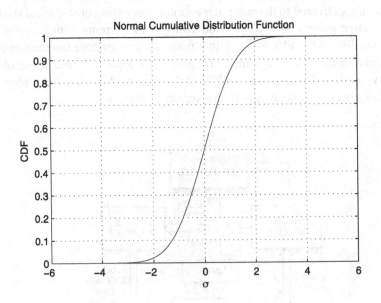

Figure 9.12. Cumulative distribution function versus the variance.

9.7.1   Code Edge Measurement

The code edges are the input voltages which give rise to the same probability of having two contiguous codes. The simplest way to obtain them is to search by trimming the input voltage: the source, possibly controlled digitally, is changed according to a suitable search algorithm; for every point the system collects a given number of samples to verify whether or not half of them fall in one code and the other half in the next one.

The method is fairly time consuming even if an effective search method (like the binary tree) and an automatic tester are used. Indeed, if the resolution in measuring Δ is $\Delta/16$ and the tested 12-*bit DAC* is intrinsically monotonic with conversion speed 2 *MHz*, with a binary search requiring five iterations per edge and *100* samples per points, then the time required to explore the entire range is $100 \cdot 2^{12} \cdot 5 = 2.048 \cdot 10^6$ corresponding to a too long test time (about 1 *sec*) and test cost for components whose selling cost should be low or medium.

Another possible method is the servo-loop test which is faster than the above described technique because it does not use data averaging in the search process. The possible measurement setup of Fig. 9.13 uses, as its analog input, the integration of a 1-*bit DAC* whose control is reversed when the digital output crosses a code. Because of the loop delay and the feedback the voltage experiences triangular transitions that continue until the output of the integrator is within a small fraction of the quantization interval. The digital voltmeter (*DVM*) measures the code edge when the voltage at the output of the integrator is stabilized.

Actually the method causes the output of the integrator to track the threshold that fluctuates because of input referred noise. However, the loop filtering rejects the high frequency spur making the measurement accurate enough for low offset and low $1/f$ noise devices.

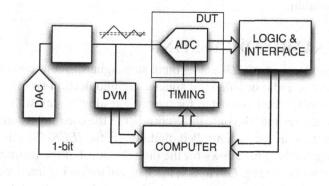

Figure 9.13.   Servo test setup for code edge measurement.

The histogram technique is another possibility that, in fact, is quite appropriate for the code edge or the code mid-point measurement. The input signal, as already studied, can be a linear ramp or a sine wave. Since collecting a relatively small number of samples per bin obtains good accuracy, then using for example an average of 20 samples per bin requires $20 \cdot 2^{12} = 81,920$ measurements for a 12-*bit*, 2 *MS/s* converter; 24 times less than the first discussed method.

The histogram made by an average of M samples per bin measures the relative amplitudes of the bins as the ratio between the samples in the *i-th* bin M_{C_i} and the expected number M. The absolute value of the code edge requires knowing the possible gain error that alters the step size from Δ to Δ_{av}. The use of the end-point measurements $V_{in}(0)$ and $V_{in}(2^n - 1)$, made by a binary search or a servo-loop test, obtains the following equation

$$\Delta_{av} = \frac{V_{in}(2^n - 1) - V_{in}(0)}{2^n - 1}. \tag{9.41}$$

The *INL* curve is given by the running sum of the error on the actual quantization steps

$$INL(k) = \Delta'_{id} \left[\sum_1^k \frac{M_{C_i}}{M} - k \right] \tag{9.42}$$

Observe that a relatively small number of samples per histogram channel does not cause a disruptive accumulation of errors in measuring the *INL*: the error in one of the bins is compensated by the error of successive bins. Thus, for example, if the error would lead to 31.3, 32.7 and 31.4 samples in three successive channels instead of a nominal 32 the measured number of samples are 31, 33, and 31. The *DNL* accuracy is $1/32$ but the $-0.3/32$, $0.7/32$ and $-0.4/32$ rounding compensate each other with no disruptive accumulation in the *INL* estimation.

9.8   DYNAMIC ADC TESTING

The dynamic testing of *ADC*s uses input step signals, sine waves (or multitone sine waves), audio or video test signals, communication-specific test signals and so forth. The basic setup for dynamic testing is almost the same as is used for static testing; the only difference being the used input source. The analog generators are often controlled digitally via the *IEEE-488* bus or other equivalent standards. This allows for the digital control of the instruments facilitating automatic testing. Obviously, the speed of the digital interface and the digital processing capabilities must be adequate for the speed and the number of bits used.

The derived specifications are typically determined by signal processing using the set of digital output data. However, in some cases it can be convenient

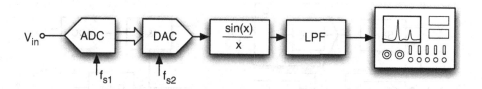

Figure 9.14. Back-to-back setup for dynamic ADC testing.

to return back to the analog domain using the so called back-to-back method that uses a *DAC* after the *ADC* under test as shown in Fig. 9.14. In some cases the sampling frequency of the *ADC* and the *DAC* conversion frequency are different to provide the benefits of under-sampling: the dynamic features of the *ADC* are tested by a high frequency repetitive input waveform; the down sampling transfers the result to a more convenient frequency range for successive characterization in the analog domain after the *DAC*.

9.8.1   Time Domain Parameters

The time response of an *ADC* determines the most relevant time-domain paramenter: the maximum sampling frequency. This specification depends on the rise and fall times, the settling, and the recovery from overdrive of *S&H* and, possibly, the first analog stage of the *ADC*. The testing of the mentioned features and identifying the bottlenecks is important in the characterization phase. For this, it is a common practice to make key blocks more observable by adding extra pins, test points accessible via needle probes or making the circuit reconfigurable for testing stand alone blocks.

Even for the routinely production it can be convenient to have extra pins for testing purposes that are not used during the normal operation. After testing they can be isolated or powered down by blowing a fuse. Another strategy is integrating a replica of the critical analog sections for making them observable through extra pins and, for a more flexible test, also using an extra clock.

Fig. 9.15 (a) shows a circuit configuration that uses a second *S&H* for testing the main *S&H* and the first analog stage of the *ADC*. The sampling frequency of the second *S&H* is scaled down with respect to f_{s1} ($f_{s2} = f_{s1}/k$, k integer) for a down-sampling of the measured analog signal. Since the result contains a beat frequency at $f_{in} - f_{s2}$, then using an input frequency close to f_{s2} obtains an almost *dc* measure with relaxed speed requirements for the second *S&H* and minimum timing errors. Moreover, because of the low beat frequency the second *S&H* sampling capacitance can be high leading to low *kT/C* noise.

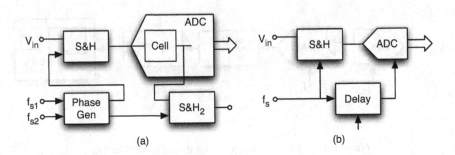

Figure 9.15. (a) Use of an extra S&H for testing analog sections. (b) Test of the S&H acquisition time.

The time required for a proper *S&H* operation can be measured with the extra circuit of Fig. 9.15 (b). The variable delay between *S&H* and *ADC* reduces the settling of the *S&H* and, at a given point, the *SNR* or the *SFDR* drop by a given amount (for example 1 *dB*). This point gives the time needed for proper sampling and, in turn, the maximum operating frequency.

9.8.2   Improving the Spectral Purity of Sine Waves

It is evident that the accuracy of the source determines the accuracy of the measurements. Namely, for a dynamic test that uses sine wave generators (like, for example, measuring the harmonic distortion) the purity of the source must be better than the desired accuracy; say, it must be at least 10 *dB* better. Accordingly, for measuring demanding specifications it can be necessary to use sources with a spectral purity higher than the ones of the generators available in the laboratory, making it necessary to purchase new accurate generators with a significant cost. Instead however, it is advisable to use methods which improve the spectral purity of the generators already available.

A method that enhances the spectral purity foresees the use of band-pass filters for removing the harmonic terms present at the output of a medium quality generator. However, since the spectrum is not the only limit but the noise floor is another important element, the choice of source to be used with filters is critical. Observe that the noise floor is caused by the phase noise; therefore, when choosing the generator the test engineer should look primarily at the phase noise performance because filtering tones is an achievable task while reducing wide band noise caused by jitter is much more difficult.

Fig. 9.16 shows typical schemes used to improve the spectral purity of sine wave generators. The schematic of Fig. 9.16 (a), used for low frequency, utilizes

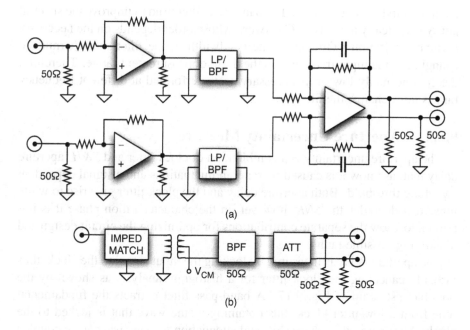

Figure 9.16. (a) Circuit for improving the spectral purity of medium frequency sine waves. (b) same function for high-frequency.

a pseudo-differential input to provide a differential output. The gain of the input amplifier is set to provide the optimal amplitude of the signal generator that gives rise to minimum phase noise and distortion. The input network also matches the 50 Ω source impedance. The low-pass or the band-pass filters used for removing the harmonics must provide high linearity to ensure that the

Be Aware

Extra circuitry used for improving the spectral purity must be very linear. The distortion contribution of the extra circuit used must be less than the desired purity level!

spur rejection of the filters is not cancelled out by the non-linear responses of the filters themselves. The fully differential amplifier must also be linear. It provides a low-pass filtering action in addition to a possible common-mode rejection. The op-amp used must ensure low output impedance for driving the matching impedances at the output.

The circuit of Fig. 9.16 (b) is a much simpler solution suitable for high frequency applications. After the input matching impedance network the *RF* transformer generates the differential signals symmetrical with respect to the common mode voltage, V_{CM}. The band-pass filter used to improve the spectral purity is typically a passive LC network whose order depends on the necessary harmonic rejection. Obviously, the bandwidth of the filter must permit the changing of the input frequency in the desired measurement range. Therefore, the Q is normally low or, if necessary, tests performed at different frequency ranges must change the filter.

9.8.3   Aperture Uncertainty Measure

The aperture uncertainty is a combination of clock jitter and $S\&H$ aperture delay that, as known, is caused by random fluctuations and a signal dependent switching threshold. Both aperture delay and the clock jitter give rise to white noise that degrades the SNR; however, in the characterization phase it is important to know the separate contributions for optimizing the circuit design and identifying possible faults.

The aperture delay is measured by locking the input signal to the clock, thus virtually canceling the clock jitter for a histogram analysis as shown by the possible test setup of Fig. 9.17. A band-pass filter extracts the fundamental tone from a low-jitter clock thus obtaining a sine wave that is locked to the clock. After a possible phase shift and attenuation the sine wave is ac coupled with a dc signal giving rise to an ADC input

$$V_{in,ADC}(t) = V_{DC} + A_1 sin(\omega_s t + \phi_1) \tag{9.43}$$

where the amplitude A_1 depends on the attenuator and the frequency response of the RC filter at the clock frequency.

The dc part corresponds to a given output code and the ac component determines a range of variation around that code. Varying the phase shift changes

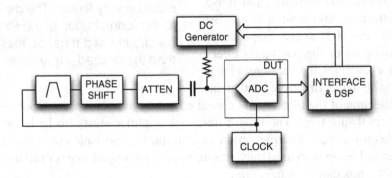

Figure 9.17.   Test setup for the measure of the aperture uncertainty.

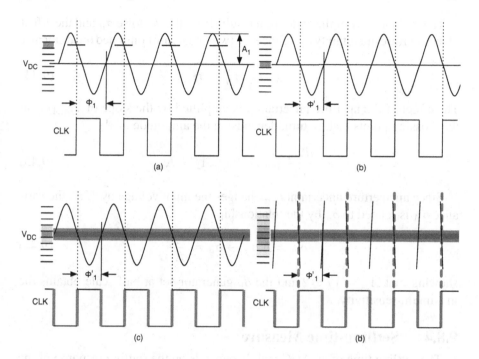

Figure 9.18. (a) and (b) Phase shift that exercises the outlined channels without dc noise and zero aperture uncertainty. (c) Noise on dc level and aperture uncertainty. (d) Amplified sine wave for making negligible the effect of the dc noise.

the sampling instant of the sine wave thus exercising all the possible codes corresponding to the $2A_1$ range.

The phase shift Φ_1 shown in Fig. 9.18 (a) gives rise to a sampling in the upper part of the sine wave above the *dc* code. However, varying the phase to Φ'_1 gives rise to an input voltage at the *dc* level. Without input noise and aperture delay all the samples fall in the same channel and the histogram is like a delta. With small sine wave amplitude the noise on the *dc* signal or in the front end of the converter (that can be referred to the dc level) gives rise to a histogram substantially dominated by the variance of the *dc* equivalent noise. The result looks like the situation shown in Fig. 9.18 (c) where the gray line representing the noise extends over three channels.

Instead, if the amplitude A_1 increases to a large value, as shown in Fig. 9.18 (d) (which only depicts the parts of the sine wave near the crossing points) the effect of the aperture delay becomes evident as the sampling time error moves the *ADC* input by many channels.

The variance σ_H of the histogram results from the *dc* noise σ_n and the effect of the aperture uncertainty σ_A. Since the two effects are combined quadratically

$$\sigma_A = \sqrt{\sigma_H^2 - \sigma_n^2}. \tag{9.44}$$

The effect of the aperture uncertainty is amplified by the slope of $V_{in,ADC}$ at the crossing points that, in turn, depends on the amplitude of A_1

$$\frac{dV_{in}}{dt}\Big|_{max} = 2\pi f_s A_1 = K \tag{9.45}$$

Since an aperture uncertainty τ_a changes the input voltage by $K\tau_a$, the variance σ_A is related to σ_a by the relationship

$$\sigma_A = K\sigma_a \rightarrow \sigma_a = \frac{\sigma_A}{K} \tag{9.46}$$

showing that $A_1 = V_{FS}/2$ and the *dc* generator set at half scale obtains the maximum sensitivity.

9.8.4   Settling-time Measure

The settling time of an *ADC* mainly depends on the settling response of the sample and hold. If the sampling time is not sufficient the signal captured by the *S&H* is not stabilized and causes an error. Therefore, for the settling-time measure it is necessary to control the time allowed for the sampling phase.

The non-ideal response of an *ADC* with finite settling-time can be decomposed into the blocks shown in Fig. 9.19 (a). The first is a continuous time unity gain amplifier with finite settling followed by an ideal *S&H*. Assume that the

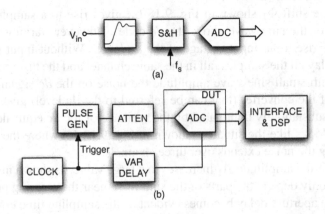

Figure 9.19. (a) Model of an ADC with finite settling response. (b) Settling time test setup.

input is a step voltage that the first block modifies according to its step response $h(t)$. If the action of the ideal *S&H* is delayed by a given amount δ with respect to the input step $U(t)$ the sampled voltage is $Uh(\delta)$. Therefore, by changing the delay it is possible to measure the settling waveform.

The test setup of Fig. 9.19 (b) obtains the above described operation by triggering a pulse generator with the clock. The delayed version of the clock is used for the *ADC* under test while the amplitude of the step input is possibly regulated with an attenuator. Since the output of the *ADC* is the conversion of the input at the time established by the delay, increasing the delay and storing the digital results enables the plotting of the settling response.

9.8.5   Use of FFT for Testing

The *FFT* is an effective way to determine the spectrum of the digital output but is also useful to relate spectral features with specifications. The setup used for the *FFT* testing is made of the input generator, such as a single sine wave (or two or multiple tones), the device under test and a buffer memory which stores the data sequence for off line processing on the *pc*. The acquisition of the digital data is done with a dedicated buffer memory or a logic analyzer.

Obviously the acquisition board must ensure the necessary quality of input signal, references and clock by using buffer amplifiers stable oscillators, filters, decoupling and matching networks. For the case of testing with two or multiple tones the signals generated by single sine wave sources, possibly filtered to ensure spectral purity, must be combined by a matching network that avoids possible reflections and cross modulation, as shown in Fig. 9.20.

The type of specification under test determines the length of the sequence which, as known, grants a required processing gain. For example, since $16,384$ samples gives a $39.1\,dB$ process gain, the testing of a *12 bit ADC* produces a noise floor at $73.8 + 39.1 = 112.9\,dB$. Therefore, the test can measure spurs even below $-100\,dB_{FS}$.

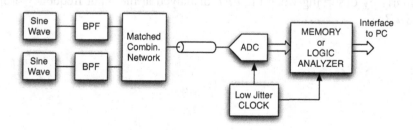

Figure 9.20.   Setup for the FFT test.

The software used on the *pc* enables the measure of various static and dynamic specifications. With a single-tone input it is possible to estimate, among others, *SNR, SINAD, SFDR,* harmonics, and *THD*. A two-tones input obtains key features like *IMD*, and as discussed shortly *NPR*.

The use of the *FFT* can also determine testing parameters by comparison of effects. It is known that some limits cause noise affecting the *SNR*. If it is possible to estimate the *SNR* without that limit and to repeat the *SNR* measure when that (and only that) limit affects the measure, a suitable processing can extract the specification. For example, since the effect of the clock jitter depends on the signal frequency, an indirect estimation of the aperture jitter is done by measuring the *SNR* at low (SNR_L) and high input frequency (SNR_H). If the band of the converter is much larger than the high frequency used and the jitter of the generator is much less than the expected measured value, the *SNR* reduction is caused by the aperture jitter.

Recall that the noise power caused by a τ_{ji} jitter in sampling a sine wave with amplitude $V_{FS}/2$ and frequency f_{in} is equal to $(V_{FS}\pi f_{in}\tau_{ij})^2/2$. Therefore

$$SNR_L = 10 \cdot log\left[\frac{V_{FS}^2/8}{V_n^2}\right];$$

$$SNR_H = 10 \cdot log\left[\frac{V_{FS}^2/8}{V_n^2 + (V_{FS}\pi f_{in}\tau_{ij})^2/2}\right] \quad (9.47)$$

where V_n^2 is the power of the output noise with negligible jitter contribution.

The solution of equation (9.47) yields

$$\tau_{ij}^2 = \frac{1}{2\pi f_{in}}\left[10^{-SNR_H/10} - 10^{-SNR_L/10}\right] \quad (9.48)$$

Another example of the use of the *FFT* for testing is for measuring the analog bandwidth. For an analog circuit the bandwidth limit is when the output amplitude drops by $3\,dB$. The measure of the *ADC* bandwidth can be done similarly by observing when the *FFT* displayed at the input frequency drops by $3\,dB$.

PROBLEMS

9.1 Draft the printed circuit board for the dynamic testing a fast A/D 12-bit converter. The IC has separate analog and digital supplies and has external the reference voltages V_{low} and V_{high}. The input signal is differential and can range from 0.1 MHz to 120 MHz. Use the recommendation given in this chapter as check list.

9.2 Make a search on the Web and find the layout of a pcb recommended in a data converter data sheet for the testing. Examine the ground plane and the placement of components used to filter the supply voltages.

9.3 The static testing of an A/D converter provides the output in binary format. The input is a slow ramp. Write a computer code or find the solution with some software available for determining gain and offset with the best-fit-line method.

9.4 Repeat the previous problem but assume that the input signal is a sine wave at a suitable frequency. Determine the used frequency that enables 10^{14} different input levels.

9.5 The static testing of a 12-bit D/A converter is done with a digital input sine wave. The output of the DAC is converter with a 16-bit ADC. Write a computer code or find the solution with some software available for determining the harmonic distortion terms. Determine the best sine wave frequency, the number of points to be used and the achievable test accuracy.

9.6 Determine the test procedure for a 12-bit current steering DAC. The circuit uses a 4-4-4 segmentation. Assume that the operation of the scheme is linear.

9.7 Determine the testing time for the case discussed in Example 9.1 but suppose that the measure is done changing by steps the input frequency and that for each point of the spectrum 1024 samples are collected for performing the DFT. The time required to automatically change the input frequency is 1 ms.

REFERENCES

Books and Monographs

M. Mahoney: *Tutorial DSP-Based Testing of Analog and Mixed-Signal Circuits*, Computer Society, IEEE, Washington D.C., 1987.

M. Burns and G. W. Roberts: *An Introduction to Mixed-Signal IN Test and Measurement*, Oxford University Press, New York, 2001.

Journals and Conference Proceedings

J. Doernberg, H. Lee, and D. A. Hodges: *Full-speed testing of A/D converters*, IEEE Journal of Solid-State Circuits, vol. 19, pp. 820–827, 1984.

E. J. McCluskey and F. Buelow: *IC Quality and Test Transparency*, IEEE Transaction on Industrial Electronics, vol. 36, pp. 197–202, 1989.

M. Burns: *High Speed measurements Using Undersampled Delta Modulation*, Teradyne User's Group Proceedings, Teradyne, 1997.

Y. Sun: *Analogue and mixed-signal test for systems on chip*, Special Session Introduction, IEE Proceedings, Part G, vol. 151, pp 335–336, 2004.

E. Truebenbach: *Instruments for Automatic test*, IEEE Instrumentation and Measurement Magazine, vol. 8, pp. 27–34, 2005.

Index